Food Engineering Series

Springer's *Food Engineering Series* is essential to the Food Engineering profession, providing exceptional texts in areas that are necessary for the understanding and development of this constantly evolving discipline. The titles are primarily reference-oriented, targeted to a wide audience including food, mechanical, chemical, and electrical engineers, as well as food scientists and technologists working in the food industry, academia, regulatory industry, or in the design of food manufacturing plants or specialized equipment.

More information about this series at http://www.springer.com/series/5996

Mohammed Al-Khusaibi
Nasser Al-Habsi
Mohammad Shafiur Rahman
Editors

Traditional Foods

History, Preparation, Processing and Safety

Springer

Editors
Mohammed Al-Khusaibi
Department of Food Science and Nutrition
College of Agricultural and Marine Sciences
Sultan Qaboos University
Muscat, Oman

Nasser Al-Habsi
Department of Food Science and Nutrition
College of Agricultural and Marine Sciences
Sultan Qaboos University
Muscat, Oman

Mohammad Shafiur Rahman
Department of Food Science and Nutrition
College of Agricultural and Marine Sciences
Sultan Qaboos University
Muscat, Oman

ISSN 1571-0297
Food Engineering Series
ISBN 978-3-030-24622-8 ISBN 978-3-030-24620-4 (eBook)
https://doi.org/10.1007/978-3-030-24620-4

This Springer imprint is published by the registered company Springer Nature Switzerland AG
The registered company address is: Gewerbestrasse 11, 6330 Cham, Switzerland

Dedicated to Asma Al-Asmi, Jamila Al-Habsi and Sabina Akhter.

Preface

Global industrialization significantly introduced the wide range of preserved foods, such as dried, canned, and frozen. This trend was followed by take-away shops that sold fast foods (such as sandwiches, pizzas, and burgers). Recently, consumers want meals which are easy and quick to prepare, are healthy, and possess ethnic or traditional flavors with varied shelf life. Thus, the new products, such as minimally processed foods and ready-to-eat, ready-to-cook, cook-chill, cook-freeze, and part-bread, part-pizza meals, are being introduced in the market. More recently, there is a great interest on the different aspects of the traditional foods, since it can fulfill and foster additional dimension of culture and tradition. This book provides a comprehensive overview on traditional foods from different regions of the world. Ten chapters in this book explore the preparation, processing, marketing, safety, and nutritional aspects of traditional foods around the world and present different chapters explaining the traditional foods of different cultures, such as Arab, Indian, and Chinese foods. It will also contain chapters to discuss the safety, nutritional, and consumer sensory perception of traditional foods as well as laws and regulations.

Chapter 1 gives an overview on the traditional foods around the word. It briefly discusses how traditional foods are defined and presents the brief summary of the traditional foods of Arabs, Indians, Chinese, African, Native Australians, Native Americans, and Europeans. An overview of sensory characteristics and needs of laws and regulations related to traditional foods is also presented.

Chapter 2 discusses the Arab traditional foods. It gives an insight into the culture of hospitality in Arab countries and its relation to traditional foods. Different traditional foods/dishes are reported in this chapter. The ingredients and preparation of plant-based, animal-based, and rice dishes, dairy products, and bread are listed. Herbs and spices used in Arab cuisines are also explained in this chapter.

Chapter 3 gives an insight into the Gulf Cooperation Council (GCC) countries' traditional foods. This chapter includes an overview about the eating habits of the countries between the past and present. This chapter gives examples of traditional foods in the GCC countries according to their ingredients and preparation methods. Selected plant-based, meat-based, dairy, and sweet food products are also discussed in this chapter.

Chapter 4 includes traditional foods from Maghreb. Maghreb countries are those countries in the Northwest Africa. Although they are Arab countries, they have different traditional dishes and staple foods. This chapter discusses different traditional dishes and raw materials that contribute to traditional foods, technology of processing, and nutritional values. It categorizes foods into oils, cereals, dairy products, meat products, and sweets.

Chapter 5 highlights Chinese traditional foods. It gives a brief history about Chinese traditional foods and the differences among different regions in China. Then, it reports the different classes of Chinese traditional foods and their preparation methods, such as rice-based and fish-based foods, poultry, fermented foods, liquors, and bread.

Chapter 6 discusses the Indian traditional foods. This chapter gives an insight into the rich variation of Indian culture, religions, climate, and communities and how those factors resulted in different food preferences and varied traditional foods. It reports several traditional foods based on regions, religious beliefs, and shelf life. Preparation, processing, safety, and sensory attributes of the Indian traditional foods are also discussed. This chapter also highlights research that has been progressed in this area.

Chapter 7 discusses Mediterranean foods and their positive health effects. It discusses the cardioprotective, anti-inflammatory, chemopreventive, pleiotropic, antidiabetic effects of olive oil and highlights the Mediterranean diet components and discusses the health effects of nuts and fish as part of this traditional diet.

Chapter 8 deals with the safety, nutrition, and functionality of traditional foods. This chapter categorizes traditional foods based on commodities. The national characteristics and functionality of the commodities are highlighted, and examples of bioactive compounds are presented. This chapter also discusses the safety of traditional foods in each commodity.

Chapter 9 highlights the laws and regulations that control traditional foods. The importance of such legislations and their evolution through the last decades are discussed. Legislations that protect the identity and integrity of traditional foods, such as intellectual property and geographical indications, are also reported. This chapter also discusses the food safety legislations and how they are integrated with the World Trade Organization (WTO) regulations, which emphasizes the General Agreement on Tariffs and Trade (GATT) framework. This chapter gives examples of legislations from the European Union, the Middle East, Africa, China, South East Asia, and Latin America.

Chapter 10 discusses the sensory characteristics of traditional foods and their role in the excitement and pleasure perceived by consumer and the motivational factors that affect the perception of tradition foods and their willingness to accept traditional foods from other communities. In addition, the effects of spices on the sensory properties of traditional foods are highlighted in the chapter. This chapter also discusses the sensory evaluation of traditional foods and the sensory aspects of selected traditional foods.

The contributing authors were selected from different parts of the world. Individually, each chapter represents a unique perspective of specific groups of traditional foods. We are confident that the academics, higher-level undergraduate and graduate students, professionals from industry, and authorities working related to food product development, food quality, and safety will find a valuable compilation and rare volume on the traditional foods.

Muscat, Oman Mohammed Al-Khusaibi
Muscat, Oman Nasser Al-Habsi
Muscat, Oman Mohammad Shafiur Rahman

Acknowledgment

We would like to express our gratitude to Sultan Qaboos University, Sultanate of Oman, for their support to execute this book project. Special thanks to our colleagues in the Department of Food Science and Nutrition at the College of Agricultural and Marine Sciences for their continued support and encouragement. We would also like to appreciate Mr. Deepak Ravi, Mr. Anthony Dunlap, and Mr. Daniel Falatko of Springer Publisher for their assistance during the progress of the project. Finally, we would like to thank our family members for their encouragements to complete this book project.

Contents

Contributors

Nasser Al-Habsi Department of Food Science and Nutrition, College of Agricultural and Marine Sciences, Sultan Qaboos University, Muscat, Oman

Mohammed Al-Khusaibi Department of Food Science and Nutrition, College of Agricultural and Marine Sciences, Sultan Qaboos University, Muscat, Oman

Laxmi Ananthanarayan Food Engineering and Technology Department, Institute of Chemical Technology, Mumbai, Maharashtra, India

Ferdaous Boughattas Université d'Artois, EA 7394, Institut Charles VIOLLETTE, Lens, France

Kriti Kumari Dubey Food Engineering and Technology Department, Institute of Chemical Technology, Mumbai, Maharashtra, India

Jamie Gallagher School of Agriculture and Food Science, University College Dublin, Dublin 4, Ireland

School of Biomedical Science, Ulster University, Coleraine, Northern Ireland

Abul Hossain Department of Biochemistry, Memorial University of Newfoundland, St. John's, NL, Canada

Romdhane Karoui Université d'Artois, EA 7394, Institut Charles VIOLLETTE, Lens, France

ISA Lille, EA 7394, Institut Charles VIOLLETTE, Lille, France

Ulco, EA 7394, Institut Charles VIOLLETTE, Boulogne sur Mer, France

Université de Lille, EA 7394, Institut Charles VIOLLETTE, Lille, France

Zaixiang Lou State Key Laboratory of Food Science & Technology, School of Food Science, Jiangnan University, Wuxi, China

Aideen McKevitt School of Agriculture and Food Science, University College Dublin, Dublin 4, Ireland

Abhijeet B. Muley Food Engineering and Technology Department, Institute of Chemical Technology, Mumbai, Maharashtra, India

Md. Jiaur Rahman Department of Food Processing and Preservation, Hajee Mohammad Danesh Science and Technology University, Dinajpur, Bangladesh

Md. Ramim Tanver Rahman State Key Laboratory of Food Science & Technology, School of Food Science, Jiangnan University, Wuxi, China

Mohammad Shafiur Rahman Department of Food Science and Nutrition, College of Agricultural and Marine Sciences, Sultan Qaboos University, Muscat, Oman

Rekha S. Singhal Food Engineering and Technology Department, Institute of Chemical Technology, Mumbai, Maharashtra, India

Mostafa I. Waly Department of Food Science and Nutrition, College of Agricultural and Marine Sciences, Sultan Qaboos University, Muscat, Oman

Zhang Yu Institute of Biology, Qilu University of Technology (Shandong Academy of Sciences), Jinan, China

Hicham Zaroual Université d'Artois, EA 7394, Institut Charles VIOLLETTE, Lens, France

Chapter 1
Traditional Foods: Overview

Mohammed Al-Khusaibi and Mohammad Shafiur Rahman

1 Introduction

Tradition is "a long-established custom or belief that has been passed on from one generation to another" (Oxford Dictionary, 2018). It is handed down from age to age, following or conforming to tradition: adhering to past practices or established conventions (Marriam-Webster Dictionary, 2018). According to the definition of the European Commission, a food is said to be traditional if the usage is proven to be transmitted between generations considering that one human generation of at least 25 years (EU, 2006). Eating habits also contribute to the concept of traditional food. The definition of traditional foods is also applied to traditional ingredients and traditional preparation methods. The traditional consumption methods also varied with culture, for example Arabs and Indians use their hand for eating, while Chinese and European use chop sticks and spoons. From the literature, a traditional food must be lined to a territory (Bertozzi, 1998; Jordana, 2000), according to Guerrero et al. (2009), a definition of traditional food from consumer perspective can be:

> *a product frequently consumed or associated with specific celebrations and/or seasons, normally transmitted from one generation to another, made accurately in a specific way according to the gastronomic heritage, with little or no processing/manipulation, distinguished and known because of its sensory properties and associated with a certain local area, region or country.*

Traditional foods are the expressions of culture, identity, heritage, and lifestyle. The quality level of traditional foods (i.e. safety, processing and preparation, and health) is a key to secure and expand the market share (Guerrero et al., 2009).

M. Al-Khusaibi (✉) · M. S. Rahman
Department of Food Science and Nutrition, College of Agricultural and Marine Sciences, Sultan Qaboos University, Muscat, Oman
e-mail: mohamedk@squ.edu.om

M. Al-Khusaibi et al. (eds.), *Traditional Foods*, Food Engineering Series,
https://doi.org/10.1007/978-3-030-24620-4_1

2 Arab Traditional Foods

Arab traditional foods represent their specific traditions and culture. The food, especially traditional food, is a symbol of hospitality. The Arab traditional foods cooking methods vary according to the types of foods. The ingredients and nutritional quality of a food also play an important role in selecting traditional foods. The cooking methods start with stir frying followed by boiling or simmering. Arab traditional cooking methods also involve baking or roasting in a traditional oven called *Tannor*. Part of the Arab traditional foods is influenced by Mediterranean food which is high in nuts, cereals, fruits and vegetables, fish and olive; and olive oil is the frequently used ingredient.

Among the Arab region, the Maghreb is a distinctive geographical area in the Northwest Africa. It includes Morocco, Algeria, Libya and Tunisia and Mauritania. It is also known as the Berber world. In the Maghreb, food is prepared by different processes including fermentation, salting, drying, and pickling. Just similar to Mediterranean people, The Maghren people use olive oil in their diet and as a medicine. *Alouana* is a special oil prepared from roasted olives. *Argan* oil is another oil which is popular in the Maghreb. Couscous is the staple food in the Maghreb. It is the carbohydrate-based food that substitute rice in the other Arab countries diet. Couscous is served with meat, chicken or it can be prepared as vegetarian. In Algeria, a sweet couscous dish called "*Mesfouf*" is popular. Other traditional dishes in the Maghreb are made from pasta, chickpeas, barley, etc. Fermented dairy products have their own place in the list of the Maghreb traditional food. Fermented camel milk and its derivatives are examples. *Tagine* is a culinary preparation from Northwest Africa. When the word "*Tagine*" is heard or read, it has the sound of the Maghreb. *Tagine* is cooked in specific utensil and may contain a mixture of meat, poultry or fish, vegetables and spices. There are small differences in the preparation of *Tagine* within the Maghreb countries. Maghreb cuisine is flavoured with different types of sweets and dessert dishes, such as samet, raisins, pastilla, baklawa and other traditional food products.

3 Indian Traditional Foods

The different cultures and religions in Indian have greatly influenced the traditions and Indian cuisines. The vegetarian diet is popular in India, but some regions, especially in the coastal areas are known for their non-vegetarian cuisines. Staple foods in India are based on rice and wheat. However, different regions in India, such as Punjab, Southern India, West Bengal, Maharashtra and Gujarat are known for specific unique dishes.

Bread is an important part in Indian cuisine. Naan, *paratha* and *rooti* (flat bread) are the very popular forms of bread. Naan is cooked in a special oven called *tandoor* over. *Parathas* are served either plain or stuffed with potatoes, called *aloo* in India.

The second important part of the meals is lentil. It is served with rice dishes or with various types of bread.

In Indian cuisines, snacks are served with tea or coffee and accompanied with mint or tamarind sauce. During festivals and religious occasions, different types of confections; such as *mithai, halwa, kheer, payasam, sevai*, are prepared. Ingredients that are involved include condensed milk, grain flours, lentils, semolina, chickpea flour and vegetables such as carrots and pumpkins.

Different beverages are known in different regions of Indian. Tea and coffee are popular across India. A special tea is called *masala* tea in which tea leaves are boiled with spices such as cardamom, ginger, and clove. India is known of its spices and herbs. Christopher Columbus discovered the new world while trying to sail to Indian looking for spices. Indian traditional foods have strong flavor owed to the combination of spices and herbs.

4 Chinese Tradition Foods

There are different regions in China, namely Inner Mongolia, Guizhou, Hunan, Yunnan, Xinjiang, Gansu, and Ningxia, and Chengdu. Each area is known for its unique traditional style of food preparation. However, rice remains the important staple food. Rice is consumed in different forms, such as rice cereals, noodles, puffed rice, rice cake, fermented rice, and extruded rice snackes. The second important product is sea food especially jelly fish, which has a very important place in the traditions of Chines and important meal in the social occasions. Poultry has also an important role in the Chines cuisines. Among the popular traditional foods, Chines vinegar and fermented dairy products like tofu are popular. Different types of soups are also common in Chinese traditional foods.

5 African Traditional Foods

Africa is the second-biggest continent in land area and the second-largest in population after Asia. While it is mostly tropical, some parts are savannas. It can be divided into 4 regions: North Africa which is made up by the Maghreb countries; Sahel which is made by Chad, Mali, Mauritania, Niger and Sudan; West Africa which includes Benin, Burkina Faso, Cameroon, Cape Verde, Cote d'Ivoire, Gambia, Ghana, Guinea, Guinea-Bissau, Liberia, Nigeria, Senegal, Sierra Leone and Togo; and Central Africa which includes Angola, Central African Republic, Republic of the Congo, Democratic Republic of the Congo, Equatorial Guinea, Gabon, São Tomé and Príncipe, South Sudan. This gives diversity in crops and livestock that contribute to different types of food. According to Bakunda and Otengei (2013), the African cuisine are not internationally popular and there is a lack of literature published in this topic. Cereals, legumes and tuber roots are the major raw materials in

Africa. The continent has more than two thousands of grains, roots and tubers, fruits, vegetables, legumes, and oil crops. Leafy green vegetables are also an important part of diets across most of Africa. In addition, alcoholic or non-alcoholic fermented foods and beverages are major parts of the African cuisine. These foods are produced at domestic level and also in small industrial scale (Jespersen, 2003). There is a variety of products that are fermented. Starch-based food products depends on maize such as '*kenkey*' and '*mawe*' or on rice such as "*masa*" or depends on Plantain such as '*agadagigi*' (Jespersen, 2003). Fermented milk products include, *amasi*, *mbanik* and *nono*. Each African country has its own way of preparation and also naming of the products.

6 Red Indian and Australian Aboriginal Traditional Foods

Australian Aboriginal has never been engaged in intensive agriculture, instead they moved from hunting and gathering to industrial system of mechanical farming, processing, transporting. Symons (1982) believes that engagement in intensive agriculture is a key role to have good cuisines. This is why plant food is considered as supplement with meat-based diets being the predominant diet (Lee, 1996). Meat products include *kangaroo*, *abalone*, *wallaby*, *emu* and *crocodile*.

The plant food eaten by aboriginal Australians is called bush food. Those foods are still popular in Australia and there is a strong believe with Australians that knowledge about bush food is important and needs to be maintained (Walsh & Mitchell, 2002). There are around 5000 native food species that were used by the Aboriginal people. According to Pont (1984) (cited in Bannerman, 2011), Australian food was also influenced by the peasants from the Mediterranean countries who immigrated to Australia after the Second World War.

Akatyerr and different Acacias are the main commercialized bush foods (Merne Altyerre-ipenhe, Douglas, & Walsh, 2011). Example of spices is lemon myrtle, Tasmannia pepper, Anise Myrtle, and Wattleseed. Examples of indigenous fruits and vegetables are bush tomatoes (*Solanurn chippendalei*), raisins, quandong (*Santalum acuminatum*), native fig (*Ficus platypoda*), Wild oranges (*Capparis* spp), desert banana (*Leichhardtia australis*) and kakadu plum (Higgins-Desbiolles, Vilkinas, Wijesinghe, Akbar, & Gifford, 2016 Konczak et al., 2009). The root and tubers include yams (*Dioscorea species*), wild potato (*Ipornoea costata*), native onion (*Cyperus* species) and the yam daisy (*Microseris scapigeris*). Coconut milk is a very popular beverage with the indigenous Australians. In addition, flowers and other parts of plant are steeped in water to make beverages (Brand-Miller & Holt, 1998).

Native Americans are called Indian or Red Indians because Christopher Columbus was supposed to be sailing to India when he arrived to the new World. This has resulted in the introduction of new food to the world. The traditional native American cuisines depends on corn, potatoes, beans, and squash. Examples of native American dishes are *succotash*, *bean bread*, *fry bread*, *baked acorn squash*

(Park, Hongu, & Daily III, 2016). Other popular foods which have originated from native American include turkey, mashed potatoes, baked beans and pumpkin pie. The animal-based foods come from hunting (bear and deer) and fishing which were a routine practice for native Americans. Preparation methods used by native Americans include boiling, steaming, oven cooking, roasting, smoking and drying.

7 European Traditional Foods

There are several definitions for traditional foods. The recent definition of traditional food by EuroFIR in 2008 considered the food practices and preparation methods prior to the Second World War and remain in use (Weichselbaum, Benelam, & Costa, 2009). There are differences in the cultures and hence the traditional foods of the European countries. Several countries had been occupied by different nations and different ethnic groups across the history that influenced the history of traditional foods. The report by the EuroFIR Traditional Foods (Weichselbaum et al., 2009) identified several traditional foods for the thirteen countries participated in the study, as shown in the Table 1.1. French and Italian cuisine were the most top rated and internationally popular when compared to other internationally recognized cuisines (Bakunda and Otengei, 2013).

8 Sensory Characteristics of Traditional Foods

In any types of foods, safety needs to be considered first followed by sensory characteristics. Distinct sensory characteristics of traditional foods is one of the major criteria for selecting traditional foods. It is important to keep the traditional ingredients and processing or preparation methods as original to produce specific

Table 1.1 Selected traditional foods in Europe

Country	Selected traditional food
Italy	Ricotta stuffed roll, Vicentina cod, Pizza Napoletana Margharita, Braised beef with Barolo wine, Tuscan costagnaccio
Poland	Cold soup “Chlodnik”, Pork chop, Stew made of sauerkraut, meat and dried mushrooms, Tree cake, Smoked ewe’s milk cheese
Austria	Gemüsesuppe (Vegetable Soup), Wiener Schnitzel (a thin, breaded, fried meat cutlet, traditionally calve meet), Cabbage noodles, Potato dumplings, Apple strudel
Belgium	Shrimp croquette, Flemish stew, Meat loaf, meat balls, endives with ham and cheese sauce, Belgian waffles
Germany	Black Forest smoked ham, Thuringian fried sausage, Swabian ravioli, Dresden fruit loaf, Pumpernickel bread
Greece	Leek sausages, Rabbit stew, Chickpea soup, Must jelly, Cherry tomato of Santorini
Turkey	Baklava, Kebab with yogurt, Anchovy stew, Dried fermented soup, Pastirma

traditional foods. The flavor of traditional foods depends on different factors. Ingredients are very important especially when spices, herbs, condiments are essential part in the culture. Secondly, the preparation methods that initiate the interaction of ingredients and/or create new flavors. For example, frying cause the liberation of volatile compounds of spices and herbs. In addition, Maillard reactions at the high frying temperatures results in the formation of new flavors and development of color. On the other hand, fermentation produce mild to vey source food products according to the degree of fermentation, which can influence people's preferences.

9 Laws and Regulations

Concerns of food safety are being noticed since Middle Ages and regulatory measures are enforced to prevent the sale of adulterated or contaminated food. Ancient Roman law reflected this concern for adulteration of food with punishment that could result in condemnation to the mines or temporary exile (Fortin, 2009). The main purpose of food laws is to avoid adulteration and economic fraud. In the case of traditional foods, more criteria need to be included, for examples authentication, and protection of the originality and sources of the traditional foods. In order to do that, first detailed characteristics of the traditional foods need to be documented, for examples ingredients, methods of preparations, flavor, texture, and taste. More emphasis is now given on the source of the traditional product. For example, the label must show clearly where the food has come from, otherwise it would be misleading not to show it. (i.e. a tub of 'Greek Yogurt' which was made in France) (Food a Fact of Life, 2012). The European Union has created three systems to promote and protect regional food products. These include the Protected Designation of Origin (PDO), the Protected Geographical Indication (PGI), and the Traditional Specialty Guaranteed (TSG). Protected Designation of Origin (PDO) is used for food produced, processed and prepared in a given geographical area using recognized know-how (e.g. West Country farmhouse Cheddar cheese and Jersey Royal potatoes). Protected Geographical Indication (PGI) (the geographical link) must occur in at least one of the stages of production, processing or preparation (e.g. Melton Mowbray Pork pie, Scottish farmed salmon and Welsh lamb). Traditional Specialty Guaranteed (TSG) highlights traditional character, either in composition or means of production (e.g. traditional farm fresh turkey in the UK). In addition, traditional foods are also regulated by the codex standards. Since Codex Alimentarius Commission adopt what is called "Horizontal" standards (i.e. standards that are not applied to categories of foods but general standards that can be applied to different categories of foods), traditional foods can be regulated by such standards.

Food safety is a big concern in traditional food trade. The risk is due to preparation of food using contaminated water, poor personnel and utensil hygiene, low level of food safety knowledge and the lack of food safety legislations. Food safety can be a boundary that prevent traditional foods from international trade. According to the World Trade Organization (WTO) and the General Agreement on Tariffs and

Trade (GATT), Codex standards are the basic which has to be implemented in national standards in order for a county to be benefited from global trading. Hence, traditional food safety is ensured in the countries that are member in the GATT through implementation of Codex standards (Garrett, Jahncke, & Cole, 1998).

10 Conclusion

Traditional foods include ingredients, preparation methods and custom of eating, which have been transferred through generations and linked to a territory. These reflect the differences in nations cultures, believes, religions, food preferences, habits and availability of resources. The traditional foods of different countries vary according to the ethnicity and also geographical areas. When they cross the boarders, these gain more value due to uniqueness. Hence, the safety of traditional food must be enhanced by good manufacturing practices (GMP) and increased food safety knowledge among producers. To protect them from adulterations and to keep their identity, different countries have set legislations that regulate traditional foods. Examples are the three systems of European Union, i.e. the Protected Designation of Origin (PDO), the Protected Geographical Indication (PGI), and the Traditional Specialty Guaranteed (TSG). In the future there would be great market for the traditional foods as well as fusion foods combining different traditional foods from different regions. Therefore, it is important first to document the characteristics of traditional foods in terms of their ingredients, preparation or processing methods, distinct sensory properties and storage. Specific laws and regulations need to be developed to determine authentication and to ensure safety of traditional foods available locally and globally.

References

Altyerre-ipenhe, M., Douglas, J., & Walsh, F. (2011). *Aboriginal people, bush foods knowledge and products from central Australia: Ethical guidelines for commercial bush food research, industry and enterprises*. DKCRC Report 71. Alice Springs: Ninti One Limited.

Bakunda, G., & Otengei, S. O. (2013). Internationalization of African ethnic cuisine: A situation analysis. *Journal of Research in International Business and Management, 3*(2), 66–72.

Bannerman, C. (2011). Making Australian food history. *Australian Humanities Review, 51*, 49–63.

Bertozzi, L. (1998). Tipicidad alimentaria y dieta mediterránea. In *El color de la alimentación mediterránea. Elementos sensoriales y culturales de la nutrición* (pp. 15–41). Barcelona: Icaria.

Brand-Miller, J. C., & Holt, S. H. (1998). Australian Aboriginal plant foods: A consideration of their nutritional composition and health implications. *Nutrition Research Reviews, 11*(1), 5–23.

EU. (2006). Council Regulation (EC) No509/2006 of 20 March 2006 on agricultural products and foodstuffs as traditional specialities guaranteed. *Official Journal of the European Union L, 93*(1), 1–11.

Food a Fact of Life. (2012). PowerPoint presentation. Food Product Labelling. Retrieved from http://www.foodafactoflife.org.uk.

Fortin, N. D. (2009). *Food regulation: Law, science, policy, and practice*. Hoboken, NJ: Wiley.
Garrett, E. S., Jahncke, M. L., & Cole, E. A. (1998). Effects of codex and GATT. *Food Control, 9*(2–3), 177–182.
Guerrero, L., Guàrdia, M. D., Xicola, J., Verbeke, W., Vanhonacker, F., Zakowska-Biemans, S., … Scalvedi, M. L. (2009). Consumer-driven definition of traditional food products and innovation in traditional foods. A qualitative cross-cultural study. *Appetite, 52*(2), 345–354.
Higgins-Desbiolles, F., Vilkinas, T., Wijesinghe, G., Akbar, S., & Gifford, S. (2016). Indigenous foods benefiting indigenous Australians. In *Proceedings of the 26th annual CAUTHE conference* (pp. 527–542).
Jespersen, L. (2003). Occurrence and taxonomic characteristics of strains of Saccharomyces cerevisiae predominant in African indigenous fermented foods and beverages. *FEMS Yeast Research, 3*(2), 191–200.
Jordana, J. (2000). Traditional foods: Challenges facing the European food industry. *Food Research International, 33*(3–4), 147–152.
Konczak, I., Zabaras, D., Dunstan, M., Aguas, P., Roulfe, P., & Pavan, A. (2009). *Health benefits of Australian native foods: An evaluation of health-enhancing compounds*. Canberra: Rural Industries Research and Development Corporation.
Lee, A. (1996). The transition of Australian Aboriginal diet and nutritional health. In *Metabolic consequences of changing dietary patterns* (Vol. 79, pp. 1–52). Basel: Karger.
Marriam-Webster Dictionary. (2018). Retrieved November 14, 2018, from https://www.merriam-Webster.com/dictionary/traditional.
Oxford Dictionary. (2018). Retrieved November 14, 2018, from https://en.oxforddictionaries.com/definition/tradition.
Park, S., Hongu, N., & Daily, J. W., III. (2016). Native American foods: History, culture, and influence on modern diets. *Journal of Ethnic Foods, 3*(3), 171–177.
Pont, G. (1984). Review of goody. *op cit Mankind, 14*(3), 225–226.
Symons, M. (1982). *One continuous picnic: A history of eating in Australia*. Adelaide: Duck Press.
Walsh, F., & Mitchell, P. (Eds.). (2002). *Planning for country: Cross-cultural approaches to decision making on Aboriginal lands*. Alice Springs: Jukurrpa Books/IAD Press.
Weichselbaum, E., Benelam, B., & Costa, H. S. (2009). Synthesis report No. 6: Traditional foods in Europe. EuroFIR Project Management Office/British Nutrition Foundation. United Kingdom. Retrieved November 18, 2018, from http://www.eurofir.net.

Chapter 2
Arab Traditional Foods: Preparation, Processing and Nutrition

Mohammed Al-Khusaibi

1 Introduction

Traditional foods represent the geographical distinct environment, resources and culture of people. According to the EUROFIR "Traditional food is a food of a specific feature or features, which distinguishes it clearly from other similar products of the same category in terms of the use of "traditional ingredients" (raw materials or primary products) or "traditional composition" or "traditional type of production and/or processing method" (Tajkarimi, Ibrahim, & Fraser, 2013). It can also be simply defined as products with specific raw materials, and/or with a recipe known for a long time, and/or with a specific process (Cayot, 2007). Food is a traditional symbol of Arab hospitality. Rosenberger (1999) wrote that even when people was suffering from scarcity of food and poverty, the traditional foods was always served to travelers and guests. Nowadays, special types of traditional cuisine are usually prepared to honor guests.

Historically, there are several factors caused the emergence of traditional foods. For examples, a special dish named "*tharid*" is prepared by soaking pieces of flat bread in a meat stock. The simplicity of this dish might represent the arid climate and scarcity of raw materials. On the other hand, abundance of some crops with the limited market in the ancient time caused farmers to think about traditional preparation methods to preserve the food. It is not only the environment and resources contributed to the nature of traditional foods in the Arab countries, but religious beliefs have also influenced to flourish of different dishes through permitted and restricted ingredients (Tajkarimi et al., 2013). For example, Muslims are prohibited from eating pork and drinking alcohol (Forum, 1999; Tannahill, 1989). This prohibition is extended to

M. Al-Khusaibi (✉)
Department of Food Science and Nutrition, College of Agricultural and Marine Sciences, Sultan Qaboos University, Muscat, Oman
e-mail: mohamedk@squ.edu.om

M. Al-Khusaibi et al. (eds.), *Traditional Foods*, Food Engineering Series,
https://doi.org/10.1007/978-3-030-24620-4_2

the use of ingredients obtained from pig and the use of alcohols in food preparation. The Arabic culture is rich in traditional food dishes that are prepared domestically and these foods are nowadays available in restaurants and sometime in the street food vendors. This chapter will give an insight view on the different plant-based and meat-based food dishes. The chapter will discuss the preparation methods, ingredients, and nutritional quality of Arab traditional foods. It must be noted that in term of the name of the traditional foods or dishes, there are always minor variations in the spelling and phonation. The English spelling of Arab traditional foods vary from country to country. In addition, different names of same dishes are noticed among different countries.

2 Food Preparation of Arab Traditional Foods

Food preparation methods play an important role in the nutritional quality and organoleptic properties of food. The cooking process in Middle Eastern food generally starts with stir frying, followed by boiling and fat enrichment. One of the cooking methods is using an oven called *Tannoor* or *Tandoor*. It is an oven traditionally made from clay but nowadays it is also made from stainless steel. The heat is traditionally generated by burning charcoal, dry tree branches and trunks inside the oven. There is a combination of radiant from the flame, radiation from the refractory service of the oven, air-forced convection and conduction heating (Saxena, Rao, & Rao, 1995). The oven is used to bake bread and pastries, cook rice dishes and grilled meat. These foods also absorbs smokes, especially when fatty foods (e.g. pieces of meats) are grilled and the fat drips into the fire making smoke.

3 Mediterranean Food

Mediterranean food refers to the food prepared by nations in the Mediterranean Basin, which is the region around the Mediterranean Sea. The Arab countries, which are among the Mediterranean region, are Jordan, Lebanon and Syria. Part of the Mediterranean food is associated to the ancient Greek who anticipated a relationship between healthy diet and diseases (Aboul-Enein, 2016). Since some of the Mediterranean countries are olive-growing countries, olive oil is one of the main culinary lipid in the Mediterranean diet. In addition, the Mediterranean diet is high in nuts, cereals, fruits and vegetables, moderate to high in fish and seafood and low in meat (Martínez-González & Gea, 2012). It must be noted that the supply (per capita) of fish and seafood in the Arab countries is less than the global average (Hwalla, Bahn, & El Labban, 2015). In general, this could cause a decrease in the consumption of fish and seafood in the Arab countries. Johnston, Fanzo, and Cogill (2014) reported a dietary shift in the Arab countries away from traditional food and more to what they called "westernized" diet, which is protein-based energy-rich diet.

4 Plant Based Arab Foods

4.1 Salads and Dips

Tahinah: *Tahinah* is also called *tahini*, which is a condiment made from sesame seeds after being toasted and ground. In some references, it is named sesame butter because it is a thick paste. Sesame seeds are soaked in water and then crushed and soaked in water to separate the bran from the kernels. The kernels are then toasted and ground to produce the paste. *Tahina* is served as a dip and it is also one of the essential components in several dishes and dips as shown in Table 2.1. Sesame seeds are good protein source, and hence the addition of *tahinan* to several dishes increase the protein quantity. The addition of *tahinah* is used in the preparation of *hummus* (chickpeas dip) and it results an increase of amino acid scores by supplementing two sulphur amino acids; methionine and cysteine (Faris & Takruri, 2003). Sesame seed contain 44–58% oil, 18–25% protein, and 5% ash (Kahyaoglu & Kaya, 2006;

Table 2.1 List of Arab traditional foods and ingredients

Type	Name	Main ingredients	Countries
Plant based			
	Fattoush or fattoosh	Lettuce, parsley, cucumber, tomato, onion, fried bread, olive oil, bread (Arabic), mint, lemon, salt, sumac	Jordan, Lebanon, Palestine, Syria
	Hummus (also Hommus, houmous)	Chickpeas, tahina (sesame butter), olive oil, garlic, lemon	Jordan, Lebanon, Palestine, Syria and Egypt
	Falafel	Broad (fava) beans (in Egypt) or chickpeas (in other countries), parsley coriander, cumin, salt, garlic, onion (green and bulb), spices	Jordan, Lebanon, Palestine, Syria and Egypt
	Ful Medames	Broad beans, tomato, garlic, cumin, salt, olive oil, lemon juice.	Jordan, Lebanon, Palestine, Syria and Egypt
	Rice dishes		
	Mansaf	Rice, lamb, fermented dried yogurt	Jourdan
	Koshari	Rice, macaroni, lentils, tomato sauce, chickpeas and fried onion strips	Egypt
	Mujaddara	rice, lentils, onion, olive oil	Iraq, Jordan, Lebanon, Palestine, Syria
	Magloba	Rice, chicken/lamb, tomatoes, potatoes, cauliflower, eggplant	Iraq, Jordan, Lebanon, Palestine, Syria
	Biryani (Arab version)	Rice, rice, chicken/lamb, potato, onions	Iraq

(continued)

Table 2.1 (continued)

Type	Name	Main ingredients	Countries
Animal based			
	Hawashi (sandwich)	Meat, bread, onion, tomato, vegetable oil, spices	Jordan, Lebanon, Palestine, Syria, Egypt
	Shawerma	Meat, spices Usually made as a sandwich with bread	Jordan, Lebanon, Palestine, Syria, Egypt
	Kibbeh (baked or fired meat balls)	bulgur, minced onions, and finely ground meat (beef or lamb)	Jordan, Lebanon, Palestine, Syria, Egypt
	Musakhan	Chicken, onions, sumac, saffron, fried pine nuts	Palestine
	Kofta (meat balls)	Minced meat/chicken, onion, garlic, parsley and spices	Jordan, Lebanon, Palestine, Syria, Egypt
	Fesikh	Fermented fish	Egypt
	Kebab	Marinated, skewered and char-coal grilled minced meat or meat cubes	Jordan, Lebanon, Egypt, Syria

Kamal-Eldin & Appelqvist, 1994; Yoshida, 1994). It was also reported that sesame paste has 42% oleic acid followed by 37% linoleic acid (Borchani, Besbes, Blecker, & Attia, 2010).

Hummus (chickpea dip): *Hummus* (Fig. 2.1, item 1) is commonly consumed during breakfast and it is a side dish during lunch or dinner. It is a creamy paste-like food, which is made from chickpeas (garbanzo beans, *Cicer aretinum* L.). The preparation of *hummus* involves the addition of sesame butter (*Tahini*), garlic, and lemon juice (Al-Holy, Al-Qadiri, Lin, & Rasco, 2006; Yamani & Al-Dababseh, 1994). The boiled chickpeas and the other ingredients are placed in a food processor and blended to get the creamy pate. Olive oil is added at the top of the paste. Sumac, cumin, pomegranate or paprika are used to decorate the creamy paste (see Fig. 2.1). Lupine beans are also prepared similarly. However, these are again soaked after boiling in order to remove the bitter taste resulting from alkaloids in bitter lupines. This traditional soaking step that takes from 2 to 5 days (with frequent change of soaking water) is not only used to remove the bitterness. Lupines alkaloids are reported to be anticholinergic and cause toxicity called lupines poisoning (Di Grande et al., 2004). Data from the National Health and Nutrition Examination Survey 2003–2010 showed that consumers who frequently consumed chickpeas or *hummus* had higher intake of dietary fiber; polyunsaturated fatty acids; vitamins A, vitamin E, vitamin C, folate; magnesium; potassium; iron and Healthy Eating Index-2015 (O'Neil, Nicklas, & Fulgoni III, 2014). In addition, the study showed that chickpeas/*hummus* consumers had lower total fat, saturated fatty acids and cholesterol. The total dietary fiber in *hummus* dish is 1.63% and the ratio of soluble

Fig. 2.1 Photographs present: (*1*) *Hummus*; (*2*) *Tabolah*; and (*3*) *Fattoush*

to insoluble dietary fiber is 1:2, which is a recommended ratio. During cooking of chickpeas in a pressure cooker (14.7 psi) for 40 min, there was no significant difference in the chemical composition was reported (de Almeida Costa, da Silva Queiroz-Monici, Reis, & de Oliveira, 2006).

Hummos dish is usually prepared fresh at homes or restaurants. However, it is currently available commercially as ready-to-eat dish, which is stored chilled at hypermarkets and it can be found canned which gives a longer shelf life. The preparation of *hummus* with a short shelf life has been linked to what is called "ingredient-driven" outbreaks. Generally, there is an association between ready-to-eat foods, which are stored chilled and foodborne botulism. A botulism poisoning case was reported in 2004 in UK linked to improper storage of *hummus* (Peck, Goodburn, Betts, & Stringer, 2008). In USA, food poisoning from *Salmonella enterica* (serotype Bovismorbificans which is rare in USA) infections was due to consumption of hummos was reported in three states (Centers for Disease Control and Prevention (CDC), 2012).

Several studies examined the effect of additives or treatments to extend the shelf life of *hummus* at chilling temperature. A combination of 1000 IU/g nisin and 0.3% citric acid was somewhat effective (~1.5-log reduction) in controlling the concentration of *Listeria innocua* and aerobic plate count for up to 6 days at 4°C (Al-Holy et al., 2006). Yamani & Mehyar (2011) reported the use of potassium sorbate, sodium benzoate and sodium metabisulfite as additives to extend the shelf life of *hummos* at 5°C. The most effective additive was Sodium metabisulfite, which extended the shelf life of *hummus* to 63 and 77 days when 0.05% and 0.1% was used, respectively. Recall of food products including *hummus* was also reported due to *Listeria monocytogenes* (Gray & Killinger, 1966).

Hummus is also commercially pasteurized which gives an extended shelf life compared to chilled ready-to-eat *hummus*. The industrial process is very similar to the conventional process. Chick peas are soaked in stainless steel tanks with water and sodium bicarbonate for around 10–14 h (Mchugh, 2016). Soaking will facilitate the removal of skin either manually or mechanically. Cooking can be done by boiling the water of soaked chickpeas for 30 min. However, according to Mchugh (2016), this results in inferior texture of *hummus*. The second option is to place the chickpeas in trays inside a cooker and cooked at 124°C for 30 min. Following cooking, the chickpeas will be drained and then grinded until becoming smooth (Mchugh, 2016). Next steps are mixing ingredients of *hummus* and then the mixture is pasteurized and packaged.

Tabolah (or *Tabbouleh*): *Tabolah* (Fig. 2.1, item 2) is a dish made mainly from finely chopped parsley and bulgur (couscous can be used instead of bulgur). Bulgur is a type of whole wheat durum grain. It is believed to have several nutrition advantages because it is unrefined and keep its content of bran and germ. This dish contains other ingredients, such as lettuce, tomatoes, mint, and onion and it is seasoned by olive oil, salt and lemon juice. According to Dashti, Al-Awadi, Sawaya, Al-Otaibi, & Al-Sayegh (2003), the ratio of saturated to polyunsaturated fatty acids is 1.3, which is similar to olive oil. Total dietary fiber in the dish is around 7.6%. The name of the dish can be misnamed with another dish called "Mutabal" which is made from eggplant.

Fattoush or *Fattoosh*: *Fattoush* salad (Fig. 2.1, item 3) is a mixture of different vegetables in addition to toasted or fried pieces of bread. Vegetables usually included are lettuces, cucumber, capsicum, radishes, tomatoes, mint leaves, and onion. However, the dish is not limited to these items. The salad could also contain feta cheese and sumac powder. Dashti et al. (2003) reported 1.8 mg/100 g cholesterol in *Fattoush* dish. The ratio of saturated to polyunsaturated fatty acids is 3.5. The difference between *Tabolah* and *Fattoush* is that the later contain Arabic bread, which had been fried in vegetable oil. The total dietary fiber is 6.8%.

Mutabal: *Mutabal* is a dip called "*baba ghannoush*". The eggplant is oven grilled and then it is peeled. It is blended in a food processor with sesame butter, garlic, and lemon juice. The creamy paste is then served in small dishes after being decorated with parsley, pomegranate, or mint leaves. The dietary fiber according to Dashti et al (2003) is around 4.4%. The dish can be found in multi-section hypermarkets and it has a limited shelf life at 4°C.

Legumes Dishes: Legumes have a good nutritional value and these are sources of many traditional foods in the Middle East. These are rich sources of carbohydrates, protein, vitamins and minerals (de Almeida Costa et al., 2006; Gowen et al., 2007). Many dishes, such as *hummus*, *falafel* and "*fol medames*" are prepared from legume. These are either cooked and served directly or used to make other dishes mentioned above. Fava beans and chickpeas are soaked overnight to rehydrates and then boiled until becoming tender. These are then served after the addition of salt, lemon, cumin or hot sauce according to individual's preferences.

Falafel: *Falafel* (Fig. 2.2) is a very popular dish in several Arab countries such as Jordan, Syria, Iraq, Lebanon, and Palestine. In some countries like Egypt and Sudan,

Fig. 2.2 A dish of *falafel*

it is known as *Taamia*. It can be made from chickpeas (Mediterranean countries) or fava beans (Egypt) or both. Vegetable burgers including *falafel* can be good alternatives to beef burgers (Kägi & Wüthrich, 1993). Ground beans are made into a dough with addition of other ingredients such as parsley, scallions, garlic, cumin, coriander and seam seeds. The dough is shaped into balls, which are deep-fried at 160–180°C for 6–8 min. Dashti et al. (2003) reported a cholesterol content of 3.57 mg/100 g of *falafel*, but no justification for the cholesterol content was given. Due to frying process, acrylamide was detected in *falafel* with a concentration of 3500 µg/kg (Al-Dmoor, Humeid, & Alawi, 2004). The dietary fiber is 5.07%. Another use of fava beans is in the *Ful Medames* dish, which is made from cooked and mashed fava beans topped with olive oil and cumin. Spices and herbs like garlic, parsley and coriander are added during mashing of the beans. In Iraq, the beans are called "*baqilla*" and there are some dished which are similar to the Egyptian mashed fava bean and can be cooked with rice. Dashti et al. (2003) reported a high ratio of saturated to polyunsaturated fatty acids in *Ful Medames*. The total dietary fiber is 6.81% (Dashti et al., 2003).

Lentil Soup: Lentil soup is a traditional soup especially in Mediterranean cuisine. The lentil is boiled in water and chili pepper, tomato, ginger, garlic, turmeric and salt are added. The mixture is simmered over medium-low heat with frequent stirring until it is thickened. Lentil soup is a good source of dietary fiber. Lentil is high in carbohydrates (56.4%) and protein (20.6%) (de Almeida Costa et al., 2006). It contains minerals such as iron, cobalt and iodine. The total dietary fiber was reported to be 4.9–6.83%, which is made from equal proportions of soluble and insoluble dietary fiber (Dashti, Al-Awadi, Khalafawi, Sawaya, & Al-Amiri, 2003).

During cooking of lentil beans for 20 min, the protein and carbohydrates content was not affected while the fiber content slightly decreased from 6.83 to 5.69% (de Almeida Costa et al., 2006). According to Abdel-Haleem and Omran (2014), a vegetarian soup mixture composed from lentil, green pea, chickpea, potato and barley; proved to be a good source of protein, iron and zinc and with good in vitro digestibility and availability.

Moussaka: The dish is very popular in Egypt and Levant countries but there are slight differences in the ingredients and preparation of this dish within the countries. The main ingredient of the m*oussaka* is eggplant. Eggplant and potato slices are fried in vegetable oil. Seasoned cooked ground beef is added between the layers of eggplant and potato slices and then backed in an oven. In Levant countries, the dish is cooked without meat but tomatoes is added and it may include chickpeas. The Egyptian form of this dish is left in the fridge for one day and then served. It is called "*Moussaka Barda*" because the word "*Barda*" literally mean "cold". *Moussaka* is also popular in Greek with similar ingredients and preparation procedures.

4.2 Spices, Seeds and Herbs (Condiments)

Orient is known for its aromatic flavoring. According to Roden (2008), different Arab countries have different spices/seeds mixtures that result in a specific and unique flavor. Spices can be sprinkled on dishes before serving such as salads or dips or used as a seasoning for dishes from meat and fish. Traditionally, the spices are bought as a whole and ground when needed in the house. Ground or crushed spices does not keep their flavor intensity for a long time compared to whole spices (Rathore & Shekhawat, 2008). Ground spices are stored in air-tight containers until needed. During preparation of dishes, the ground spices are toasted or roasted to increase the intensity of flavor (Raghavan, 2006). In some cases, the whole spices are toasted or roasted and ground and kept in the air-tight containers for further use (Arndt, 2017). In many instances, aroma or flavor to food is used during preparation of food dishes (Roden, 2008).

Sumac powder: Sumac is a general name for around 35 species of a plant in the genus *Rhus* of the family *Anacardiaceae*. The fruit of the species *Rhus coriaria* is famous in Middle East for its culinary and medicinal use (Sezik, Tabata, & Yesilada, 1991). Sumac powder is purple powder prepared by drying and grinding the fruits with addition of salt (Fazeli et al., 2007). It gives a tart and lemony taste. It is used for salads, dips, and some meat based dishes. Sumac powder extract showed antimicrobial activities against Gram-positive and Gram-negative bacteria (*Staphylococcus aureus*, *Bacillus cereus*, *Escherichia coli*, *Salmonella typhi*, *Proteus vulgaris*, and *Shigella flexneri*) at concentrations as low as 0.05% for Gram-positive bacteria. Gram negative bacteria needed higher concentrations of 0.2 % (Fazeli et al., 2007). Antimicrobial effect was also reported by Nasar-Abbas, Halkman, and Al-Haq (2004) and they mentioned that high sensitivity of gram-positive bacteria compared to gram-negative. The extract of *Rhus coriaria* was also reported to have antioxidant properties (Kosar, Bozan, Temelli, & Baser, 2007; Van Loo, De Bruyn, & Verzele, 1988).

Za'atar (in Arabic): In many Arab countries, the Arabic word refer to thyme (*Thymus vulgaris*). Thyme is one of the traditional herbs used for seasoning food and it is used as therapeutic. There is a popular drink called "*Za'atar* drink or *Za'ater* tea" although tea leaves are not part of the drink. It is made by simply boiling thyme leaves in water. In Arab cuisine, the word *Za'atar* also refer to a mixture made from wild thyme, sumac and sesame seeds (Kadi, 1994). Thyme volatile components were reported to have antioxidant activities comparable to those of known antioxidants such as α-tocopherol and butylated hydroxy toluene (BHT) (Lee, Umano, Shibamoto, & Lee, 2005). Thyme also has an antibacterial and antifungal activity (Paster et al., 1990). Addition of essential oil of thyme or thymol cause a significant decrease in the viable counts of *Salmonella typhimurium* (Juven, Kanner, Schved, & Weisslowicz, 1994). Thyme essential oils assist in increasing the shelf life of perishable foods such as chicken (Chouliara & Kontominas, 2006) and lamb meat (Karabagias, Badeka, & Kontominas, 2011). Thyme essential oil was found to have strong fungicidal effect (Rasooli & Abyaneh, 2004). Aqueous extracts were also found to have antibacterial effect against gram-positive and gram-negative bacteria and displayed high radical scavenging activity and reducing power (Martins et al., 2015).

Cumin: Cumin (*Cuminum cyminum*) is a flowering plant that is used as a spice for its distinctive flavour and aroma. Ground Cumin has an earthy, nutty, spicy taste with a hint of lemon (Spices INC, 2018). Some people might get confused between cumin and black cumin (*Nigella sativa*), caraway (*Carum carvi*) and nigella (*Nigella* spp). Cumin is used as a whole seed or as a powder. The unique aroma of cumin seeds is due to the volatile oil, which is at a concentration of 2–5%. The oil has an antihydrolytic effect and found to have strong antioxidant properties as compared to conventional synthetic antioxidants (Behera, Nagarajan, & Rao, 2004). The major volatile compounds in the cumin oil are cuminaldehyde (*p*-isopropyl benzaldehyde, *p*-cuminaldehyde), p-mentha-1,4-dien-7-al, cumin aldehyde, γ-terpinene, and β-pinene (Iacobelliset al., 2005; Varo & Heinz, 1970). Processing of cumin (such as roasting) affects the volatile compounds concentration. Behera et al. (2004) recommended an optimum roasting temperature of 125°C for 10 min. Antifungal and antiaflatoxigenic properties of *Cuminum cyminum* (L.) was reported by Kedia, Prakash, Mishra, and Dubey (2014). Cumin has antibacterial activity against several bacteria such as *Vibrio* spp. Strains (Hajlaoui et al., 2010), *Clavibacter*, *Curtobacterium*, *Rhodococcus*, *Erwinia*, *Xanthomonas*, *Ralstonia*, and *Agrobacterium* (Iacobellis, Lo Cantore, Capasso, & Senatore, 2005), *E. coli*, *S. aureus*, and *S. faecalis* (Allahghadri et al., 2010). According to Dhandapani, Subramanian, Rajagopal, and Namasivayam (2002), aqueous extract of *C. cyminum* can lower the blood glucose and plasma and tissue lipids in alloxan diabetic rats at low concentration doses. Antioxidant effect of cumin has been also reported (Allahghadri et al., 2010; Einafshar, Poorazrang, Farhoosh, & Seiedi, 2012).

Cinnamon: Cinnamon is obtained from several trees that belong to the genus Cinnamomum in the family *Lauraceous*. There are about 300 species of *Cinnamomum* but four species are used to obtain cinnamon and *Cinnamomum zeylanicum* is the most widely used (Jayaprakasha & Rao, 2011). When used in cuisine, it produce a pungent taste and smell. The major volatile compound is cinnamaldehyde and there are around 80 volatile compounds detected in the cinnamon essential oil. It has antioxidant, antimicrobial, and antidiabetic activities (Badei, Faheid, El-Akel, & Mahmoud, 2002; Jayaprakasha & Rao, 2011). Antioxidant activity of cinnamon from *Cinnamomum verum* (Mathew & Abraham, 2006) and from *Cinnamomum zeylanicum* (Mancini-Filho, Van-Koiij, Mancini, Cozzolino, & Torres, 1998) was also reported. *Cinnamomum zeylanicum* demonstrated beneficial effects as potential therapeutic agent for diabetes when studied in vitro or in vivo (Ranasinghe et al., 2012).

Cardamom (*Elettaria cardamomum*): In Arab tradition, this is a luxury spices and one of the most expensive spices. It gives the Arab Traditional coffee its distinct flavor and name: “Cardamom coffee”. It is also used in tea and rice dishes. Twenty three volatiles were reported by Husain & Ali (2014) with predominance of 1,8–cineole. Ten volatile components from *Elettaria cardamomum* were tested for their antibacterial effect and found effective against *Bacillus subtilis*, *Brevibacterium ammoniagenes*, *Staphylococcus aureus*, *Staphylococcus mutans*, *Propionibacterium acnés*, *Escherichia coli*, *Pseudomonas aeruginosa*, *Enterobacter aerogenes*, *Proteus vulgaris*, *Saccharomyces cerevisiae*, *Candida utilis*, *Plasmodium ovale*, *Trichophyton mentagrophytes*, and *Penicillium chrysogenum* (Kubo, Himejima, & Muroi, 1991). Immunomodulatory effects of *Elettaria cardamomum* aqueous extracts was reported by Majdalawieh and Carr (2010). They concluded that cardamom could be used as a therapeutic tool to regulate inflammatory responses and prevent/attenuate carcinogenesis.

Nutmeg: This seed or the powder obtained from *Myristica fragrans*. It has a warm, spicy aroma and flavor used in both sweet and savory cooking. Nutmeg extracts has several bioactive properties such as antioxidant (Piaru, Mahmud, Majid, & Nassar, 2012; Saeed et al., 2016), aphrodisiac activity (Tajuddin, Ahmad, Latif, & Qasmi, 2003), anticariogenic activity (Chung, Choo, Lee, & Hwang, 2006) and antiinflammatory activity (Ozaki, Soedigdo, Wattimena, & Suganda, 1989). Sonavane, Sarveiya, Kasture, and Kasture (2002) reported an anxiogenic activity of *M. Fragrans* seeds. The aqueous extracts of the plant was found to free radicals and has protection ability against lipid peroxidation. Antibacterial effect of *M. fragrans* against *Streptococcus mutans* was reported by Chung et al. (2006). Several extracts (acetone, ethanol, methanol, butanol and water extracts) has been evaluated for antioxidant activity, antimicrobial activities against *Bacillus subtilis*, *Staphylococcus aureus*, Pseudomonas putida, Pseudomonas aeruginosa and antifungal activity against *Aspergillus fumigatus*, *Aspergillus niger*, *Aspergillus flavus* (Gupta, Bansal, Babu, & Maithil, 2013). All extracts showed antioxidant and antimicrobial activity. The highest antioxoidant and antimicrobial activity was found with acetone extract. The possible compounds that contributed to this activity include Sabinene, b-Pinene, a-pinene, myristicin, isoeugenol, p-cymene, carvacrol, eugenol and b-caryophellene.

Nutmeg also found to possesses anti-inflammatory and anti-cancer activity (Olajide et al., 1999).

Turmeric (*Curcuma longa*): The Arabic word for this spice is "Curcum" which is similar to the active compounds in turmeric; curcuminoids. It is mainly used in powder form in rice dishes and curries for both its yellow colour and for its taste. It imparts a mustard-like, earthy aroma and pungent flavor (Chaturvedi, 2009). Curcuminoids from *C. Longa* such as curcumin, 4-hydroxycinnamoyl (feruloyl) methane and bis (4-hydroxycinnamoyl) methane were found to have antioxidant activity with the highest activity reported for curcumin (Toda, Miyase, Arichi, Tanizawa, & Takino, 1985). *Curcuma longoa* was found to exhibit anti-inflammatory (Ammon & Wahl, 1991), anti-cancer activities (Kuttan, Bhanumathy, Nirmala, & George, 1985), antioxidant (Reddy & Lokesh, 1992; Unnikrishnan & Rao, 1995), antiprotozoal activity (Araujo et al., 1999; Rasmussen, Christensen, Kvist, & Karazmi, 2000) and antibacterial activity (Araujo & Leon, 2001).

Fenugreek seeds: The scientific name for fenugreek is *Trigonella foenum-graecum*. The seeds have a strange bitter taste and a strong smell. The seeds are very popular for their functional and therapeutic properties. Hence, a dish called "*Hilbeh*" is prepared by boiling the seeds in water. Either the extract of the water or the water with the seeds is consumed. A paste also is made that contain fenugreek seeds. Fenugreek dishes are usually sweetened with honey to reduce the bitterness. It is one of main dishes that lactating women are given after giving birth (Yadav & Baquer, 2014).

In different cultures, food has always been used for its nutritive value and in many cases for its medicinal values (Halberstein, 2005). When looking at the traditional heritages, it might be difficult to find a sharp dividing line between food and drugs. In Arab countries, there are around 200–250 plant species are being used currently (Azaizeh, Saad, Khalil, & Said, 2006; Lev, 2002). Those substances are used for their content of the bioactive ingredients. Among them, 62% of those plants are spices commonly used in traditional food (Heneidy & Bidak, 2004). Azaizeh et al. (2006) reviewed the use of herbs as traditional medicine in the eastern region of the Mediterranean area and gave a thorough survey of plants, fruits, and seeds and their different uses in therapeutic remedies. However, the safety of the medicinal herbs is of a great concern especially with the lack of governmental regulation on the production and claims of herbal medication. The negative effects of the misuse of herbal medicines include hepatotoxicity, carcinogenicity, dermatologic effects, and effects on kidney, nervous system and blood (Saad, Azaizeh, Abu-Hijleh, & Said, 2006). In addition, some spices and herbs are known for their antimicrobial effect mainly for their essential oils (Arora & Kaur, 1999; Brandi, Amagliani, Schiavano, De Santi, & Sisti, 2006; Ceylan & Fung, 2004).

Ginger: Ginger is a flowering plant with binomial name *Zingiber officinale*. The rhizome or root (commonly named ginger) is widely used as a spice or for medicinal uses. It can be found commercially as powder or it can be used as fresh from grocery stores. Ginger is used worldwide to produce different food such as ginger drink (ginger bear), ginger tea (by steeping ginger in boiling water), and zinger candy. According to He, Li, Li, Guo, and Guo (2001), there are around 44 hexanextracted compounds in the essential oil of ginger rhizomes with terpenes being as the major constituent. The essential oil of ginger is constituted by high percentage of sesquiterpenes (67%), followed by monoterpenes (17 %) and aliphatic compounds (14%) (Sasidharan & Menon, 2010). Ginger is known for its antioxidant properties due to the presence of beta-carotene, ascorbic acid, terpenoids, alkaloids, and polyphenols such as flavonoids, flavones glycosides and rutin (Aruoma et al., 1997). Aqueous extract of ginger was found to poses anti-inflammatory and anti-thrombotic activity (Ali, Blunden, Tanira, & Nemmar, 2008). The antimicrobial activity of ginger against *Pseudomonas aeruginosa*, *Salmonella typhimurium*, *Escherichia coli* and *Candida albicans* was reported by Jagetia, Baliga, Venkatesh, & Ulloor (2003).

4.3 Arabian Cardamom Coffee: **Qahwah**

Coffee is one of the traditional hot drinks in the Arab countries and it is considered as a symbol of warm hospitality. The traditional method of making Arabic coffee start with roasting of coffee beans under wood fire. After toasting, the beans are ground to have coffee grounds. At this step, several spices can be added such as saffron, cinnamon, and cardamom. The latter is the most common spices added and from which the name "Arabic cardamom coffee" came. The ground spices are added to a hot water, which then is brought to boiling. Boiling is usually done in a special metal coffee pot which will be used later to serve the coffee. The suspension is boiled until it begin to forth. After boiling, it is left for another few minutes so coffee powder is settled down and coffee drink can be served. As it can be noticed from the preparation method, the coffee drink produced is strong and bitter. Sugar is not commonly added and if so, it is not with a quantity that can be perceived. Turkish coffee is prepared is a similar way but in two-steps boiling and sugar can be added significantly to have a sweet coffee drink. Coffee reading or coffee tasseography is type of fortune-telling that depends on the sediment left in a coffee cup after coffee being consumed. This is similar to tea-leaf reading which is practiced in different countries around the word. Coffee reading is popular in the Arabian in the popular heritage.

4.4 Rice Dishes

Rice is the staple food in the majority of Arab countries. It is not grown in the Arab countries and instead imported from India, Pakistan and Thailand. There are different ways to cook rice and accordingly there are different names for the rice based dishes. In Maghreb countries (Morocco, Algeria, Tunisia and Libya), the staple food is couscous which is derived from durum wheat. Both rice and couscous are served with meat or fish. Some rice dishes include lentil as one of the main ingredients.

Mansaf: *Mansaf* is a Jordanian cuisine from rice and spices served with lamb. Rice can be substituted with bulgur. The uniqueness of Mansaf is that the lamb is cooked in fermented dried yogurt, which is called *Jameed* (see: 6. dairy products). A broth is made from *Jameed* in which the pieces of lamb are cooked.

Koshari: In Egypt, a dish called *Koshari* is very popular and it is made from a mixture of rice, macaroni and lentil. The mixture is topped with chickpeas, tomato sauce, and crispy fried onion strips. The dish contain no animal derived ingredient and hence can be describe as vegan dish as far as no animal oil is used during preparation.

Mujaddara: *Mujaddara* (Fig. 2.3) is similar to the Egyptian *Koshari*, but from Levant countries. It is made from rice and lentil and topped with crispy fried onion strips. The dish is seasoned with cumin, coriander, or mint. Like in *Koshari*, the dish is vegan as far as no animal oil is used during preparation.

Magloba: *Magloba* is a rice dish in Levant countries where the name means "upside-down". The dish originated from Palestine and spreaded to the Levant countries. The cooking of *Magloba* involves cooking the meat or chicken and vegetables in a cooking pot until these are fully cooked, and then adding a half-cooked rice on top and leave all under heat until rice is fully cooked. When it is ready, the pot is flipped upside down so the rice will be at the bottom of the dish and the meat/chicken is at the top.

Biryani: *Biryani* rice is a popular dish in several countries in Asia and each country/region has its own preparation method and ingredients. In addition to rice, meat used in the form of either chicken, mutton, beef, prawn or fish. *Biryani* is also found in India, Pakistan, Afghanistan, and Bangladesh. In Arab countries, Iraqi *biryani* is

Fig. 2.3 Mujaddara dish decorated with capsicum and reddish

Fig. 2.4 Stuffed grape leave "*Mahshi*"

popular where lamb or chicken is used with the rice. The dish it topped with fried onion strips, fried potato cubes, almonds and raisins.

The preparation of some rice dishes involve cooking the rice and the meet together in a traditional oven called *tandoor*. Traditionally, *tandoor* is a cylindrical clay oven that is submerged in the ground. The heat source is charcoal or wood fire in a specific cavity at the bottom of the oven. The rice, meat and any other ingredients (such as vegetables) are stacked in layers at which the meet is in top of the rice. This allows the meat to cook and cause its juices to drip on the rice giving it a unique taste. *Quzi* (also called Qoozi or Ghozi) is a popular example in the Levant countries. The preparation method involves stuffing a whole lamb with rice, vegetables and spices. The whole lamb is then cooked slowly at moderate heat in an oven. In Yemen, the dish is called *Haneeth* where rice is cooked under pressure using a young-small size lamb.

Rice is also used in Dolma, a family of stuffed foods. The name in Arabic is "*Mahshi*" as singular and "*Mahashi*" as plural (Fig. 2.4). Types of food used in stuffing include rice-stuffed grape leaves, rice-stuffed sweet pepper, rice-stuffed cabbage, rice-stuffed Zucchini. In such type of foods, it must be noted that bulgur replace rice is some dishes. Figure shows one type of *Mahshi* which is grape leaves *Mahshi* (stuffed grape leaves).

4.5 Bread

Bread is an important food item in human diet as a source of protein and carbohydrates. It can be considered as a staple food in the Arab region. Flat bread is the common type of bread in the Arab countries and it is called "*Khebz*" in Arabic. Different countries have their own different style of bread preparation either unleavened or leavened with baking powder. Hence, there are different names for bread, which is made basically form the same main ingredients. Some types of bread take their names from the originating country, like Yemeni bread from Yamen and Lebanese bread form Lebanon. Other types are named according to the way of cooking like the *Tannoor* bread. The bread is usually baked in a traditionally made oven called "*Tandor*". One of the unique methods of bread preparation is spreading a thin sheet of dough on a heated metal surface. Another type of bread is the *Saj* bread or the *Markook* or *Shrak* bread. The naming—*Saj*—came from the way of preparation. The bread is cooked in a domed or convex metal griddle, called Saj. A popular bread is called "*Taboon* bread". The name came from the oven used to bake the bread, which is called "*Taboon* oven". The oven is truncated cone oven made from clay. There is an opening at the bottom to stoke the fire. Traditionally, dried tree branches, tree trimmings, wood chips are used as a fuel. Nowadays, Arabic bread or as called in some countries "*lebanes* bread" is very common in restaurants and coffee shops. It is made in automated or semi-automated bakeries that supply the catering service their needs of bread.

5 Meat and Fish Dishes

Meat (especially lamb) is very important in the diet of this region for families that can afford it. Chicken is generally less expensive (Roger, 2000). The most popular types of meat used in the Arab countries are lamb, goat and chicken. Beef and camel meet is occasionally used while pork is not used because of religious restrictions. Slaughtering of animals needs to follow specific procedures and conditions based on Islamic law. The meat that is obtained following those procedures and conditions is called Halal meat, which has the same meaning of permissible and lawful.

Hawawshy: *Hawawshy* is a type of sandwich that is mainly prepared from minced meat with other ingredients like onion, spices, green pepper; stuffed in pita bread and then oven baked (Ismail, 2006).

Shawerma: *Shawerma* is a very popular street food made from either chicken, lamb or beef. It is not commonly prepared in homes. Fig. 2.5 shows chicken (left) and meat shawerma (right). The food is special in the way of preparation. The marinated meat cuts are skewered in a vertical rotating roasting fork (Spit) and roasted by radiant heat over time. This method is called Rotisserie. The meat is skewered in a large diameter (around 30–60 cm) so the meat is cooked slowly. The temperature is controlled so the outer layer is roasted while the inner part stay uncooked. The

Fig. 2.5 *Shawerma* meat skewed in a vertical rotating roasting fork

outer layer which has been roasted is shaved and shredded (Ahmed, El-Hakem, & Ibrahim, 2015) and made as a sandwich with lettuce and tomatoes and tahini (see Sect. 4.1) or served with *hummos* in a dish called *Hummus shawerma*. The sanitary quality of shawerma is a concern. In a study by Elfaki and Elhakim (2011), the source of shawerma samples (from Sudan market) showed significant effect in term of total counts, coliforms and pathogenic bacteria which confirm the importance of hygiene and good handling of meat during storage and preparation. *Escherichia coli*, *Salmonella* spp, *Klebsiella* spp, *Citrobacter* spp and *Proteus* spp were isolated from 20%, 4%, 12%, and 8% of shawerma samples from Egyptian market (Al-Mutairi, 2011). According to Ahmed et al. (2015), samples shawerma from Egyptian market had 45 % moisture, 25 % carbohydrates, 18% protein and 10% fat.

Kibbeh (or *kebbah*): A type of food made by mixing ground bulgur and minced lamb or beef meat, which are made as a dough. The dough is then made into small balls or oval shapes and stuffed with stirred fried ground meat, onion, nuts and spices (cinnamon, nutmeg, clove). The balls then deep fried at 160–180 °C for 8–10 min in vegetable oil resulting in gritty crust and brown surface. Acrylamide present in *kibbeh* was observed as of 3600 $\mu g\ kg^{-1}$ (Al-Dmoor et al., 2004). In Iraq, the *kibbeh* is made from rice instead of bulgur. There are also variations from one Arab country to another in term of the ingredients of the stuffing. The *kibbeh* balls are usually served with yogurt or Tahinah as a dip. There is a very popular *kibbeh* dish in Syria and Lebanon called "*kibbeh nayeh*" which can be translated to raw/uncooked kibbeh. The fresh meat is minced twice in a mincer and then mixed with minced onion and spices. The mixture looks like a paste of meat and it is served without any further cooking. This poses a food safety hazard from pathogenic bacteria that can be contaminated meat. Raw *kibbeh* sampled from Uberlândia, Brazi, was found to contain thermotolerant coliforms (40 % of samples) and *Salmonella* spp. (30 % of samples). *Klebsiella*, *Enterobacter*, *Serratia liquefaciens* and *Hafnia alvei* were also isolated from the samples (Carvalho, Queiroz, & Silva, 2017).

Musakhan: Musakhan is considered as the national dish of Palestine and consists of roasted chicken baked with onions, sumac, saffron, and decorated with fried pine nuts. Each piece of chicken is then placed in a piece of traditional bread called taboon bread, which is a type of flatbread wrap.

Kofta: Kofta is another dish made with minced meat or chicken. Minced onion, garlic, parsley and spices are added to the minced meat and then mixed together to make a dough. The dough is then formed into small balls or rods. The balls are roasted for a short period. The balls are then cooked in a tomato sauce. Kofta is usually served with tahinah as a dip. The safety of dishes prepared from minced meat is a global concern. *Escherichia coli*, *Salmonella* spp, *Klebsiella* spp, *Citrobacter* spp, and *Proteus* spp were isolated from kofta samples from Egyptian market (Al-Mutairi, 2011).

Fesikh: One of the very traditional foods in Egypt is the fish dish called "*Fesikh*". It is characterized as a wet and soft fish product with a strong pungent smell and a shiny silvery appearance (Essuman, 1992; Mostafa & Salem, 2015). The food is a celebrity dish from fermented and salted fish. The most popular fish used for this dish is Mullet of the genus *Mugil*. The fish preparation is done by families who are specialized in this dish and who pass the preparation experience through generations (e.g., from father to sun). The person prepares this food is called "*Fasakhani*" which means the one who prepare the *fesikh*". *Fesikh* is consumed during a festival called "*Sham AlNasim*" which marks the beginning of the spring. *Fesikh* can be prepared with low salt content and hence its shelf life is 15–20 days, or can have high salt content with an increased shelf life of 2 – 3 months (Mostafa & Salem, 2015). There are concerns about the safety of *fesikh* following different report of outbreaks linked to the consumption of the food. In April 24, 2012, the Canadian Food Inspection Agency (CFIA) warned people not to consume *fesikh* which had being sold in the local markets of Toronto, Canada (Walton et al., 2014, accessed March 4, 2018). In 17th and 18th of April 2012, an outbreak of type E botulism was reported in Toronto, Canada as a result of consumption of *fesikh* (Walton et al., 2014).

Kebab and Tikka: The word kebab refers to a variety of meat dishes that are grilled. The dish originated from the Middle East and now popular in many counties even with slight difference names such as *Shashlik* (in Europe) or *Chuanr* (in China). In Arabic counties, *Kebab* refers to marinated minced meat, skewered in wood or metal skewers and charcoal grilled. In some countries, *Kebab* also refer to meat cubes that are prepared same way. In Arabic countries, the meat cubes skewers are called *Tikka*. The common meat used is mutton, beef and chicken. When

Fig. 2.6 A dish of kebab (*1*) Beef or mutton tikka, (*2*) Chicken tikka, (*3*) Beef or mutton kebab, and (4) Chicken kebab

ordering *kebab* or *tikka* in an Arabic restaurant, meat always understood as mutton or beef (i.e. ordering meat *kebab* means a dish of beef or mutton, whatever available). If prefer to have chicken *kebab*, then costumer needs to specify. Fig. 2.6 shows meat *kebab and tikka* served with French fries and Arabic flat bread. Traditionally flat bread and fresh sliced onion are served with *Kebab*, however recent trend is to serve with potato fries.

It worth mentioning that the word kebab is used in other counties in the world to denote either minced meat or meat cubes. In some countries, *Kebab* is not grilled, it is prepared by stir frying followed by simmering in a curry sauce. Because of the widespread of kebab around the word, it is common to see a prefix before the word "*kebab*" which represent the type of kebab according to the country such as *kofta kebab*, *lyulya kebab, chopan kebab, kakori kebab,* etc. (Kraig & Sen, 2013).

6 Dairy Products

Arab people depend on milk and milk products, which are obtained from livestock they raise. Milk is obtained from cows, goats or sheeps. In many countries, milk is traditionally consumed in its raw condition without any heat treatment to destroy bacteria, which poses food poisoning risk. There are also several types of Arab traditional food products produced by fermentation. Arab people produce various types of cheese from cow or goat milk. *Karish* or *Kareish* cheese is a type of cheese made from skimmed milk via lactic acid bacteria fermentation. Traditionally, the milk to be used for cheese making is not pasteurized. The result is unique flavor produced due to the present of several species of lactic acid bacteria.

Labneh (concentrated yogurt): Labneh is consumed widely in Jordan, Middle East and Mediterranean diet (Ibrahim, 1996; Mazahreh, 2009). It is considered as soft cheese made from strained yogurt, which make it thicker than yogurt.

Nabulshi cheese: *Nabulshi* cheese is produced by renneting raw sheep milk or a blend of sheep and goat milk without starter culture (Mazahreh, 2009).

Jameed: *Jameed* is a hard and dry fermented yogurt. It is prepared by boiling of sheep or goat's milk, which is then left to dry and ferment for few days with daily addition of salt. The mixture becomes thick and it is formed into round balls. The drying should be enough to make a shelf stable product and this is why the balls are very hard and looks like cheese balls.

Milk cream: Milk cream is also prepared traditional in different Arab countries like Syria, Lebanon and Egypt. One popular cream is called *Halabi* cream (*Halab* literally means Aleppo, a Syrian city).

Ghee: Ghee is a clear, still and solid fat, which is made from cow or goat milk. The Arabic name for the ghee is "*semneh*". Preparation involves production of butter by a traditional butter churn. Butter is then simmered to obtain the ghee. During simmering, impurities are skimmed from the surface. Simmering is what make ghee different from butter. It results in a golden brown color, nutty taste and unique aromatic flavor. After simmering, the liquid part (which will later solidify at room temperature) is collected and the residues that settled down the bottom are discarded. It is reported that ghee made from buffalo milk is more prone to oxidative deterioration than that made from cow's milk.

7 Sweet of the Arab Countries

In Arabic language, sweet is called "*Helo*" and the plural is "*Halawyat*". Sweets are very common in the Arab nation meals. During occasions, parties, or ceremonies, sweets have a special place and attendance. Sweets are also part of the hospitality with tea, coffee or a cold drink. Sweets are also presented as gifts when visiting a family at their house.

Ghuraiba: The sweet is common in the Lavent countries. The same sweet is called "*Shakrulma*" in Iraq. It is made by wheat flour, icing sugar, cooking oil and cardamom, which give the sweets its unique flavor. All ingredients are mixed to make a dough. The dough is then formed into small balls and each ball is pressed gently by hand to form a disc of 1 cm high and 2 cm diameter. Each discs can be decorated by an almond half, or a piece of pistachio. The discs are baked in an oven at around 180°C and they are left until a slight change in the colour can be noticed.

Zalabia: *Zalbia* is a very popular sweet in all Arab countries. It has the shape of a twisted knot, pretzel. It is made from flour, starch, butter and sugar which are mixed together and deep fried into pretzels. After frying, the pretzels are coated with a syrup. The texture of *Zalabia* is chewy and that might be because of the outer layer of crystalized sugar.

Fig. 2.7 *Konafa* sweet

Konafa (shredded pastry with cheese): *Konafa* (Fig. 2.7) is a very popular sweet in Levant countries. It is made from shredded dough or vermicelli noodles. The pastry is heated in butter or margarine. Traditionally, ghee from cow milk (called *semneh* in Arabic) is used and it gives the food a unique flavor. A layer of cheese is sandwiched between two layers of pastry and then baked in an oven. After baking, a sugar-based syrup is poured at the top of the *konafa* and it is decorated with pistachio powder.

Baglawa: It is a sweet made from Filo (or phyllo) dough, which is a dough that is used to make some pastries. A paste is made from grinded roasted pistachios, sugar and rose water. The paste is then sandwiched between two layers of filo. The top layer is smeared with butte or Ghee (*Semneh*) and then backed. After baking, syrup is added to the top layer of the baklawa and then cut into small cubes.

Samsamia: *Samsamia* is a sweet based on sesame. The sesame is roasted until it turns golden brown. Sugar and honey or molasses are caramelized by heating with ginger and/or cinnamon and left. The roasted sesame is added to the caramelized sugar and formed into a thin sheet that is cut into small slabs.

Basbosa or *Basbousa*: *Basbosa* is a very popular sweet cake. It is made from the dough of semolina with sugar, milk, butter (or ghee). The dough is spread in an oven tray and sliced into small slabs and each slab is decorated with sliced almond. The dough is then baked in an oven. After baking, basbosa is sweetened with a vanilla-flavored syrup.

Balah Ash-sham: The name literally mean "Date fruit of Levant". *Balah* literally means date fruit and *Ash-sham* means Levant. The sweet names came from its shape, which looks like date fruits and originated from Levant countries. In Iraq, it is known as "*Datli*". It is made from flour and starch, which are used to make a paste in addition to water and vegetable oil. The ingredients of the paste are mixed at high temperature and then left to cool down when the paste has the desired consistency. Then egg and vanilla flavor are added. The dough is then formed into the previously mentioned shape (date fruit shape) and fried in frying oil. After frying, they are dipped in syrup and they are ready for serving.

8 Safety of Arab Traditional Foods

Food safety is a major challenge in the Arab counties. Several factors can make the food safety systems ineffective. One of the challenges is to control both locally produced foods and imported foods under the same standards (Al-Kandari & Jukes, 2009). It has also been reported that food from street vendors is a major issue (Al-Kandari & Jukes, 2009; Idriss & El-Habbab, 2014). In addition, food safety inspection covers major cities more than rural areas and hence the safety of food is not comparable. In relation to traditional foods, the safety challenges can more pronounced. In general, the practices, knowledge and awareness about food safety is low in the Arab countries. Traditional foods are commonly made by family-owned small business that try to help the family to get some income. There is no proper link between the business and other food chain elements such as ingredient suppliers, retailer, etc.

In many cases where the food business is run from home, food safety authorities have no access to inspect and control the quality and safety of the production. Public awareness about food safety is crucial in order to achieve high level of food safety and this is true when talking about production of traditional foods at home levels. The level of awareness depends on several factors including the level of income and the level of education. Due to low awareness or the neglect of food safety control measures, several infractions may happen during preparation of traditional food at homes: the quality of raw materials (Ismail, 2006), Lack of overall hygiene, improper food handling, Improper transportation of food (e.g., unlicensed vehicles), Unsuitable storage facilities, lack if insects and pests control systems) (Alsayeqh, 2010).

Several outbreaks have been reported as a result of traditional food consumption such as pickled vegetables (Barari & Kalantar, 2010) and soup from sour yogurt (Keramat, 2000) in Iran, chicken noodles in Lebanon (Hanna, Adib, & Daoud, 2009). A massive outbreak of 91 patients was reported in Egypt as result of salted fish. Patients suffered from botulism, which caused the death of 20% of patients (Weber et al., 1993). Several foodborne outbreak and/or recall issues has been linked to infractions in handling and processing of food products (Hamlin, 1997; Liu, 2007). FAO and WHO (Codex Alimentarius Commission) has issued recommendations for street vendors that cover different issues such as ingredients, cooking and handling of food, transportation, food storage and hygiene control, water and also personal health. (FAO/WHO, 1999). During festivals and fairs, food poisoning can results from bacteria that can contaminate the food through raw materials, cooking utensils, poor people hygiene. Organizers of such events need to pay a special attention to this issue because it is of a great challenge (Chan, 2014).

Acknowledgment The author would like to thank the instagram account *"#oman_restaurant_guide"* for providing the photos of the traditional foods. Special thanks to Mr. Salem Al-Uraimi.

References

Abdel-Haleem, A. M. H., & Omran, A. A. (2014). Preparation of dried vegetarian soup supplemented with some legumes. *Food and Nutrition Sciences, 5*(22), 2274.

Aboul-Enein, B. H. (2016). The Mediterranean-Style diet: Historical and contemporary perspectives. *Journal of Nutritional Biology, 2*(1), 73–74. Retrieved from http://researchonline.lshtm.ac.uk/2965173/

Ahmed, A. M., El-Hakem, N. A. B., & Ibrahim, G. A. (2015). Chemical and microbial assessment of beef and chicken shawarma sandwiches in Ismailia governorate and its impact on consumer health. *Egyptian Journal of Chemistry and Environmental Health, 1*(1), 686–693.

Al-Dmoor, H. M., Humeid, M. A., & Alawi, M. A. (2004). Investigation of acrylamide levels in selected fried and baked foods in Jordan. *Journal of Food Agriculture and Environment, 2*, 157–165.

Al-Holy, M., Al-Qadiri, H., Lin, M., & Rasco, B. (2006). Inhibition of Listeria innocua in hummus by a combination of nisin and citric acid. *Journal of Food Protection, 69*(6), 1322–1327.

Ali, B. H., Blunden, G., Tanira, M. O., & Nemmar, A. (2008). Some phytochemical, pharmacological and toxicological properties of ginger (Zingiber officinale Roscoe): a review of recent research. *Food and Chemical Toxicology, 46*(2), 409–420.

Al-Kandari, D., & Jukes, D. J. (2009). A situation analysis of the food control systems in Arab Gulf Cooperation Council (GCC) countries. *Food Control, 20*(12), 1112–1118.

Allahghadri, T., Rasooli, I., Owlia, P., Nadooshan, M. J., Ghazanfari, T., Taghizadeh, M., & Astaneh, S. D. A. (2010). Antimicrobial property, antioxidant capacity, and cytotoxicity of essential oil from cumin produced in Iran. *Journal of Food Science, 75*(2), H54–H61.

Al-Mutairi, M. F. (2011). The incidence of Enterobacteriaceae causing food poisoning in some meat products. *Advance Journal of Food Science and Technology, 3*(2), 116–121.

Alsayeqh, A. F. (2010). *Possible factors for food safety infraction and fraud continuity in restaurants in Saudi Arabia*. Retrieved SFDA's website http://www.sfda.gov.sa from 11.

Ammon, H. P. T., & Wahl, M. A. (1991). Pharmacology of Curcuma longa. *Planta Medica, 57*(01), 1–7.

Araujo, C. A. C., Alegrio, L. V., Gomes, D. C. F., Lima, M. E. F., Gomes-Cardoso, L., & Leon, L. L. (1999). Studies on the effectiveness of diarylheptanoids derivatives against Leishmania amazonensis. *Memórias do Instituto Oswaldo Cruz, 94*(6), 791–794.

Araujo, C. A. C., & Leon, L. L. (2001). Biological activities of Curcuma longa L. *Memórias do Instituto Oswaldo Cruz, 96*(5), 723–728.

Arndt, A. (2017). *Seasoning savvy: How to cook with herbs, spices, and other flavorings*. Boca Raton: CRC Press.

Arora, D. S., & Kaur, J. (1999). Antimicrobial activity of spices. *International Journal of Antimicrobial Agents, 12*(3), 257–262.

Aruoma, O. I., Spencer, J. P. E., Warren, D., Jenner, P., Butler, J., & Halliwell, B. (1997). Characterization of food antioxidants, illustrated using commercial garlic and ginger preparations. *Food Chemistry, 60*(2), 149–156.

Azaizeh, H., Saad, B., Khalil, K., & Said, O. (2006). The state of the art of traditional Arab herbal medicine in the Eastern region of the Mediterranean: A review. *Evidence-based Complementary and Alternative Medicine, 3*(2), 229–235.

Badei, A. Z. M., Faheid, S. M. M., El-Akel, A. T. M., & Mahmoud, B. S. M. (2002). Application of some spices in flavoring and preservation of cookies: 2-Antimicrobial and sensory properties of cardamom, cinnamon and clove. *Deutsche Lebensmittel Rundschau, 98*(7), 261–265.

Barari, M., & Kalantar, E. (2010). An outbreak of type A and B botulism associated with traditional vegetable pickle in Sanandaj. *Iranian Journal of Clinical Infectious Diseases, 5*, 111–112.

Behera, S., Nagarajan, S., & Rao, L. J. M. (2004). Microwave heating and conventional roasting of cumin seeds (Cuminum cyminum L.) and effect on chemical composition of volatiles. *Food Chemistry, 87*(1), 25–29.

Borchani, C., Besbes, S., Blecker, C. H., & Attia, H. (2010). Chemical characteristics and oxidative stability of sesame seed, sesame paste, and olive oils. *Journal of Agricultural Science and Technology, 12*, 585–596.

Brandi, G., Amagliani, G., Schiavano, G. F., De Santi, M., & Sisti, M. (2006). Activity of Brassica oleracea leaf juice on foodborne pathogenic bacteria. *Journal of Food Protection, 69*(9), 2274–2279.

Carvalho, P. R., Queiroz, L. L., & Silva, H. A. (2017). Contamination of raw kibbeh as important public. Paper presented at the 29th Brazilian Congress of Microbiology, 22nd–25th October 2017, Foz do Iguaçu, Brazil.

Cayot, N. (2007). Sensory quality of traditional foods. *Food Chemistry, 102*(2), 445–453.

Centers for Disease Control and Prevention (CDC). (2012). Multistate outbreak of Salmonella serotype Bovismorbificans infections associated with hummus and tahini—United States, 2011. *Morbidity and Mortality Weekly Report (MMWR), 61*(46), 944.

Ceylan, E., & Fung, D. Y. C. (2004). Antimicrobial activity of spices. *Journal of Rapid Methods & Automation in Microbiology, 12*(1), 1–55.

Chan, W. E. I. L. (2014). Food safety at fairs and festivals. In R. Ghiselli, B. Almanza, & M. Khan (Eds.), *Food safety: Researching the hazard in hazardous foods* (p. 341). Boca Raton: CRC Press.

Chaturvedi, T. P. (2009). Uses of turmeric in dentistry: An update. *Indian Journal of Dental Research, 20*(1), 107.

Chouliara, I., & Kontominas, M. G. (2006). Combined effect of thyme essential oil and modified atmosphere packaging to extend shelf life of fresh chicken meat. *Recent Progress in Medicinal Plants: Natural Products, 15*, 423–442.

Chung, J. Y., Choo, J. H., Lee, M. H., & Hwang, J. K. (2006). Anticariogenic activity of macelignan isolated from Myristica fragrans (nutmeg) against Streptococcus mutans. *Phytomedicine, 13*(4), 261–266.

Dashti, B., Al-Awadi, F., Khalafawi, M. S., Sawaya, W., & Al-Amiri, H. (2003). Soluble and insoluble dietary fibre in thirty-two Kuwaiti dishes. *Food Chemistry, 83*(4), 557–561.

Dashti, B., Al-Awadi, F., Sawaya, W., Al-Otaibi, J., & Al-Sayegh, A. (2003). Fatty acid profile and cholesterol content of 32 selected dishes in the state of Kuwait. *Food Chemistry, 80*(3), 377–386.

de Almeida Costa, G. E., da Silva Queiroz-Monici, K., Reis, S. M. P. M., & de Oliveira, A. C. (2006). Chemical composition, dietary fibre and resistant starch contents of raw and cooked pea, common bean, chickpea and lentil legumes. *Food Chemistry, 94*(3), 327–330.

Dhandapani, S., Subramanian, V. R., Rajagopal, S., & Namasivayam, N. (2002). Hypolipidemic effect of Cuminum cyminum L. on alloxan-induced diabetic rats. *Pharmacological Research, 46*(3), 251–255.

Di Grande, A., Paradiso, R., Amico, S., Fulco, G., Fantauzza, B., & Noto, P. (2004). Anticholinergic toxicity associated with lupin seed ingestion: Case report. *European Journal of Emergency Medicine, 11*(2), 119–120.

Einafshar, S., Poorazrang, H., Farhoosh, R., & Seiedi, S. M. (2012). Antioxidant activity of the essential oil and methanolic extract of cumin seed (Cuminum cyminum). *European Journal of Lipid Science and Technology, 114*(2), 168–174.

Elfaki, A. E., & Elhakim, S. A. (2011). Quality evaluation of two Sudanese street foods of animal origin. *Advance Journal of Food Science and Technology, 3*, 219–223.

Essuman, K. M. (1992). *Fermented fish in Africa: A study on processing, marketing and consumptione* (FAO Fisheries Technical Paper No.329). FAO: Rome.

FAO/WHO. (1999). Revised regional guidelines for the design of control measures for StreetVended foods in Africa, codex Alimentarius commission, CAC/GL-22 - Rev. 1 (1999).

Faris, M. A.-I. E., & Takruri, H. R. (2003). Study of the effect of using different levels of tahinah (sesame butter) on the protein digestibility-corrected amino acid score (PDCAAS) of chickpea dip. *Journal of the Science of Food and Agriculture, 83*(1), 7–12.

Fazeli, M. R., Amin, G., Attari, M. M. A., Ashtiani, H., Jamalifar, H., & Samadi, N. (2007). Antimicrobial activities of Iranian sumac and avishan-e shirazi (Zataria multiflora) against some food-borne bacteria. *Food Control, 18*(6), 646–649.

Forum. (1999). Pew Research Center's Forum on Religion & Public Life, MAPPING THE GLOBAL MUSLIM POPULATION: A Report on the Size and Distribution of the World's Muslim Population, October 2009.

Gowen, A., Abu-Ghannam, N., Frias, J., & Oliveira, J. (2007). Modelling the water absorption process in chickpeas (*Cicer arietinum* L.)—The effect of blanching pre-treatment on water intake and texture kinetics. *Journal of Food Engineering, 78*(3), 810–819.

Gray, M. L., & Killinger, A. H. (1966). Listeria monocytogenes and listeric infections. *Bacteriological Reviews, 30*(2), 309.

Gupta, A. D., Bansal, V. K., Babu, V., & Maithil, N. (2013). Chemistry, antioxidant and antimicrobial potential of nutmeg (Myristica fragrans Houtt). *Journal of Genetic Engineering and Biotechnology, 11*(1), 25–31.

Hajlaoui, H., Mighri, H., Noumi, E., Snoussi, M., Trabelsi, N., Ksouri, R., & Bakhrouf, A. (2010). Chemical composition and biological activities of Tunisian Cuminum cyminum L. essential oil: A high effectiveness against Vibrio spp. strains. *Food and Chemical Toxicology, 48*(8–9), 2186–2192.

Halberstein, R. A. (2005). Medicinal plants: Historical and cross-cultural usage patterns. *Annals of Epidemiology, 15*(9), 686–699.

Hamlin, S. (1997). Hummus maker's recall widened. *The New York Times*, New York, June 25, 1997. https://www.nytimes.com/1997/06/25/garden/hummusmaker-s-recall-widened.html, accessed 18 Mar 2018.

Hanna, N. M., Adib, S. M., & Daoud, Z. (2009). Food-borne salmonella outbreak at a bank cafeteria: An investigation in an Arab country in transition. *Eastern Mediterranean Health Journal, 15*, 470–474.

He, W., Li, L., Li, Y., Guo, S., & Guo, B. (2001). GC-MS analysis of different solvent extracts of ginger. *Journal of Tropical and Subtropical Botany, 9*(2), 154–158.

Heneidy, S. Z., & Bidak, L. M. (2004). Potential uses of plant species of the coastal Mediterranean region, Egypt. *Pakistan Journal of Biological Sciences, 7*, 1010–1023.

Husain, S. S., & Ali, M. (2014). Analysis of volatile oil of the fruits of Elettaria cardamomum (L.) maton and its antimicrobial activity. *World Journal of Pharmacy and Pharmaceutical Sciences, 3*(2), 1798–1808.

Hwalla, N., Bahn, R., & El Labban, S. I. (2015, Nov 16). Sustainable food consumption in Arab countries. Arab environment 8: Sustainable consumption for better resource management. In A. Gelil & Saab (Eds.), *2015 report of the Arab forum for environment and development* (pp. 134–171).

Iacobellis, N. S., Lo Cantore, P., Capasso, F., & Senatore, F. (2005). Antibacterial activity of Cuminum cyminum L. and Carum carvi L. essential oils. *Journal of Agricultural and Food Chemistry, 53*(1), 57–61.

Ibrahim, S. A., Al-Khatib, K. A., Al-Haik, S. A., & Daguri, M. H. (1996). Labneh 'Concentrated yogurt': A traditional fermented dairy product in Jordan. *Dairy, Food and Environmental Sanitation, 16*, 296–298.

Idriss, A. W., & El-Habbab, M. S. (2014). Middle East food safety perspectives. *Journal of the Science of Food and Agriculture, 94*(10), 1922–1927.

Ismail, S. A. (2006). Microbiological quality of hawawshy consumed in Ismailia, Egypt. *Journal of Food Safety, 26*(4), 251–263.

Jagetia, G. C., Baliga, M. S., Venkatesh, P., & Ulloor, J. N. (2003). Influence of ginger rhizome (Zingiber officinale Rosc) on survival, glutathione and lipid peroxidation in mice after whole-body exposure to gamma radiation. *Radiation Research, 160*(5), 584–592.

Jayaprakasha, G. K., & Rao, L. J. M. (2011). Chemistry, biogenesis, and biological activities of Cinnamomum zeylanicum. *Critical Reviews in Food Science and Nutrition, 51*(6), 547–562.

Johnston, J. L., Fanzo, J. C., & Cogill, B. (2014). Understanding sustainable diets: a descriptive analysis of the determinants and processes that influence diets and their impact on health, food security, and environmental sustainability. *Advances in Nutrition, 5*(4), 418–429.

Juven, B. J., Kanner, J., Schved, F., & Weisslowicz, H. (1994). Factors that interact with the antibacterial action of thyme essential oil and its active constituents. *Journal of Applied Microbiology, 76*(6), 626–631.

Kadi, J. (1994). *Food for our grandmothers: Writings by Arab-American and Arab-Canadian feminists*. Boston, MA: South End Press.

Kägi, M. K., & Wüthrich, B. (1993). Falafel burger anaphylaxis due to sesame seed allergy. *Annals of Allergy, 71*(2), 127–129.

Kahyaoglu, T., & Kaya, S. (2006). Modeling of moisture, color and texture changes in sesame seeds during the conventional roasting. *Journal of Food Engineering, 75*(2), 167–177.

Kamal-Eldin, A., & Appelqvist, L. Å. (1994). Variation in fatty acid composition of the different acyl lipids in seed oils from four Sesamum species. *Journal of the American Oil Chemists' Society, 71*(2), 135–139.

Karabagias, I., Badeka, A., & Kontominas, M. G. (2011). Shelf life extension of lamb meat using thyme or oregano essential oils and modified atmosphere packaging. *Meat Science, 88*(1), 109–116.

Kedia, A., Prakash, B., Mishra, P. K., & Dubey, N. K. (2014). Antifungal and antiaflatoxigenic properties of Cuminum cyminum (L.) seed essential oil and its efficacy as a preservative in stored commodities. *International Journal of Food Microbiology, 168*, 1–7.

Keramat, F. (2000). Outbreak of food borne botulism in a family in Hamedan (Iran) due to consumption of soup prepared with traditional kashk. In *8th congress of infective diseases*, Tehran, Iran.

Kosar, M., Bozan, B., Temelli, F., & Baser, K. H. C. (2007). Antioxidant activity and phenolic composition of sumac (Rhus coriaria L.) extracts. *Food Chemistry, 103*(3), 952–959.

Kraig, B., & Sen, C. T. (Eds.). (2013). *Street food around the world: An encyclopedia of food and culture*. Santa Barbara: ABC-CLIO.

Kubo, I., Himejima, M., & Muroi, H. (1991). Antimicrobial activity of flavor components of cardamom Elettaria cardamomum (Zingiberaceae) seed. *Journal of Agricultural and Food Chemistry, 39*(11), 1984–1986.

Kuttan, R., Bhanumathy, P., Nirmala, K., & George, M. C. (1985). Potential anticancer activity of turmeric (Curcuma longa). *Cancer Letters, 29*(2), 197–202.

Lee, S.-J., Umano, K., Shibamoto, T., & Lee, K.-G. (2005). Identification of volatile components in basil (Ocimum basilicum L.) and thyme leaves (Thymus vulgaris L.) and their antioxidant properties. *Food Chemistry, 91*(1), 131–137.

Lev, E. (2002). Reconstructed materia medica of the Medieval and Ottoman al-Sham. *Journal of Ethnopharmacology, 80*(2–3), 167–179.

Liu, D. (2007). Taste of Chicago Salmonella outbreak linked to Hummus shirazi. Foodconsumer.org. Jul 14, 2007. http://foodconsumer.org, accessed 18 Mar 2018.

Majdalawieh, A. F., & Carr, R. I. (2010). In vitro investigation of the potential immunomodulatory and anti-cancer activities of black pepper (Piper nigrum) and cardamom (Elettaria cardamomum). *Journal of Medicinal Food, 13*(2), 371–381.

Mancini-Filho, J., Van-Koiij, A., Mancini, D. A., Cozzolino, F. F., & Torres, R. P. (1998). Antioxidant activity of cinnamon (Cinnamomum Zeylanicum, Breyne) extracts. *Bollettino Chimico Farmaceutico, 137*(11), 443–447.

Martínez-González, M. A., & Gea, A. (2012). Mediterranean diet: The whole is more than the sum of its parts. *British Journal of Nutrition, 108*(4), 577–578. https://doi.org/10.1017/S0007114512001833

Martins, N., Barros, L., Santos-Buelga, C., Silva, S., Henriques, M., & Ferreira, I. C. F. R. (2015). Decoction, infusion and hydroalcoholic extract of cultivated thyme: Antioxidant and antibacterial activities, and phenolic characterisation. *Food Chemistry, 167*, 131–137.
Mathew, S., & Abraham, T. E. (2006). Studies on the antioxidant activities of cinnamon (Cinnamomum verum) bark extracts, through various in vitro models. *Food Chemistry, 94*(4), 520–528.
Mazahreh, A. S., & Ershidat, O. T. M. (2009). The benefits of lactic acid bacteria in yogurt on the gastrointestinal function and health. *Pakistan Journal of Nutrition, 8*(9), 1404–1410.
Mchugh, T. (2016). How hummus is processed. *Food Technology, 70*(1), 91–94.
Mostafa, A. A., & Salem, R. H. (2015). Characterization of microbiological and nutritional variations in processed mullet (Mugil cephalus) Fish. *International Journal of Microbiological Research, 6*(2), 108–122.
Nasar-Abbas, S. M., Halkman, A. K., & Al-Haq, M. I. (2004). Inhibition of some foodborne bacteria by alcohol extract of sumac (Rhus coriaria L.). *Journal of Food Safety, 24*(4), 257–267.
O'Neil, E., Nicklas, A., & Fulgoni, V. L., III. (2014). Chickpeas and hummus are associated with better nutrient intake, diet quality, and levels of some cardiovascular risk factors: National health and nutrition examination survey 2003–2010. *Journal of Nutrition & Food Sciences, 4*, 254.
Olajide, O. A., Ajayi, F. F., Ekhelar, A. I., Awe, S. O., Makinde, J. M., & Alada, A. R. A. (1999). Biological effects of Myristica fragrans (nutmeg) extract. *Phytotherapy Research: An International Journal Devoted to Pharmacological and Toxicological Evaluation of Natural Product Derivatives, 13*(4), 344–345.
Ozaki, Y., Soedigdo, S., Wattimena, Y. R., & Suganda, A. G. (1989). Antiinflammatory effect of mace, aril of Myristica fragrans Houtt., and its active principles. *The Japanese Journal of Pharmacology, 49*(2), 155–163.
Paster, N., Juven, B. J., Shaaya, E., Menasherov, M., Nitzan, R., Weisslowicz, H., & Ravid, U. (1990). Inhibitory effect of oregano and thyme essential oils on moulds and foodborne bacteria. *Letters in Applied Microbiology, 11*(1), 33–37.
Peck, M. W., Goodburn, K. E., Betts, R. P., & Stringer, S. C. (2008). Assessment of the potential for growth and neurotoxin formation by non-proteolytic Clostridium botulinum in short shelf-life commercial foods designed to be stored chilled. *Trends in Food Science & Technology, 19*(4), 207–216.
Piaru, S. P., Mahmud, R., Majid, A. M. S. A., & Nassar, Z. D. M. (2012). Antioxidant and anti-angiogenic activities of the essential oils of Myristica fragrans and Morinda citrifolia. *Asian Pacific Journal of Tropical Medicine, 5*(4), 294–298.
Raghavan, S. (2006). *Handbook of spices, seasonings, and flavorings*. Boca Raton: CRC Press.
Ranasinghe, P., Jayawardana, R., Galappaththy, P., Constantine, G. R., de Vas Gunawardana, N., & Katulanda, P. (2012). Efficacy and safety of 'true' cinnamon (Cinnamomum zeylanicum) as a pharmaceutical agent in diabetes: A systematic review and meta-analysis. *Diabetic Medicine, 29*(12), 1480–1492.
Rasmussen, H. B., Christensen, S. B., Kvist, L. P., & Karazmi, A. (2000). A simple and efficient separation of the curcumins, the antiprotozoal constituents of Curcuma longa. *Planta Medica, 66*(04), 396–398.
Rasooli, I., & Abyaneh, M. R. (2004). Inhibitory effects of Thyme oils on growth and aflatoxin production by Aspergillus parasiticus. *Food Control, 15*(6), 479–483.
Rathore, M. S., & Shekhawat, N. S. (2008). Incredible spices of India: From traditions to cuisine. *American-Eurasian Journal of Botany, 1*(3), 85–89.
Reddy, A. C. P., & Lokesh, B. R. (1992). Studies on spice principles as antioxidants in the inhibition of lipid peroxidation of rat liver microsomes. *Molecular and Cellular Biochemistry, 111*(1–2), 117–124.
Roden, C. (2008). *The new book of Middle Eastern food*. New York: Knopf.
Roger, D. (2000). The history and culture of food and drink in Asia. In K. F. K. K. C. Ornelas (Ed.), *The Cambridge world history of food*. New York: Cambridge University Press.

Rosenberger, B. (1999). Arab cuisine and its contribution to European culture. In J. L. Flandrin & M. Montanari (Eds.), *Food: A culinary history* (pp. 207–223). New York: Columbia University Press.

Saad, B., Azaizeh, H., Abu-Hijleh, G., & Said, O. (2006). Safety of traditional Arab herbal medicine. *Evidence-based Complementary and Alternative Medicine, 3*(4), 433–439.

Saeed, A., Rehman, S. U., Akram, M., Bhatti, M. Z., Naz, R., Latif, A., … Saeed, A. (2016). Evaluation of antioxidant effects and inhibitory activity of medicinal plants against lipid peroxidation induced by iron and sodium nitroprusside in the mouse brain. *Journal of the Chemical Society of Pakistan, 38*(2), 333–340.

Sasidharan, I., & Menon, A. N. (2010). Comparative chemical composition and antimicrobial activity fresh & dry ginger oils (Zingiber officinale Roscoe). *International Journal of Current Pharmaceutical Research, 2*(4), 40–43.

Saxena, D. C., Rao, P. H., & Rao, K. R. (1995). Analysis of modes of heat transfer in Tandoor oven. *Journal of Food Engineering, 26*(2), 209–217.

Sezik, E., Tabata, M., & Yesilada, E. (1991). Traditional medicine in Turkey. 1. Folk medicine in Northeast Anatolia. *Journal of Ethnopharmacology, 35*, 191–196.

Sonavane, G. S., Sarveiya, V. P., Kasture, V. S., & Kasture, S. B. (2002). Anxiogenic activity of Myristica fragrans seeds. *Pharmacology Biochemistry and Behavior, 71*(1–2), 239–244.

Spices Inc. (2018). Cumin Seed: description. From www.spicesinc.com. Retrieved 15 September 2018.

Tajkarimi, M., Ibrahim, S. A., & Fraser, A. M. (2013). Food safety challenges associated with traditional foods in Arabic speaking countries of the Middle East. *Trends in Food Science & Technology, 29*(2), 116–123.

Tajuddin, T., Ahmad, S., Latif, A., & Qasmi, I. A. (2003). Aphrodisiac activity of 50% ethanolic extracts of Myristica fragrans Houtt.(nutmeg) and Syzygium aromaticum (L) Merr. & Perry.(clove) in male mice: a comparative study. *BMC Complementary and Alternative Medicine, 3*(1), 6.

Tannahill, R. (1989). *Food in history*. New York: Crown Publishers.

Toda, S., Miyase, T., Arichi, H., Tanizawa, H., & Takino, Y. (1985). Natural antioxidants. III. Antioxidative components isolated from rhizome of Curcuma longa L. *Chemical and Pharmaceutical Bulletin, 33*(4), 1725–1728.

Unnikrishnan, M. K., & Rao, M. N. (1995). Inhibition of nitrite induced oxidation of hemoglobin by curcuminoids. *Die Pharmazie, 50*(7), 490–492.

Van Loo, P., De Bruyn, A., & Verzele, M. (1988). On the liquid chromatography and identification of the flavonoids, present in the "sumach tannic acid" extracted from *Rhus coriaria*. *Chromatographia, 25*(1), 15–20.

Varo, P. T., & Heinz, D. E. (1970). Volatile components of cumin seed oil. *Journal of Agricultural and Food Chemistry, 18*(2), 234–238.

Walton, R. N., Clemens, A., Chung, J., Moore, S., Wharton, D., Haydu, L., … Austin, J. W. (2014). Outbreak of type E foodborne botulism linked to traditionally prepared salted fish in Ontario, Canada. *Foodborne Pathogens and Disease, 11*(10), 830–834.

Weber, J. T., Hibbs, R. G., Jr., Darwish, A., Mishu, B., Corwin, A. L., Rakha, M., … Tauxe, R. V. (1993). A massive outbreak of type E botulism associated with traditional salted fish in Cairo. *Journal of Infectious Diseases, 167*(2), 451–454.

Yadav, U. C. S., & Baquer, N. Z. (2014). Pharmacological effects of Trigonella foenum-graecum L. in health and disease. *Pharmaceutical Biology, 52*(2), 243–254.

Yamani, M. I., & Al-Dababseh, B. A. (1994). Microbial quality of hoummos (chickpea dip) commercially produced in Jordan. *Journal of Food Protection, 57*(5), 431–435.

Yamani, M. I., & Mehyar, G. F. (2011). Effect of chemical preservatives on the shelf life of hummus during different storage temperatures. *Jordan Journal of Agricultural Sciences, 173*(796), 1–26.

Yoshida, H. (1994). Composition and quality characteristics of sesame seed (Sesamum indicum) oil roasted at different temperatures in an electric oven. *Journal of the Science of Food and Agriculture, 65*(3), 331–336.

Chapter 3
GCC Traditional Foods: Preparation and Processing

Nasser Al-Habsi, Mohammed Al-Khusaibi, and Mohammad Shafiur Rahman

1 Introduction

Traditional food and beverage derived from land and sea, and it reflect an important trait of any nation's cultural heritage. These foods are considered as essential daily or occasional dishes by the local populace for long time. The method and recipes of preparation have been inherited from generation to generation. Furthermore, traditional foods characterize a crucial component of people's diet, health, and socioeconomic status (Trichopoulou, Soukara, & Vasilopoulou, 2007).

The traditional foods in the Arabian Gulf countries (Gulf Council for Cooperation, GCC), which comprises Oman, Kuwait, Qatar, Bahrain United Arab Emirates (UAE) and Kingdom of Saudi Arabia (KSA), are almost common. This is due to the similarity in social, religious and economic characteristics. Islam influences the culture of GCC countries and it even impact Arab's eating habit. Muslims in general are obliged to eat permissible (halal) food. The lifestyle of the GCC countries has been greatly experienced by the rapid growth, which has helped to shape their nutritional needs and patterns in life style (Musaiger, 1993). Standard foods, such as milk, dates, bread, fish and seafood have been the basis of the GCC diets and cuisines. However, the modern diets of the GCC also contains a mixture of dishes from around the world. Breads such as date bread (*Khubz al-tamer*), thin pan bread (*al-rigag*), and flat bread (*lanoor*) are commonly consumed in GCC countries. Also, rice is considered a staple food for the entire region and it can be cooked for both lunch and dinner. Lamb and mutton are consumed more than beef. One of the most common dishes, served on many different occasion, is *al mansaf.* It consists of

N. Al-Habsi (✉) · M. Al-Khusaibi · M. S. Rahman
Department of Food Science and Nutrition, College of Agricultural and Marine Sciences, Sultan Qaboos University, Muscat, Oman
e-mail: habsin@squ.edu.om

M. Al-Khusaibi et al. (eds.), *Traditional Foods*, Food Engineering Series,
https://doi.org/10.1007/978-3-030-24620-4_3

roasted lamb and rice with herbs. Religious occasions and celebrations have also shaped the Arab in general and GCC countries cuisine. There are special dishes (meals and sweets) which are prepared only during *Ramadhan* month, a holy month in which Muslims fast from sunrise to sunset. Muslims have two religious celebrations which are called *Eid Alfitr* (after *Ramadhan*) and *Eid Aladha* (two month after *Ramadhan*) in which special traditional dishes are prepared such as Omani *halwa* and *shuwa.*

The GCC cuisine is known for its aromatic food from the Arab merchants who controlled the spice trade. Food preparation methods and particular food ingredients were incorporated and modified to suite the people taste over many decades. The incorporation of spices and curry powder into the traditional local cuisine, for example spices originated from India, is an indication of the influence of the Indian culture on the food practice in the GCC (Musaiqer, 1993). The GCC countries traditional foods have not been marketed internationally and this can be noticed from the lack of literature about the foods. The content of this chapter will mainly emphasis on the traditional cuisine of the GCC countries. Moreover, it will also focus on the preparation methods, ingredient, nutritional value, composition and proximate analysis of traditional composite dishes.

2 Food Consumption Pattern in GCC

Food consumption pattern in the GCC countries has changed rapidly due to the economic growth and change in the life-style of the peoples. Factors that have contributed to the changes are food subsidy, food preferences, education and womens' employment (Musaiqer, 1993). Traditional foods such as date, milk rice, high fiber bread and fish have been substituted by a more diversified diet. The consumption of fish and mutton is more frequent compared to red and poultry meat. Lamb became more popular than beef. Rice has not lost its popularity and is the most staple cereal and is considered the main dish in a daily bases a combined by other complements. Wheat is also widely consumed as a bread and macaroni, however, the preparation from home has reduced dramatically and replaced with out-of-home foods. Milk and dairy product especially yoghurt, cheese, *laban*, butter and raw milk have also gain popularity and become an essential meal items with the new generation. Many widely-consumed foods in GCC are maid mainly from cereals, legumes and oilseed crops. Consumption of traditional sweet such as *rehash and halwah* and dates, which are normal taken between meals. These are now replaced with nuts, seeds, carbonated beverage, potato-chips, corn-puffs, chocolates and candies. Nevertheless, tea has held its popularity and is consumed sweetened with or without milk. Moreover, Arabian or Gulf bitter coffee (*gahwah khalijia*) is a hospitable traditional drink available on all occasions. It is served from a pot (*dallah*) in small handle less cups with no sugar added. Coffee is usually served with dates. Other traditional

Table 3.1 Common traditional beverages in the GCC countries

Local name	Common name	Ingredients
Qahwah	Brown-sugar coffee	Brown sugar, ghee, cinnamon, ginger, black pepper and water
Al-Liqah	Date palm pollen	Date palm pollen, sugar, and saffron
Al-Ward	Rose water	Rose water, cardamom, and sugar. Rose water is mixed with the other ingredients for taste.
Loomi	Dried lemon extract	Dried lemon, water, and sugar
Zanjabil	Ginger extract	Ginger, sugar, and milk
Darseen	Cinnamon extract	Cinnamon, sugar, and water
Tamar hindi	Tamarind	Tamarind pulp, water, and sugar.
Karkade	Rosella extract	Roselle calyces, sugar, and water
Zaafaran	Saffron extract	Saffron, cardamom, sugar, and rose water
Khawaja Ibrahim	Basil seed extract	Basil seeds, date palm pollen extract, and saffron
Hail	Cardamom extract	Cardamom pods, water, and sugar
Shamandar	Beetroot	Beetroot, sugar, and water

Sources: Alalwan, Mandeel, & AL-Sarhani (2017) and Musaiger et al. (2000)

beverages are listed in Table 3.1. Despite that fast foods are very popular nowadays, especially among the younger generation (Musaiger, 1993), traditional dishes are still attracting consumers and what are called "traditional restaurants" that serve traditional dishes are popular. Table 3.2 shows different traditional foods in the GCC countries.

3 Plant Based Products

3.1 Wheat

Khubz Al-Rigag is a traditional bread and popular in the GCC and is usually prepared at home. The name of this type of bread could differ among the GCC country. *Khubz Al-Rigag* is prepared by mixing flour with salt and water with continuous kneading until the dough become elastic. The dough is then spread by hand on the top of iron pan. The thickness of the bread should be very thin. When the bread is cooked a knife is used to remove the bread from the pan. Sometimes egg, spreadable chocolate, sugar and cardamom are added to the bread when it is on the on the pan. Table 3.3 shows different type of bread consumed in the GCC with their local names, ingredient and preparation methods.

Table 3.2 Popular traditional dishes in the GCC countries

Local name	Common name	Ingredients	Country
Legumes, vegetables, wheat dishes			
Dahl	Lentil stew	Lentils, onion, tomato, vegetable oil, mixed spices and water	Oman
Skanah	Fenugreek soup	Fenugreek seeds, wheat flour, sugar, whole milk, ghee, cardamom, black pepper, salt and water	Oman
Shorba	Soup	Meat, blanched wheat, onion, ghee, mixed spices, salt and water	Oman
Muhogo	Cassava soup	Cassava, coconut milk, chili, salt and water	Oman
Mooz matbokh	Banana stew	Green banana, meat, coconut milk, mixed spices, salt and water	Oman
Hareese	Beaten wheat with beef, Wheat porridge	Blanched whole wheat, meat (beef), ghee, salt and water, spices	Oman, Bahrain, UAE, KSA
Gerish		Whole wheat grits, laban (sour milk) butter, onion, corn oil and salt	KSA
Asseda	Beaten flour	Wheat flour, sugar, powdered cardamom, oil, powdered cinnamon, saffron with rose water, margarine	All GCC countries
Jola bil-assal	Bread with honey/date syrup	Wheat flour, black pepper, ghee, honey, salt and water	Oman
Nakhee	Chickpea	Chickpeas, salt and water	Bahrain
Bajelah	Broad bean	Broad beans, salt and water	Bahrain
Sago		Sago, sugar, rose water with saffron, powdered cardamom, margarine, and nuts.	All GCC countries
Rice, meat and fish dishes			
Qabooli	Rice with meat/chicken	Rice, beef, tomato garlic, ginger, vegetable oil, mixed spices, salt and water	Oman
Kabsa (also called Machbous)	Rice with meat/fish	Rice, meat or fish, onion, tomato, garlic, green chili, vegetable oil, mixed spices, salt and water	Oman, UAE
Aish abaid	Plain rice	Rice, onion, ghee, salt and water	Oman
Baranuosh	Sweet rice	Rice, date-syrup, fish, onion, vegetable oil, mixed spices, salt and water	Oman
Kalambah	Rice with fish	Fish, rice, lentil, onion, mixed spices, ghee, salt and water	Oman
Momowash	Rice with shrimps	Rice, dried shrimp, mung beans, fried onion and mixed spices	Bahrain
Muaddas	Rice with lentils	Rice, lentils, onion and mixed spices	Bahrain
Mahamer	Sweet rice	Rice, date molasses, saffron and rose water	Bahrain
Mbezar	Turmeric rice	Rice, turmeric and chili powder	Bahrain

(continued)

Table 3.2 (continued)

Local name	Common name	Ingredients	Country
Madroba	Beaten rice and meat	meat, rice, water, parsley, dill, coriander, onions, garlic, tomatoes, tomato paste, turmeric, powder cardamom, cinnamon, salt, sunflower oil	Qatar
laham mufour	Boiled meat	Meat, tomato, onion, garlic, black pepper, salt and water	Oman
Babloh smak	Fish stew	Fish, onion, tomato, mixed spices, garlic, coriander, wheat four, salt and water	Oman
Marag	Meat/fish curry	Meat/fish, onion, tomato, potato, eggplant, okra, squash, garlic, mixed spices, salt and water	Oman
Bahraini Kebab	Vegetable kebab Deep-frying	Chickpea flour mixed with onion, dill, bell pepper, and leek	Bahrain
Mbasal	Sauteed onion	Onion and spices	Bahrain
Thareed Laham		Meat with bone and fat (lamb), water, squash, carrots, onions, potato, spicy green pepper, garlic, tomatoes, tomato paste, turmeric, cardamom, salt, capsicum, traditional flat thin beard	Qatar
Margoog Laham		Meat with bone and fat (lamb), water, potato, onions, capsicum, squash, parsley, coriander, tomato paste, Iranian bread, salt, spice	Qatar
Maleh	dried, salted fish	Fish, salt	Oman, UAE
Gurus	Doughnut cake	Wheat flour, milk powder (or yogurt), egg, sugar, oil, salt, yeast	Oman, UAE
Balaleet	Fried vermicelli	Vermicelli, sugar, egg, water	Oman, UAE
Marassia		Whole wheat flour, egg, whole milk, butter, honey, or syrup	KSA
Mataziz		Whole wheat flour, mutton, carrots, peas, onion, tomato, corn oil, salt and spices	KSA

Sources: Musaiger et al. (1998), Alalwan et al. (2017), Habib et al. (2011), Al-Kanhal, Al-Mohizea, Al-Othaimeen, and Khan (1994), Elobeid, Phoboo, and Magdad (2015)
UAE United Arab Emirates, *KSA* Kingdom of Saudi Arabia

Table 3.3 Different types of bread in the GCC countries

Local name	Common name	Ingredients
Khubez tamer	Date bread	Wheat flour, date syrup, cardamom, and depitted dates
Khubez Halibi	Milk bread	Wheat flour, milk, sugar, and sesame seeds
Khubez tanoor	Clay oven bread	Wheat flour, salt, water, and yeast
Khubez khamira	Fennel enriched bread	Wheat flour, yeast, water, and fennel seeds
Khubez rigag	Crispy flat bread	Wheat flour, salt, and water
Maghlaq	Oatmeal date bread	Wild oats, dates, mustard seeds, eggs, and ghee

Sources: Alalwan et al. (2017), Musaiger et al. (2000), ElObeid et al. (2015), Habib et al. (2011)

This simple mixture of flour and water with a pinch of salt, smeared on a sizzling-hot pan and chipped free with a pastry knife or a paint chipper. It takes some practice to get down, but it is a fun and easy bread once you get the hang of it. Some countries are popular for their own type of bread. For example, Oman is known by its very thin round-flat unleavened bread. There are traditionally two version called "*Khubz Omani*" and *rikhal"* or *"reqaq"*. The word *Khubz* literally mean "bread". These two types are prepared by a batter made from whole wheat flour. The viscosity of the batter (influenced by the water content) determine the final product *"Khubz Omani"* or *"rikhal"*. The bread are baked in a round metal grid. The breads are smeared with traditional ghee (*Samnah*) made from goat milk. Ali, Al-Nassri, Al-Rasasi, Akhtar, and Al-Belushi (2010) studied the glycemic index (GI) of eight different types of Omani traditional bread. It was concluded that the GI of the breads were within small to medium values and no correlation was found with the GI and the chemical composition. Figure 3.1 shows the preparation of Omani bread "*rikhal*".

Parathas (*Chapatis*). The word *chapati* is of the South Asian (Indian-subcontinent) and Africa. The *chapatis* in the Gulf are referring to unleavened bread made with oil which called *parathas*. However, distinctly Omani variation of *chapatis* has an African spin. This type of traditional food is generally produced and consumed fresh in homes and restaurants. Since this type of food is adopted from other cultures,

Fig. 3.1 Preparation f Omani bread "*Khubz Omani*"

it is commonly eaten with Indian dishes, which is popular in the Gulf. *Chapatis* (Fig. 3.2) are prepared by mixing 100 g of wheat flour with 55 g water and 0.7 g salt. A thorough mixing of the ingredient in a bowl until dough of firm consistency is achieved. The dough is then portioned into a small ball, flattened and grilled on a greased griddle over a stove at 183–205 °C. The nutritional value per 100 g edible portion for protein, fat, carbohydrate and energy is 8.2%, 3.6%, 55% and 285 kcal, respectively (Musaiger, 1997).

Tharid. This dish is a very popular dish in the Arab countries and especially in GCC countries. It has been characterized by its simple preparation. It is made from pieces of bread that are immersed in meat (mutton or chicken) stew. The meat should be cooked enough to be pulled from the bones easily. Traditionally, the GCC countries traditional bread (*khubz rigag*) is used. The amount of bread pieces should be enough to absorb most of (sop) the liquid in the stew. Traditional, the *tharid* dish is prepared and then served ready to be consumed (i.e., the bread pieces have sop the broth of the stew, Fig. 3.3). Alternatively, the meat stew and the bread can be served separately and consumer would mix them in a small dish/bowl.

Hareese: It is also called *Jareesh* in some countries. The dish is made from boiled, cracked, or coarsely-ground wheat. The wheat is soaked overnight to rehydrate. The meat is cooked until it becomes tender and then the wheat is added to cook. When they are ready, the mixture is beaten to form a consistent thick meal. The dish is served topped with traditional ghee. No condiments other than salt are used in this dish.

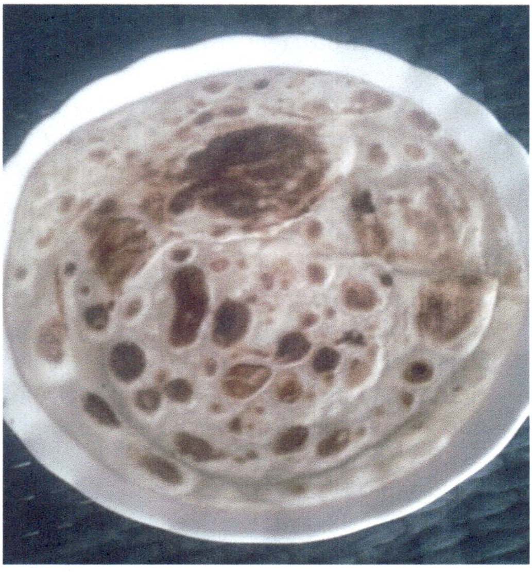

Fig. 3.2 A photo represent *Chapati* bread

Fig. 3.3 Tharid dish

3.2 Rice and Grain

In GCC counties, rice and grains form the staple food. Rice is consumed mainly at lunch time and sometimes at dinner time. Milled rice is used widely in the gulf compared to un-milled. The preparation method of most the rice dishes among the Gulf countries is common the only difference is the spices. Rise is first washed several times followed by boiling and straining the water before cooking. However, the straining of water is not practiced in some of the houses to maintain the water soluble nutrients (such as vitamin B) (Alalwan et al., 2017; Al-Kanhal et al., 1994). Rice is cooked in different dishes such as white rice with curry or *Kabsa* (also called *Makboos*) wich is a spiced rice. There are different names for the *kabsa* rice. It is called *Qabooli* in Oman. *Mandi* is word originated from Yemen but commonly used in the GCC countries to denote a type of *kabsa*. In Kuwait, the popular rice dish is called *Ghosi*. The dish is a different form of *kabsa* and decorated with nuts and stir fried onion and potato strips. All rice dishes are served with meat, commonly mutton and chicken. Table 3.1 shows different rice dishes consumed in the GCC with their local names, ingredient and preparation methods. Figure 3.4 shows an example of *kabsa* dish.

*Ursia***:** The dish is very popular in Oman and UAE and served in special occasions and celebrations of religious days. The dish preparing is similar to the *Harees* preparation and the only different is that rice is used instead of wheat. The dish is served with a traditional sauce called "*tursha*" which is made from raisins, chickpeas, garlic, tomatoes and salt.

Dry legumes represent the common legumes in the GCC countries. This include lentil which is called "*dahl*", dried bean, dried chickpeas and dried fava beans. They are used – individually – to prepare stews which is consumed with different types of traditional bread. Figure 3.5, a. shows the *ursia* preparation and b. the final dish.

Fig. 3.4 Kabsa rice dish with meat

Fig. 3.5 The preparation of ursea (left) and final dish (right)

4 Animal (Meat and Fish) Based

Shuwa, The unique Omani specialty for which meat is slathered in a thick blend of oil and spices, wrapped in palm fronds (Fig. 3.6), and cooked over hot embers underground. This type of dish is an occasional and is consumed during *Eid Alfitr* after *Ramadhan* (Muslims holy month) and during the *Eid Aladha*. Lamb, goat, or even camel meat may be coated with the spice rub before being wrapped in banana leaves and tucked into a palm frond roasting bag. The duration of burying the *shuwa* is from 1 to 2 days. The smoky flavors permeate the meat and a nice, crisp crust of

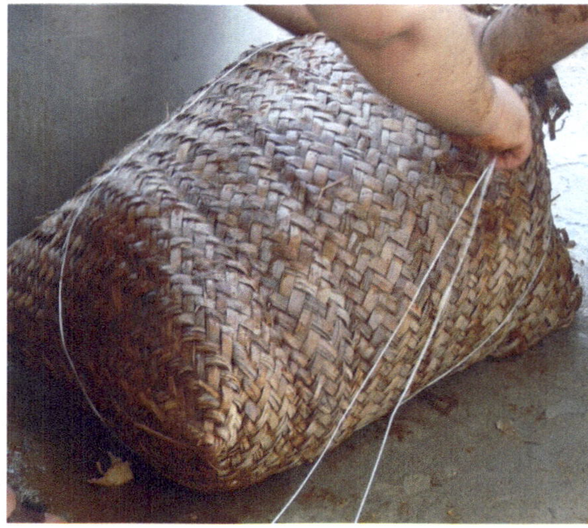

Fig. 3.6 The shuwa packed in the palm frond bag

spice developed on the outside of each piece. The meltingly tender meat is enjoyed with a large platter of rice and tangy Omani-style tomato sauce for one of the most anticipated meals of the year.

Malah (salted fish). This dish is very popular in coastal areas in Oman and UAE and commonly prepared from tuna fish. In order to preserve fish, the fish is fillet and small cuts are formed in the fillet. The fish fillet are heavily salted. The fish fillets are then packaged in a brine of high salt concentration. The fish can be preserved by this way for several months at room temperature. During preparation of meals, the fish fillet must be soaked in water which is changed twice in order to get red of the high concentration of salt.

Salted shark. This is another way to preserve fish by salting and drying. The very common type of fish is shark. Shark fishes are fillet (with skin), heavily salted and sun dried. The final product range from dry to semi-dry product. The process has no consistency because it is still practiced at fishermen level. Fish preserved by this way is locally called "*Oal*". It is used to prepare rice dishes with either plain rice or spiced rice "*Qaboli*".

5 Dairy and Sweet Products

5.1 Dairy Products

Before the discovery of oil in the GCC countries, people activities varied from oversea trading, local small scale trading, agriculture, animal husbandry. Animals were raised for its meat, eggs, and also milk. Traditionally, animals were milked on a daily basis due the lack of processing plants and due to the small quantity. Milk was also used to produce butter. Milk is churned in a traditional churn made from treated animal skin (a goat skin bag) (Guizani, Kasapis, & Al-Ruzeiki, 2001). Churning produce butter and buttermilk which is called *Laban* in the GCC countries. Nowadays, *Laban* also refer to a fermented milk drink.

Samn (Samnah). This refers to the ghee made from goat, sheep or cow milk. The ghee is prepared by simmering the butter. Impurities are removed from the surface and the product retain a clear liquid fat. Unique flavor is obtained by the process and also by adding herbal spices.

5.2 Sweet

Halwa is a dessert like sweet jelly (Fig. 3.1), which is made up mainly starch, sugar, water, ghee, nuts and flavored and/or colored with *saffon* and/or rose water (Rahman et al., 2012). Omani *halwa* is commonly made with sugar (50%), water (25%), mill flour (10%) and ghee (i.e. clarified butter) (Al-Shamsi, Rahman, Abdullah, & Claereboudt, 2013). Traditional *halwa* is prepared by heating water first followed by adding different ingredients one by one (Al-Shamsi et al., 2013) (Fig. 3.2). First, sugar is added to boiling water in an open pan and starch is added to the sugar syrup after dissolving all sugar. The heating process continues with manual mixing with special spoon (i.e. handle with wood and square metal plate at the other end to make it easy for continuous scraping so that gel does not stick on the surface and caused burning) until the mixture turns into a thick, jelly-like substance. At this stage, fats, nuts, and spices are added, and gels are poured into dishes of varying shapes, designs. There are mainly 3 types of Omani halwa, based on their final color: white, yellow and black. They differ in the sweetener, flavor and texture. Al-Shamsi et al. (2013) measured the chemical compositions (moisture, protein, fat, ash, carbohydrate, color, pH, and fatty acid profiles) of locally available traditional *halwa* prepared commercially. The main variations were observed in protein (0.51–2.27 g/100 g *halwa*), fat (0.8–14.4 g/100 g *halwa*), sugar (14.9–56.3 g/100 g *halwa*), and color (L-value: 24.5–35.7), while other compositions remained similar. The type and amount of fat and sugar content in halwa could be considered high fat and high sugar diet. The successful development of healthier *halwa* with lower fat and sugar with the incorporation of health functionality would interesting. *Halwa* is also known in UAE and Bahrain with minor differences in the ingredients.

Maamoul (pronounced [ma'-mul]). The sweet is a date-filled cookie which are served in special occasions and celebrations. The sweet is made from white flour but semolina can be used instead. The ingredients also include ghee, yeast and rose water, cardamom and saffron. The filling is mainly date fruit but it can also be fig. This is similar to *Kleicha* sweet which is popular in Iraq and KSA. Different types of sweet are shown in Table 3.4.

Table 3.4 Traditional sweets of the GCC countries

Local name	Common name	Ingredients
Halwa	Sweet	Corn flour, fat, brown sugar or dates syrup, nuts, cardamom and saffron
Maho	Coconut sweet	Coconut, sugar and milk
Qeshat	Dried corn flour sweet	Corn flour, fat and sugar
Assidah	Wheat flour porridge	Wheat flour, sugar or date syrup, ghee, and water
Khabees	Flour and oil	Wheat flour, date syrup, ghee, cardamom, and saffron
Mamrus	Flour and oil	Wheat flour, date syrup, ghee, sesame, and almonds
Hesso	Fenugreek seeds and eggs	Fenugreek seeds, flour, sugar, and water
Rashoofa or (Gellab)	Garden cress seeds and flour	Garden cress seeds, sugar, flour, and spices
Legaimaat	Sweet dumplings	Wheat flour, sugar, dry yeast, and water
Rangina	Dates in butter sauce	Rutab dates, flour, butter, and cinnamon
Madquq bisr	Pounded dates	Khalal (bisr) dates, date syrup, cardamom, and saffron
Nashab	Filo cinnamon and sugar roll	Wheat flour, date syrup, ground cinnamon, and sugar
Rahash	Sesame paste	Sesame seeds, sugar syrup, and egg whites
Mahala	Sweet thin pancake	Wheat flour, sugar, eggs, water, cardamom, and rose water
Gurs Al-taabi	Bahraini pancake	Wheat flour, sugar, eggs, water, cardamom, and rose water
Khabeesa bobar	Pumpkin sweet	Pumpkin, flour, sugar, cardamom, and saffron.
Khanfaroosh	Semolina cake	Semolina, eggs, saffron, cardamom, and sugar
Nashab	Wheat pastries	Wheat flour, sugar, oil and cardamom
Sambosa Helwah	Sweet Sambosa	Wheat flour, sugar, nuts and oil
Shaar Banat	Sugar sweet	Starch, sugar and oil
Eggbaith	Sesame sweet	Egg white, sugar and sesame

Sources: Alalwan et al. (2017), Musaiger et al. (2000), ElObeid et al. (2015), Habib et al. (2011)

References

Alalwan, T. A., Mandeel, Q. A., & AL-Sarhani, L. (2017). Traditional plant-based foods and beverages in Bahrain. *Journal of Ethnic Foods, 4*, 274–283.

Ali, A., Al-Nassri, H. A. S., Al-Rasasi, B., Akhtar, M. S., & Al-Belushi, B. S. (2010). Glycemic index and chemical composition of traditional Omani breads. *International Journal of Food Properties, 13*(1), 198–208.

Al-Kanhal, M. A., Al-Mohizea, I. S., Al-Othaimeen, A. I., & Khan, M. A. (1994). Nutritive value of some wheat based dishes consumed in the kingdom of Saudi Arabia. *Ecology of Food and Nutrition, 32*, 219–226.

Al-Shamsi, Q., Rahman, M. S., Abdullah, A., & Claereboudt, M. R. (2013). Consumers' preference mapping and their behavior with regard to traditional halwa using a structured survey questionnaire. *Journal of Food Products Marketing, 19*(4), 311–328.

Elobeid, T., Phoboo, S., & Magdad, Z. (2015). Proximate and mineral composition of indigenous Qatari dishes: Comparative study with similar Middle Eastern dishes. *Journal of Food Chemistry and Nutrition, 3*, 27–34.

Guizani, N., Kasapis, S., & Al-Ruzeiki, M. (2001). Microbial, chemical and rheological properties of laban (cultured milk). *International Journal of Food Science & Technology, 36*(2), 199–205.

Habib, H. M., Ali, H. I., Ibrahim, W. H., & Afifi, H. S. (2011). Nutritional value of 10 traditional dishes of the United Arab Emirates. *Ecology of Food and Nutrition, 50*(6), 526–553.

Musaiger, A. O. (1998). Change in dietary habits, life style and trend in diseases in the GCC countries. *Bahrain Medical Bulletin, 20*, 87–90.

Musaiger, A. O. (1993). *Traditional foods in the Arabian Gulf countries*. Cairo, Egypt: The FAO/RNEA.

Musaiger, A. O. (1997). *Tradtional dishes of Oman*. Al-Ain: Al-Ain Printing Establishment.

Musaiger, A. O. (2000). The State of Nutrition in Bahrain. *Nutrition and Health, 14*, 63–74.

Musaiqer, A. A.-R. H. A. U. (1993). *The state of food and nutrition in Bahrain*. Riyadh: UNICEF, Gulf Area Office.

Rahman, M. S., Al-Shamsi, Q., Abdullah, A., Claereboudt, M. R., Al-Belushi, B., Al-Maqbaly, R., & Al-Sabahi, J. (2012). Classification of commercial Omani hlawa by physic-chemical properties and Instrumental Texture Profile Analysis (TPA). *Italian Journal of Food Science., 24*(3), 292–304.

Trichopoulou, A., Soukara, S., & Vasilopoulou, E. (2007). Traditional foods: A science and society perspective. *Trends in Food Science & Technology, 18*, 420–427.

Chapter 4
Traditional Foods in Maghreb: Production and Research Progress

Hicham Zaroual, Ferdaous Boughattas, and Romdhane Karoui

1 Introduction

Traditional food products in the Maghreb constitute a huge element of North African culture and identity. These foods contribute significantly to maintaining and even the development of sustainability of rural economy. Thus, the production of traditional foods could be considered as an issue that assumes increasing importance within the food industry. The traditional foods are a major concern not only to rural population, and consumers, but also to producers and distributors. Indeed, regulatory authorities, food processors, retailers and consumer groups have interests in ensuring that these foods present specific quality related to the production area and the know-how of the producers.

Traditional foods in the Maghreb are mainly produced following the application of one and/or several technology processes including fermentation, salting, drying, pickling. These processes contribute significantly to maintain the quality of traditional food products for several months. These include fermented olive and camel milk, canned capers, pastilla, assida, etc. These traditional products are made mostly at

H. Zaroual · F. Boughattas
Université d'Artois, EA 7394, Institut Charles VIOLLETTE, Lens, France

R. Karoui (✉)
Université d'Artois, EA 7394, Institut Charles VIOLLETTE, Lens, France

ISA Lille, EA 7394, Institut Charles VIOLLETTE, Lille, France

Ulco, EA 7394, Institut Charles VIOLLETTE, Boulogne sur Mer, France

Université de Lille, EA 7394, Institut Charles VIOLLETTE, Lille, France
e-mail: romdhane.karoui@univ-artois.fr

M. Al-Khusaibi et al. (eds.), *Traditional Foods*, Food Engineering Series,
https://doi.org/10.1007/978-3-030-24620-4_4

home through wisdom handed down from generation to generation. The quality and shelf life of these products varied greatly according to the country and even for the same country from one region to another and even from community to community.

The production of these products is mainly performed by individual farmers at rural zones and only small enterprises are present. Traditional food products could be considered important levers on which can be based long-term strategies for the development of rural areas. In order to enter and survive on the national and international markets, agricultural quality policy should be implemented. The maintaining of the production of these productions in rural zones allows to protect them from depopulation. This chapter will provide a review on the most known traditional foods in the Maghreb including vegetable, dairy, meat, cereal and sweet products. The salted and un-salted fermented foods will also be reviewed.

2 Olive and Olive Oil Products

2.1 Olive Oil

The wild olive (*Oleaster*) appeared several tens of thousands of years ago in the Mediterranean region. Its cultivation, combined with oil production, dates back to 4000 BC, probably in Crete; with Minoan civilization; other scholars cite the Middle East and the Phoenicians as the first farmers. Its culture has gradually spread to the entire Mediterranean basin thanks to trade and conquests.

Under the Roman Empire, the olive tree contributes to the pax romana by offering the conquered countries a product with high benefit. From Greece to Spain *via* Egypt, Italy, Tunisia, France or Morocco, the culture of the olive tree was established permanently throughout the Mediterranean (Vossen, 2007). The consumption of olive oils in the Maghreb countries has seen a great increase after the arrival of Islam on their territories. The Algeria, Morocco and Tunisia countries produced 273 thousand tons of olive oil during the 2016–2017 season, with a local consumption of 190 thousand tons. Morocco and Tunisia exported to the EU, United States and China about 80 thousand tons during 2016–2017 season (COI: International trade of olive oil 2016–2017).

The Maghreb people use olive oil for food and medicine. As food, olive oil is consumed raw with bread or as an ingredient in different culinary preparations. Generally, olive oil is used with medicinal plants for medical reasons with: (1) garlic for asthma, arteriosclerosis, hypertension, (2) green anise for palpitations, (3) basil for intellectual overwork, (4) cinnamon for impotence, (5) coriander for difficult digestions, (6) juniper for diabetes and gout, (7) lavender for enteritis, infectious diseases, migraines, (8) nutmeg for asthenia, cholelithiasis, (9) onion for respiratory diseases, arteriosclerosis, diabetes, genitourinary infections, obesity, rosemary for gout, hepatitis, cholesterol.

2.1.1 Nutritional Value of Olive Oil

Olive oil is a liquid fat obtained from olives (the fruit of *Olea europaea*; family *Oleaceae*), and is the only liquid fat consumable raw and virgin. It retains in its natural state and all its biological properties including vitamins and antioxidants. Olive oil is made, as all vegetable oils, of 100% of lipids and provides 884 kcal per 100 g. It becomes rancid less quickly as compared to other oils and has a low iodine index 78/88 against 83/98 for peanut and 120/132 for sunflower oils (Olmo-García, Bajoub, Monasterio, Fernández-Gutiérrez, & Carrasco-Pancorbo, 2017).

2.1.2 Composition of Olive Oil

Fatty acids: Saturated fatty acids (palmitic, stearic), harmful to health, are present in lower proportions as compared to polyunsaturated fatty acids that break down into two categories.

Mono-unsaturated fatty acids: (73–78%) contributing to lower the LDL (Low-Density Lipoprotein). Olive oil is exceptionally rich linoleic acid (omega 9) which gives it a great dietary interest. Omega 9 in combination with the antioxidants (polyphenols) present in olive oil seem to lower blood levels of LDL cholesterol. Linoleic acid modifies biological constituents by integrating into lipoproteins and incorporating into cell membranes that are, thereby, protected from the free radicals (Boggia, Borgogni, Hysenaj, Leardi, & Zunin, 2014).

Poly-unsaturated fatty acids: that cannot be synthesized by the body, such as linoleic acid are also present in olive oil at important levels (10–15%).

Micronutrients and other phytochemicals: Olive oil contains: (1) phenolic compounds that delay the oxidation process; (2) sterols that preserve the intestine from the absorption of cholesterol, (3) terpenic alcohols, including cycloartenol, favoring the elimination of cholesterol, (4) chlorophyll that stimulates cell growth and accelerates healing processes, and (5) vitamins E that protects the body tissues and preserves vitamin A (Table 4.1). Polyphenols and vitamins are the natural antioxidants making virgin olive oil the most resistant oil to rancidity.

2.1.3 Processing Technology of Olive Oil Extraction

Maghreb's olive cultivars: In Morocco, the most recognized variety is the Moroccan Picholine with more than 96% of all olive plantations. The Moroccan Picholine is characterized by its high content of polyphenols and its yield that varies between 15 and 24% according to agronomical and industrial factors. The remaining varieties are the local ones as Haouzia, Menara, Dehbia and Meslalla, and foreign ones like Arbequina, Arbozana and Picholine of Languedoc (Bajoub et al., 2017).

In Tunisia, the Chemlali variety is present ~85% of olive plantations and accounts for more than 80% of the national production of olive oil. The other local varieties

Table 4.1 Nutritional values of olive oil (USDA Branded Food Products Database Release v0.0/ June 2018), Alouana (Gharby et al., 2018), fermented black olives (Stan Kailis & David John Harris., 2007) and argan oil (Khallouki et al., 2017; Rueda et al., 2014; Seiquer et al., 2015)

Nutritional value	For 100 g of olive oil	For 100 g of Alouana	For 100 g of fermented black olives	For 100 g of argan oil
Energy	884 kcal	884 kcal	179 kcal	899 kcal
Water	–	–	60–70 g	–
Carbohydrates	–	–	7.14 g	–
Proteins	–	–	0–1.5 g	–
Saturated fat	15.30 g	15.30 g	–	12–16 g
Polyunsaturated fat	12.10 g	12.10 g	–	29–36 g
Monounsaturated fat	72.10 g	72.10 g	–	44–51 g
Total fat	99.50 g	99.50 g	14.29 g	99 g
Fibers	–	–	5.5 g	–
Vitamins				
Vitamin-E	14.39 mg	14.39 mg	14.39 mg	70.2 mg
Vitamin-K	60.2 μg	60.2 μg	60.2 μg	–
Minerals				
Sodium	2 mg	2 mg	2 mg	–
Potassium	1 mg	1 mg	1 mg	–
Calcium	1 mg	1 mg	1 mg	–
Copper	–	–	0 mg	–
Iron	0.56 mg	0.56 mg	0.56 mg	–
Zinc	0.01 mg	0.01 mg	0.01 mg	–
Others				
Sterols	221 mg	221 mg	221 mg	26.7 mg
γ-tocopherol	–	–	–	48.3 mg
Polyphenols	20–80 mg	22.6–29.7 mg	29.7 mg	Less than 0.5 mg

are Chetoui, Oueslati, Zalmati, Chemlali, Gerboui and Chemchali. In irrigated areas, foreign varieties could be founded such as Arbiquina, Arbozana and Koroneiki (Brahmi et al., 2011; Manai-Djebali et al., 2012). In Algeria, the most prominent varieties are Chemlali, a variety native to Kabylie, with 40% of the Algerian olive orchard occupation and a yield varies from 18 to 22%. The remaining 60% of the orchards are comprised of other varieties such as Azeradj, Sigoise and Ferkani (Fig. 4.1).

Harvesting: The most common practice of olive harvesting is padding with long poles. Olives are collected on tarpaulins, jute or nylon and sometimes on the ground. Harvesting by hand is used in some areas, especially for table olives. Hand picking olives is a common practice for small growers. The introduction of mechanization involves, in most cases, intermediate steps concerning tree accessibility, size, shape and timing of harvesting (Fig. 4.2a, b). The mechanized harvest seems to be the only alternative since it contributes to the reduction of the costs of the harvest.

Fig. 4.1 Local and foreign olive cultivars in Maghreb

Fig. 4.2 Traditional (**a**, **b**) and modern (**c**, **d**) harvesting techniques

Several modern harvesting methods have recently appeared in Maghreb orchards. Pneumatic vibratory forks that appeared for about ten years are three to four times faster than padding. The olives are much less damaged and the branches are less broken (Fig. 4.2c, d). Another method of harvesting even more suitable is the use of mechanical machinery equipped with vibrator. In fact, these agricultural tractors on which is mounted a hydraulic clamp aims to shake the branches or the trunk of the olive tree. It allows the reduction of the number of shoots slaughtered and the proportion of impurities, which allows the appreciation of the quality of the olive oil.

Olive oil extraction: The extraction process goes through the following steps: Washing the olives avoids the following problems: (1) the interference of earth with color and other organoleptic properties (smell, taste) of the oil; (2) the decrease of the yield; (3) the reduction of the shelf life of olive oil because some trace metals present in the earth could be catalysts of oxidation reaction in olive oil (Fig. 4.3a). Olives are washed by immersing the olives in a water tank or, in modern installations, in washing machines that keep the water in a forced movement to reduce most of the impurity. To obtain a good quality of olive oil, it is important to use clean and potable water. Generally, a single washing is enough to obtain clean olives.

Preparation of the olive paste: Grinding is the first step of processing olives into olive oil. Olives are subjected to mechanical actions that cause the rupture of cell walls and membranes resulting in the release of cell juices and oil. This action results from the friction of the fragments of nuclei on the pulp, or shocks of the mechanical devices in rotation at high speed in the mass of the pulp. The product obtained at the end of this phase, in most installations, is a paste. The oil paste is a semi-fluid mass composed of a solid fraction (fragments of cores, skins and pulp) and a liquid fraction (emulsion of water and oil). The time between harvesting and

Fig. 4.3 Olive washing (**a**), grinding using stone mills (**b**) or mechanical grinder (**c**) and pressing of olive oil (**d**)

pressing the olives has a great influence on the final quality of the produced oil (Chimi, 2006). It is therefore necessary to squeeze the olives as quickly as possible, while respecting all the hygienic standards that are required.

Grinding mills: There are two groups of grinders. The millstone is an instrument used for the grinding by applying mechanical action following the rotation of one or more large stone wheels (usually on granite materials) on the mass olive fruits (Fig. 4.3b). In mechanical grinders, all-metal shredders are the preferred devices in modern continuous cycle plants because they fully integrate the automation requirements (Fig. 4.3c). These consist of a series of metal elements rotating rapidly and provided with sharp edges, hammers or knives varying in number and thickness; their rotation speeds varied between 1200 and 3000 revolutions per minute (rpm).

Mixing and oil extraction: It is an operation that aims to agglomerate the oil particles into larger drops, and separating spontaneously from the vegetation water. Olive oil extraction is used to separate the oil from the residual paste called "pond". The extraction methods are related to two basic types. In traditional production (i.e. discontinuous pressure extraction), olive oil extraction is carried out by hydraulic press (Fig. 4.3d). The dough coming from the kneading is distributed in layers laid on circular diaphragms made of vegetable fibers called "scourtins", which serve to drain the oil. This is the most common extraction system currently used is continuous system of extraction by centrifugation. The system works in a "dynamic" phase where the paste (diluted with hot water in the mixer) is reached by a feed pump. This paste moves gradually along horizontal axis that rotates at a high speed (6000 rpm). Under the effect of this speed, the separation of liquids and solids is ensured continuously without stopping the machine. Two continuous extraction systems are employed by two-phase and three-phase modes.

Two-phase mode that is a continuous mode where olives are grinded and then kneaded at 20 °C for 30 min. A process based on the use of centrifugation (Decanter for liquid-dough separation) to separate oil from the dough constituted by pomace and olive vegetation water by horizontal centrifuge; this system do not use the hot water for the separation of oil (Chimi, 2006).

Three-phase mode or continuous mode where olive fruits are grinded and then kneaded at 35 °C for 30 min to facilitate the liberation of oil (Chimi, 2006). A first separation by horizontal centrifuge (Decanter for solid-liquid separation) is performed allowing the separation of solid phase (pomace) from liquid phase and a second one realized by a vertical centrifuge (Decanter for liquid-liquid separation) is undertaken in order to separate oil from olive vegetation water (Chimi, 2006).

Preservation of olive oil: The extracted oils undergo a quality control before their packaging and conservation. This control is carried out with particular attention in order to avoid that the olive oils are mixed. Defective oil, mixed with good quality oils even in small proportion is forbidden, since it decrease the quality of the latter. The shelf life of the oils depends on the flow of the product. Oil mills often have a sufficient storage capacity to meet their needs in case of slump during the transformation period.

2.2 *Roasted Olive Oil (Alouana/Alwana)*

Typical of the Hyayna tribes in the Province of Taounate (Morocco), Alouana is a type of oil prepared from roasted olives. It has a very particular taste. It is known that the inhabitants of the Hyayna region prefer to consume the olive oil extracted from the fruits at their last stage of maturity (black olives) that belongs to the harvesting period of the end of December and the beginning of January. Alouana (Fig. 4.4a) is eaten accompanying with pancakes or traditional bread for breakfast, and it is characterized by its darker color compared to virgin olive oil, and its roasting taste (Guillaume, Gharby, Harhar, & Charrouf, 2018).

2.2.1 Nutritional Value of Alouana

The composition of Alouana oil is similar to virgin olive oil except some differences in vitamin E (tocopherols) and polyphenols contents resulting from the roasting of olive fruits and appearance of minor compounds that are responsible for the typical aroma of Alouana olive oil. Fatty acid (FA) composition is an essential indicator of the nutritional value of the oil. According to Gharby et al. (2018), no significant differences ($P > 0.05$) were observed between the fatty acid and sterol composition of virgin olive oil and Alouana olive oil. As for tocopherols (i.e. natural antioxidants), the content of α-tocopherol in olive oil decreased after roasting contrary to γ-tocopherol. However, there were no significant differences ($P > 0.05$) in the contents of d-tocopherol (Gharby, 2018). The level of polyphenols increased after the roasting process by 30% (passing from 226 to 297 ppm) (Table 4.1).

Fig. 4.4 Roasted olive oil (Alouana) (**a**), and picking of olive fruits (**b**), roasting of olives in the traditional oven (Lforna) and roasted olives (**c**)

2.2.2 Processing Technology of Alouana oil

The production of Alouana oil requires several manual operations to prepare it from harvesting to filtration and storage. When the olives are designated for the production of Alouana oil, olive fruits are handpicked and collected in large baskets by rural women (Fig. 4.4b). In this environment where pottery is largely produced by women, rural women build the traditional oven (Lforna) outside their home, from clay, sand and crushed pottery debris for the seating of the oven house. After baking the bread, the remains of charcoal and ash are released and the hot oven is swept. The batch of fresh olives (about 5 kg) is carefully spread on the base of the oven. The oven is tightly closed 12 h (Fig. 4.4c). The olives, without marigolds, turn a dark color and are removed from the oven with a shovel and placed in a red clay bowl (Lguessàa) (Fig. 4.4c). In a large spiral-shaped van (Lmaydouna) protected by plastic, because it is also used as a kneader and sieve on the fly (Tessyar), one or two women sitting opposite each other, crush the olives between two hot stones contained in the jatt (Fig. 4.5a). A flat stone is erected at the bottom of the Lmaydouna, the other stone, rounded (Aoudy/Awdi), is held by hand to crush the fruit (pulp and stones), by posed percussion. The pomace (pulp and stone debris), obtained by crushing, are warmed in a recipient on a small wood fire (Lkanoune) and kneaded in a clay bowl (Lguessàa), adding a little hot water (Fig. 4.5b). This operation, that takes 15 min, is followed by sitting the crushed olives in the pressing mats and pressing. Two small scourtins, round baskets of small size, braided in palmetto fiber are filled with the mixed pomace. When installed on the bottom plate of the press, the upper plate is then lowered by tightening the nuts (Fig. 4.5c).

2.3 Traditional Fermented Olives in Maghreb

Fermented olives (called also table olives) (Fig. 4.6) are an edible product of great importance in Mediterranean countries; they are also consumed worldwide. Since several decades, fermented olives constitute a main component of Algerian, Moroccan and Tunisian cuisine, and in most regions, table olives are an essential food in all meals of the day. Indeed, the black olives accompanied the tea and the bread with the olive oil during the breakfast, the lunch and the diner (Fig. 4.6).

2.3.1 Nutritional Value of Traditional Fermented Olives

The average chemical composition of the olive is as follows: Water: 50%, Protein: 1.5%, Fat: 22%, Carbohydrates: 1.8%, Polyphenols: 1.5%, cellulose: 5.5%, Minerals: 1.5%. Chlorophylls and carotenoids are also present (Table 4.1).

Fig. 4.5 Crushing (**a**), mixing (**b**) and pressing (**c**) of roasted olive

2.3.2 Processing Technology of Traditional Fermented Olives

Three types of table olives can be identified according to the color of olives: Green olives is obtained from fruits harvested during the maturation cycle, before turning totally green and when they reached their normal size. The color of the fruit can vary from green to straw yellow. Semi black olives are obtained from rose-colored, vinous-pink or brown fruits, harvested before attaining their complete maturity. Black olives is obtained from fruits harvested at their full maturity; their coloration varies, depending on the area of production and the time of harvesting, from reddish to purplish black, to dark purple black and to dark brown.

Harvesting: The harvesting process is a delicate operation since olives must be picked carefully and should not be damaged.

Sorting and sizing: The damaged olives are discarded. The olives are classified in at least two categories: large and small.

Debittering: The time between harvesting and debittering should be as short as possible. It will not exceed 24 h at 20 °C and 5 days at 5 °C. The method of debittering olives depends on the color of the olives.

Debittering of green and semi black olives: The containers where the preparation is performed should be clean. The water used must be potable. Pieces of lemon can

Fig. 4.6 Green, semi black and black fermented olives

be used to speed up the softening of olives (Fig. 4.7a). Debittering is done with drinking water. Olives are placed in containers and are submerged with the water. The water is changed every day for a period of 5–15 days until the disappearance of the bitterness of the olives. We can achieve rapid bettering by breaking the olives before putting them in water (Fig. 4.7b).

Debittering of black olives: The black olives (Fig. 4.7c) are sprinkled with salt and put in bags hermetically sealed by strings (Fig. 4.7d). The bags are placed side by side, exposed to the sun, on a height of 2 bags to avoid a high pressure on the olives. Another plastic film is used in order to protect the olives from dust, rain and moisture from the air covers the bags.

Fig. 4.7 Green and semi black olives before (**a**) and after (**b**) the bottling and black olives before (**c**) and after (**d**) sprinkling with salt

Conservation: There are two methods of preserving olives. In the cold, without changing the brine, the olives are put between 3 and 6 °C. Fermented olives can maintain their quality for 10–12 months. In salt (brine), it is prepared by adding 30 g of salt per liter and placing the olives in the coolest possible place. Olives can be kept for up to 6 months and in all cases until the first heat (late June early July) (Fig. 4.6).

Aromatization: The addition of herbs (thyme, bay leaves, basil, fennel, chili, etc.) is done a few days only before the consumption of olives by mixing some of the olives preserved with these aromatics (Fig. 4.6). Thus, it will be preferable to carry out preliminary decoctions, to allow olives to cool and then to incorporate them into the conditioning brine.

2.4 Argan Oil

2.4.1 Argan Oil in Morocco

During the tertiary era, the argan tree or *Argania spinosa* benefited from a temperate climate, which allowed it to develop and spread over vast areas from Morocco to western Algeria. Subsequently, during the Quaternary ice age, it moved to the southwest. This explains its presence in the Rabat region, in northern Morocco, near the Mediterranean coast (Beni-Snassen Mountains) and northwest of Oujda. It is also found near Tindouf in Algeria. The Phoenicians were the first to use this very luminous oil in their oil lamps.

Argan oil comes from the fruit of the argan tree (Fig. 4.8a). This fruit, yellow green, looks like an olive but is both longer, about 3 cm and rounder. The shell (Fig. 4.8a) that encloses the nut represents about a quarter of the total weight of the fruit. The walnut can contain two to three "almonds" (Fig. 4.8a) which are submitted to a pressure at low temperatures, to extract the famous and delicate argan oil. The argan oil could be used for culinary or cosmetic purposes, depending on whether or not the seeds have been roasted.

Argan oil represents for the inhabitant of Souss a real "green gold" with several medicinal properties. According to the communities, it is also part of annual or seasonal rituals. Berber women uses argan oil for its cosmetic virtues and to heal since it is known to protect their skin and hair from sometimes extremely harsh weather conditions, strong winds, sun, etc.

2.4.2 Nutritional Value of Argan Oil

Argan oil is used to prepare, cook, and add taste to the food. It enhances the flavor of dishes: amlou, tagine, couscous, (Fig. 4.8b, c) and fish, etc. It is used for traditional Moroccan cuisine but also by the great chefs of Western countries. Roasting makes the oil less bitter and more delicious.

As for the nutritional value, 100 g of edible argan oil provides 899 calories (Table 4.1). Rueda et al., (2014) reported that the edible virgin argan oils contains: myristic acid $\leq 0.2\%$, palmitic acid 11.5–15.0%, palmitoleic acid $\leq 0.2\%$, stearic

Fig. 4.8 Fruit, shell and almond of argan (**a**), couscous with argan oil (**b**), Argan oil and Amlou (**c**)

acid 4.3–7.2%, oleic acid 43.0–49.1%, linoleic acid 29.3–36.0%, arachidic acid ≤0.5%, gadoleic acid ≤0.5%, and behenic acid ≤0.2%. In contrast to olive and sunflower oils in which β-sitosterol is predominant, the major sterols detected in the argan oils are schottenol (mean 147 mg/kg) and spinasterol (mean 122 mg/kg) (Khallouki et al., 2003). The same authors reported that argan oil contains several phenolic compounds such as vanillic acid, tyrosol, ferulic acid, and syringic acid. The concentration of polyphenols in argan oil is however extremely low (<5 mg/kg) compared to olive oil (>200 mg/kg) (Khallouki et al., 2003). The major vitamin identified in argan oil is γ-tocopherol with a mean value of 483 mg/kg (Khallouki et al., 2003).

2.4.3 Processing Technology of Extraction of Edible Argan Oil

Harvesting: The harvest of Argan fruits is done manually by climbing the tree. The uses of harvesting machines or mechanical shaking are not recommended as they damage the tree. The harvested Argan fruits are picked manually from the soil are air-dried by spreading them out on the roofs the houses. This practice makes sure that the fruits can be stored for years and used for Argan oil extraction when needed (Fig. 4.9a).

Peeling: In order to separate pulp and almond, the dried fruit pulp (the pericarp) is squashed between two stones (Fig. 4.9b). The pulp is then stored for later used as premium-quality concentrated feed for the domestic animals during the dry period, whereas the extracted light-brown almonds are further processed.

Shelling: In painstaking detail work, the nuts are then carefully cracked between two stones making sure that the almonds contained in the nuts remain undamaged (Fig. 4.9c). Each nut usually contains two or three almonds, from which the Argan oil is extracted. Argan almonds are of pumpkin seed size and are called pits or seeds. The Argan nut is fifteen times harder than a hazel nut.

Roasting: Roasting argan almonds accentuate and improve the nutty oil flavour. Two roasting systems could be used. The traditional roasting takes place over a light open wood fire by occasional stirring. The second one consists in the use of modern roasting equipment.

a b c

Fig. 4.9 Harvesting (**a**), pealing (**b**) and shelling (**c**) of argan shells

Extraction: As for the roasting, two extraction methods could be used: (1) The traditional extraction method consists in cooling down the argan almonds and then placing them in a manually powered stone mill for grinding. The almond mash, which is slowly running out of the mill, is expelled of the oil with care by constant kneading and the addition of lukewarm water. After that, the unfiltered oil is decanted into vessels (Fig. 4.10a). (2) The modern extraction method consists in the using of a thermodynamic extractor. The rotation of the screw is ~100 rpm. The temperature of the extracted oil (which should be lower than 44 °C) and its organoleptic properties are controlled (Fig. 4.10b).

Decantation and filtration: The Traditional decantation consists in filling the argan oil into a decanter. Due to pressing residues, the fresh oil is rather turbid in the beginning. After a resting period of ~2 weeks, suspended solids and solid components settle to the vessel bottom to form the oil's sediment. The clear Argan oil is then carefully decanted and filtered to ensure clearly improved quality. Depending on the degree of purity required, the argan oil could be filtered up to three times. The argan oil obtained is again decanted into a vessel. The microfiltration (Fig. 4.10c) consists in the use of a filter with a porosity of 1 μm and the obtained argan oil is usually filtered on three stages. The first filtration is carried out just after the extraction, while the two remaining ones are carried out after 24 h of the first.

Fig. 4.10 Traditional extraction (**a**), modern extraction (**b**) and microfiltration of argan oil

3 Cereal Products

3.1 Couscous

Couscous is a staple food in the Maghreb that requires very little utensils for its preparation. It is an ideal food for both nomadic and agricultural peoples. The preparation of couscous symbolizes "happiness and abundance," in the words of one culinary anthropologist. In Algeria, Morocco, Libya and Tunisia couscous is a staple, and is served with meat, fish and/or vegetables and spices. Cooked simply with sour milk and melted butter, it left the hungry traveller feeling full and was the traditional food of the poorest, namely the nomadic Berbers. For centuries, black African women were employed as couscous cookers, a phenomenon that might be indicative of the Saharan origins of couscous.

The Moroccan Saharans have developed several types of couscous, and the best known since centuries is the couscous of five seeds named Couscous El-Khomassi (Fig. 4.11). This couscous, made at home by Sahara housewives from the Laayoun-Dakhla region, is considered one of the best Moroccan couscous because of its nutritional value and its wonderful taste. Couscous El-Khomassi is a mixture of five types of flour: flour, barley flour, grilled barley flour, grilled corn flour and durum wheat and a little salt.

Another type of couscous characteristic of the Souss region of Morocco is known Berkoukech (Fig. 4.11). Berkoukech is a mixture of white flour and fine and medium-quality semolina, which is worked in an excellent way to obtain thicker seeds at the end of the procedure than ordinary couscous.

Fig. 4.11 Different couscous dishes

3.2 *Nutritional Value Processing Technology of Couscous*

Table 4.2 gave the nutritional value of couscous. The methods of preparation of couscous El-khoumassi and Berkoukech are the same; the only difference is the types of flours used.

3.2.1 Mixing, Primary Sieving and Granulation

After mixing flours, the women sprinkle a little water gradually to move the flour with the fingertips after each drop of water until the flour is floured, and this is the beginning step to form the couscous. The flour must be completely rubbed between

Table 4.2 Nutritional values of couscous El-khomassi (USDA Branded Food Products Database Release v0.0/Jan 2018), Azembou (USDA Branded Food Products Database Release v0.0/April 2018), bsissa of barley and Msemmen of wheat flour (USDA Branded Food Products Database Release v0.0/ May 2018) nutritional value

Nutritional value	For 100 g of couscous	For 100 g of Azembou	For 100 g of Bsisa	For 100 g of Msemen
Energy	420 kcal	375 kcal	283 kcal	321 kcal
Water	12.67 g	12.11 g	16 g	71.43 g
Carbohydrates	72.83 g	64.42 g	50.92 g	14.21 g
Proteins	12.68 g	10.5 g	11.83 g	1.79 g
Fats	1.05 g	1.6 g	6.13 g	9 mg
Fibers	3.9 g	10.1 g	14.7 g	10.7 g
Vitamins				
Vitamin-A	0.012 mg	–	–	–
Vitamin-B1	–	0.37 mg	–	–
Vitamin-B3	–	6.27 mg	–	–
Vitamin-B6	0.12 mg	–	–	–
Vitamin-C	3 mg	–	–	–
Vitamin-K	–	2.2 μg	–	–
Minerals				
Sodium	486 mg	4 mg	101 mg	–
Potassium	166 mg	309 mg	693 mg	346 mg
Phosphorus	70 mg	296 mg	–	–
Calcium	8 mg	32 mg	–	25 mg
Magnesium	8 mg	96 mg	–	–
Iron	1.67 mg	2.68 mg	–	3.93 mg
Zinc	–	2 mg	–	–
Selenium	–	0.038 mg	–	–
Cooper	–	0.37 mg	–	–
Others				
Polyphenols	–	29.7 mg	–	–

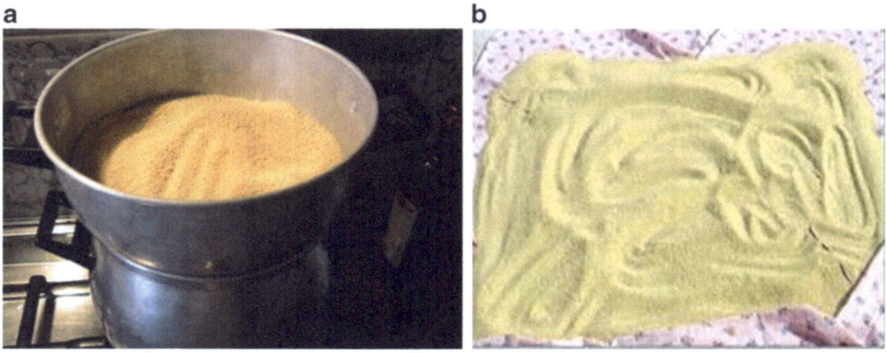

Fig. 4.12 Bleaching (a) and drying (b) of couscous

the hands. Then, the flour is put in the hot sieve and rubbed with the sieve with the palm of the hand until it comes out of the sieve as couscous with different grain sizes. The couscous is then sieved in three stages through sieves with progressively smaller holes called "ghurbal". It is sieved numerous times to form a uniform grain.

3.2.2 Bleaching and Drying

The bleaching (Fig. 4.12a) consists in placing couscous in the traditional equipment. After bleaching, the couscous is spread in a large bag beaten with a stick to separate the grains before performing a secondary sieving. The last operation consists in the drying. The couscous is spread on a clean blanket away from the dust all night (Fig. 4.12b), and the next day the fabric must be changed. The process is repeated four or five times, changing the fabric each time for the couscous to dry well and can be kept longer.

3.3 *Mesfouf in Algeria*

Mesfouf (Fig. 4.13a) is a sweet couscous dish with raisins that originates from North Africa. It is usually served in "souhour" during Ramadan due to its high nutritional value. The recipe varied from region to region since grenadine, cinnamon or other dried fruits like prunes, apricots, almonds or walnuts could be added. It is best served with milk as its sour taste complements the sweetness of fruits. To have a good dish with persistent flavor, it would be recommended to toast or roast the mixture of dried fruits (nuts, almond, etc.) in order to bring out their natural oils and to make them crispier.

a b

Fig. 4.13 Mesfouf with raisins, almond and nuts (**a**) and Mesfouf final dish (**b**)

3.3.1 Nutritional Value of Mesfouf

The nutritional value of Mesfouf is the sum of that of couscous indicated in the Sect. 3.2 and the nutrient values of the other ingredients (i.e. nuts, raisins, almond). The consumption of such dish reduces the risk of developing cancer, hypertension, diabetes, and heart disease (Goodson, 2018). In fact, Mesfouf is a good source of vitamins and fibres which are associated with a lower risk of developing diabetes and digestive problems.

3.3.2 Processing Technology of Mesfouf

Couscous preparation: A volume of water is placed in the bottom part of the couscous maker and heated until boiling. In the meantime, the couscous is placed in a bowl and olive oil is added. The couscous is then rubbed to make sure that all the couscous is coated. Then, a few quantities of water and salt are added and the couscous is left for a short time (~15 min) to allow water absorption. After that, couscous is worked between palms to break up all lumps.

Cooking: The couscous is placed in the steamer part of the couscous maker, covered with the lid and left until the steam permeates the couscous. This will take about 15 min. The couscous is then removed and placed in a bowl and left to cool for a couple of minutes. This procedure is repeated 3 times. Once the grains are ready (soft, nice and fluffy), the cooked couscous is placed in a bowl, and the grains are separated again and left to be cooled.

Addition of dried fruits: In order to garnish the dish, raisins are steamed until being soft and plump. The mix of raisins and the other dried fruits are added to the couscous and left to be cooled before enjoying (Fig. 4.13b).

3.4 *Azembou and Sellou*

Zembou or Azembou is also called Tazemmit is a typical dish of Rif (Northern region of Morocco) and Essaouira (Southern region of Morocco). The Zembou is related to the people of Al Hoceima, while Tazzemit for those of Nador and Azembou for the people of Essaouira. This product is a derivative of barley, which is a cereal that belongs to the family of Graminae. Barley has been consumed for millennia on the land, but has gradually been replaced by wheat in current consumption. The flour of Zembou is mixed with a little warm water and salt. A pressure is applied by the hand to make small piles from where it gets its particular form. It could be eaten with olive oil (Fig. 4.14a). For the Zembou consumed in the southern region, semolina is used to prepare couscous with goat meat and argan oil, a sign of hospitality, and respect for the guest (Fig. 4.14b). For flour, it can be served directly with honey and argan oil or used to prepare Sellou (Fig. 4.14c).

Fig. 4.14 Zembou with olive oil (**a**), couscous with Azembou (**b**), Sellou (**c**) and pottery couscoussier (d)

3.4.1 Nutritional Value of Azembou

The nutritional value of Azembou is given in Table 4.2. It is a very nutritious dish, low-fat and high in fiber that have been used to help regulate lazy digestive systems. Barley is recognized for the prevention of certain cancers, and helps to reduce the risk of cholesterol, diabetes and cardiovascular diseases. It is a good source of Vitamin B and various minerals such as iron, zinc, magnesium, phosphorus, copper, selenium.

3.4.2 Processing Technology of Azembou

The name Azembou refers to two products: semolina or barley flour. The production is described as follow:

Harvesting, threshing, cooking, and drying: Barley is harvested at the pastry stage during the March–May period of each season. Manual threshing of ears is performed by women's to obtain barley seeds. Cooking step consists in exposing the barley grains to steam with a pottery called "couscoussier" (Fig. 4.14d) (Kiskis) for a period of 20–30 min. Drying step consists in displaying seeds in a thin layer at ambient air-drying (Fig. 4.15a). The relative humidity of the ambient air, the cleanliness of the air and the available surface area significantly affects the quality of Azembou.

Roasting, grinding and sieving: After roasting of the seeds, manual grinding is performed using a stone grinding wheel followed by sieving/calibration, which consists in passing the mixture through a wire mesh screen. Therefore, two products, are obtained: (1) the semolina (large size retained by the first sieve); and (2) the flour (small size) that passes through the sieve (Fig. 4.15b).

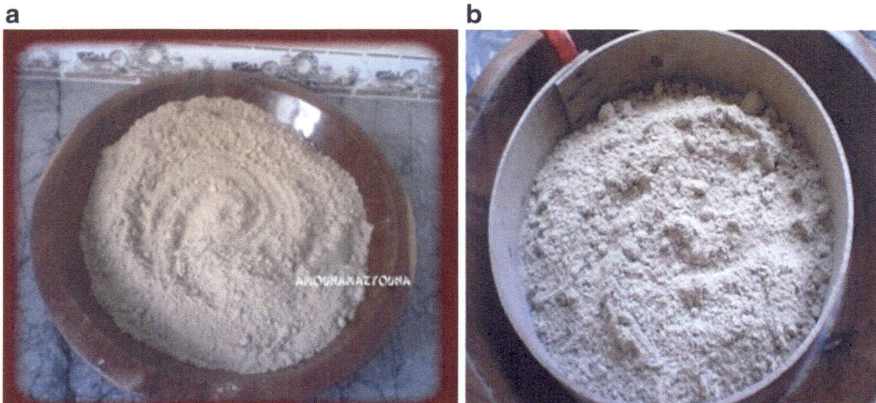

Fig. 4.15 Drying (**a**) and sieving (**b**) processes

3.5 *Hlalem*

3.5.1 History of Hlalem

The hlalem are a pasta-based dish, prepared in the form of soup, from the Tunisian culinary tradition. The pasta (Fig. 4.16a) is cooked as a soup in hot water with tomatoes, onions, starchy foods (chickpeas, beans, lentils or peas) and meat (fresh or dried) (Fig. 4.16b). Hlalem is nutritious and a good source of fibers that can be consumed to regulate the digestive system.

3.5.2 Process Technology of Hlalem

To make the pasta, 600 g of semolina are placed into a container and 1 teaspoon of salt is added. Then cold water is added until obtaining a malleable paste. The mixture is then covered with a clean cloth and let stand for 15–30 min. At the end of this time, the dough is kneaded and the pasta is made. For this, a piece of dough is taken and rolled between the hands to have a big pudding of 10 cm. Then, one side of the dough is refined between the hands and rolled and pieces of pasta were hand cut of 1 cm length over a large tray (Fig. 4.17a).

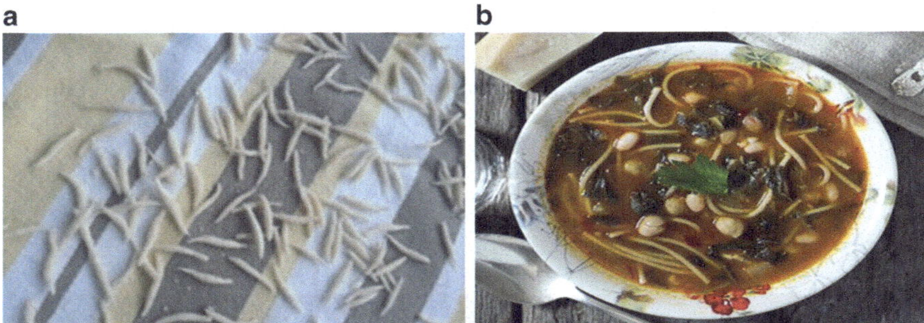

Fig. 4.16 Pasta (**a**) and dish (**b**)

Fig. 4.17 Preparation (**a**), sieving (**b**) and drying (**c**) steps of hlalem

The pasta should not overlap (Fig. 4.17b), otherwise they would stick together. Once the pasta is prepared, it is dried in the sun for one to several day(s) to get dry pasta (Fig. 4.17c). The pasta is ready when they do not stick together anymore. The pasta can be kept very well in an airtight jar for several weeks.

3.6 Bsissa

3.6.1 History of Bsissa

Bsissa (Fig. 4.18a, b) is a typical North African traditional food; Tunisian Bsissa is based on the use of flour of roasted barley which dates back to Roman times and Carthaginians. Ancient Roman, Byzantines used barley and other grains to make bsissa and when it comes to the North African, the Berbers add lentil, chick peas (hummus) and seeds to get a product with high nutritional value. Bsissa is a variety of mixtures of roasted cereals ground with fenugreek, anis seed, cumin and sugar. This kind of food is known throughout Tunisia and Libya.

Fig. 4.18 Bsissa (**a**, **b**), Rowina (**c**) and festival of bsissa in Lamta, Tunisia (**d**)

The herbs and spices added to the mixture varied according to the country and even from region to region, and the mixtures can also be used as a liquid when added to milk or water, creating a strongly floured drink called Rowina (Fig. 4.18c). The most common use for bsissa is to mix them with olive oil into a paste. This product is typically eaten with dates or figs for a quick meal which is energy-rich and healthy. It is eaten as a paste with olive oil or honey, warming in winter or as a refreshing summer drink, obtained after dilution of the powder in the water.

Bsissa, tamina and zrira are specialties of eastern Algeria whose name differs according to the region. It is prepared for birth and prophet's birthday (mouloud) or breakfast. It should be noted also that their preparations and compositions varied from one region to another.

Lamta, a Tunisian city, organizes every year its festival of bsissa (Fig. 4.18d). This event continues to provide inhabitants moments of great happiness. This competition brings together participants from several regions of Tunisia who present different varieties of bsissa. The best bsissa is chosen by a jury of experts tasting.

3.6.2 Nutritional Values of Bsissa

Bsissa is a very important source of fiber, vitamins, minerals and antioxidants. Bsissa contains important nutrients that can offer a range of health benefits such as decreasing the risk of heart disease, cancer, diabetes and obesity. Moreover, two to three spoons of bsissa are sufficient to provide high quantity of calories. It is generally given to women who have just given birth to regain strength and increase the flow of milk (Table 4.2).

3.6.3 Process Technology of Bsissa

The barley, wheat and chickpeas are grilled, separately, until obtaining golden brown color. The ingredient is stirred continuously to avoid burning and then cooled at ambient temperature. The different ingredients (barley and spices for rowina and wheat, chickpea and spices for bsissa) are ground using an industrial (Fig. 4.19a) or traditional (Fig. 4.19b) crushers allowing to obtain powder (Fig. 4.19c) which is the

Fig. 4.19 Grounding using modern (**a**) and traditional (**b**) processes getting the powder (**c**)

base for obtaining rowina or bsissa. To obtain bsissa, olive oil is added to the powder until obtaining a consistent paste; at this stage, almond and other dried fruits could be added; for rowina, the powder is diluted with a quantity of water until obtaining fairly soft dough.

3.7 Melaoui

3.7.1 Ingredients

The melaoui or msemmen (Fig. 4.20a, b) is a Moroccan culinary specialty. These pancakes have the distinction of having dough prepared from very fine semolina and a little wheat flour to which is added olive oil giving them a specific taste. Unlike other pancake recipes, melaoui do not contain milk and eggs which are replaced by warm water and olive oil. The melaoui could be served hot or cold, at breakfast or afternoon, drizzled with honey or sugar, topped with eggs, or simply plain. A stuffing with onions, tomatoes, peppers, a strong spicy and fragrant sauce is often added, to which could be added either minced meat, chicken, and peppers.

The msemmen are also used for the preparation of the rfiss (Fig. 4.20c), a dish where the puff pastries are cut roughly into pieces and served with onions and chicken or turkey meat in the occasion of a birth in honour of the mother. There is also a sweet variant of the rfiss served as a dessert, where the msemmen leaves are cut more finely and sprinkled with icing sugar and almond chips. This sweet dish, prepared in the occasion of a wedding (the day after the wedding for the newlyweds), is often accompanied by hot tea.

3.7.2 Nutritional Value and Processing Technology of Msemmen

For a preparation of the msemmen with 45% flour, 45% fine semolina, 8.5% olive oil, 1% salt and 0.5% baking powder, energy of 332 kcal is provided for 100 g of msemmen (Table 4.2). The first step corresponds to sieving the flour (Fig. 4.21a)

Fig. 4.20 different formats of melaoui (**a**, **b**) and rfiss (**c**)

Fig. 4.21 Processing technology of msemmen: (**a**) sieving, (**b**) mixing, (**c**) Partitioning, (**d**) flattening, folding (**e**) and cooking (**f**)

followed by adding water and salt little by little. All the ingredients are then kneaded in order to obtain a homogenous ball (mixing) (Fig. 4.21b). The partitioning step consists in cutting the dough into small balls (Fig. 4.21c). These latter are spread on an oiled surface as finely as possible to obtain a large round of very fine shaped dough (flattening) (Fig. 4.21d). Then, these fine shaped dough are folded on themselves to obtain squares about seven centimetres apart (Folding) (Fig. 4.21e) and the squares are flatten as thin as possible in order to obtain uniform and fine crêpe-like flatbreads. The last operation is the cooking (Fig. 4.21f) *of the msemmen* on by heating it in a heavy-bottomed pan. The pan must be very hot by adding some drops of oil. The paste is cooked on both sides, flipping over several times until obtaining a golden brown color (5–6 min).

3.8 Lablabi

Lablabi (Fig. 4.22a) is a Tunisian culinary speciality. It is the most popular dish in winter. The dish is prepared with chickpeas, bread, harissa, tuna, and pepper.

3.8.1 Nutritional Value of Lablabi

Lablabi is a good source of carbohydrates and protein. Moreover, the chickpea, the major compound in the recipe, has significant amounts of the essential amino acids and unsaturated fatty acids like linoleic and oleic acid. Calcium, magnesium, phosphorus and especially potassium are also present in lablabi. In addition, chickpea is a good source of vitamins such as riboflavin, niacin, thiamine, folate, vitamin A precursor and β-carotene. Indeed chickpea has several potential health benefits on some of the important human diseases like cardiovascular disease, type 2 diabetes, digestive diseases and some cancers (Jukanti et al., 2012).

3.8.2 Processing Technology of Lablabi

The different steps for preparing lablabi are presented as follows: (1) chickpeas are placed in a large bowl and water is added. The chickpeas, totally covered by water are soaked overnight (Fig. 4.22b), (2) the chickpeas and the water are then poured into a large pot and brought to a boil. Then the heat is lowered and the mixture is allowed to cook partially covered until a very tender product is obtained (Fig. 4.22c), and (3) salt, pepper, and cumin are added to the mixture, which is stirred before

Fig. 4.22 Lablabi dish (**a**) and different steps for producing lablabi (**b–e**)

removing the pot from heat (Fig. 4.22d), (4) lablabi is served in a bowl to which old (stale) bread is added in small piece and allowed to in the broth. A generous spoonful of harissa, tuna, and a poached egg are added to the top (Fig. 4.22e).

4 Dairy Products

4.1 Fermented Camel Milk

The land of milk production concerns the entire pastoral area of the countries of the Sahel and Sahara that includes Algeria, southern Morocco, Tunisia and; northern Mali and Chad as well as all the territory of Mauritania and Libya. Camel breeding is the Saharan type, which is subdivided into three historical breeds named Guerzeni, Marmouri and Khouari. The most adopted farming system is of the pastoral type with permanent mobility and distant movements of herds (transhumance) in search of pasture.

Camel milk is a product consumed fresh alone or with dates, and it is used in the preparation of local recipes such as "Zrig", a well-known drink in the Saharan region made by a mixture of milk, water and a little sugar. It is also traditionally appreciated for its anti-infectious, anti-cancer, antidiabetic and more generally as a restorative in convalescent patients (Dheeb et al., 2015).

The local population recognizes that camel milk has medicinal properties (antidiabetic, antituberculous and antibacterial). Traditionally, nomads served camel milk in specially made wooden bowls. The sale concerns 20% of milk production, while the remaining majority is self-consumed. Nomads have developed several products based on camel milk such as fermented milk (Lfrik in morocco) and cheese.

4.1.1 Lfrik

This derivative of camel milk is particularly popular among consumers in the southern regions of Morocco. Besides its usual preparation in the households, it is sold in creameries, artisanal dairies and certain cooperatives of the cities of the region (Fig. 4.23a). Moroccan prohibits the marketing of dairy products obtained from raw milk that did not undergo a heat treatment at least equivalent to pasteurization during their manufacture.

4.1.2 Fresh Camel Cheese

The fresh cheese of camel milk draws its typicality from the specificity of the raw material (Fig. 4.23b): Camel milk is known for its superior nutritional quality and its therapeutic virtues because of its composition, especially its richness in vitamins and trace elements.

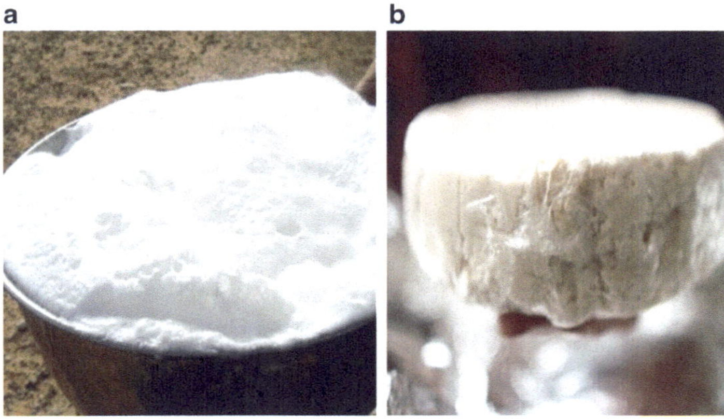

Fig. 4.23 Fermented camel milk (**a**) and fresh camel cheese (**b**)

4.1.3 Nutritional Value of Camel Milk

In camel milk, the fats representing 2.7–3.6% of the overall composition are dispersed in the form of fat globules. In addition, the fat of camel milk appears bound to proteins; this trend could explain the difficulty in churning camel milk to extract the butter. Lactose is the most important carbohydrate in milk. Its content in camel milk varies from 3.4% to 5.6%, with average levels slightly higher than those found in cow's milk (Ismaili, Saidi, Zahar, Hamama, & Ezzaier, 2016). Camel's milk is a considerable source of proteins and peptides capable of modulating various physiological functions. Nutritionally, it is of good quality since all the essential amino acids are found. The protein content in camel milk is in the 3.0–3.5% range. Camel's milk is richer in vitamins (B_3, B_6, and B_{12}) than bovine milk (Ismaili et al., 2016) (Table 4.3).

4.1.4 Processing Technology of Fermented Milk and Fresh Camel Cheese

Fermented camel milk (Lfrik): The traditional process consists in placing camel milk (Fig. 4.24a) in a goatskin (Fig. 4.24b) and allowing it to acidify milk by spontaneous fermentation at room temperature for about 12 h. The milk is then churned induce its fluidization. This traditional process is similar to that used to produce lben (or laben) from cow's milk, except that the latter is generally more acidic and partly skimmed, because of the separation of the butter during churning (Fig. 4.24c).

Camel milk cheese: The first step consists in the heating of milk at 70 °C for 15 min. The addition of other milk species (cow, ewe) and/or powder milk is performed in order to increase the dry matter content. Then the milk is allowed to coagulate by the addition of a coagulant (enzymes) and/or lactic ferments (Fig. 4.24d).

Table 4.3 Nutritional values of Dried camel meat, hump fat (Alaoui Ismaili et al., 2016), lamb meat nutritional value (Cabrera & Saadoun, 2014) and camel milk (Ismaili et al., 2016)

Nutrient value	For 100 g of dried camel meat (Tichtar)	For 100 g of lamb meat	For 100 g of hump fat	For 100 g of camel milk
Energy	236 kcal	294 kcal	857 kcal	491 kcal
Water	62 g	54 g	–	89–91 g
Carbohydrates	1.3 g	g	–	3.4–5.6 g
Protein	16 g	25 g	–	3–3.5 g
Fats	20 g	21 g	99.98 g	2.7–3.6 g
Cholesterols	–	–	10 mg	–
Vitamins				
Vitamin-A	–	–	21 μg	
Vitamin-B1	0.12 mg	–	–	
Vitamin-B2	0.18 mg	–	–	
Vitamin-B3	–	–	–	0.46 mg
Vitamin-B5	–	–	–	0.25 mg
Vitamin-B6	0.25 mg	–	–	
Vitamin-B12	–	2.6 μg	1 μg	
Vitamin-C	–	–	–	2.4–3.6 mg
Vitamin-D	–	2 mg	–	
Vitamin-E	0.61 mg	–	1 mg	0.5 mg
Minerals				
Sodium	–	–	–	36–62 mg
Potassium	–	310 mg	1 mg	60–210 mg
Phosphorus	–	–	–	58–104 mg
Calcium	6.5 mg	17 mg	–	106–157 mg
Magnesium	–	23 mg	–	7.5–16 mg
Iron	–	1.9 mg	–	0.25 mg
Zinc	–	–	–	0.5 mg
Copper	–	–	–	0.18 mg

5 Meat Products

5.1 *Tichtar and Loudeg*

The first traditional product made from the dromedary meat known in the Saharan territory of morocco is the dried dromedary meat called Tichtar (Fig. 4.25a), which can be kept for long period (2–3 months). It is prepared after the slaughter of a dromedary, and its distribution is made in equal parts between many families living in the tents commonly called "lafrik". The meat is then cut and exposed to the sun. Once fully dried, the meat is placed in special bags. According to the Saharawi culture, the fact of offering this meal represents the greatest honour that a Saharawi could make to his guest.

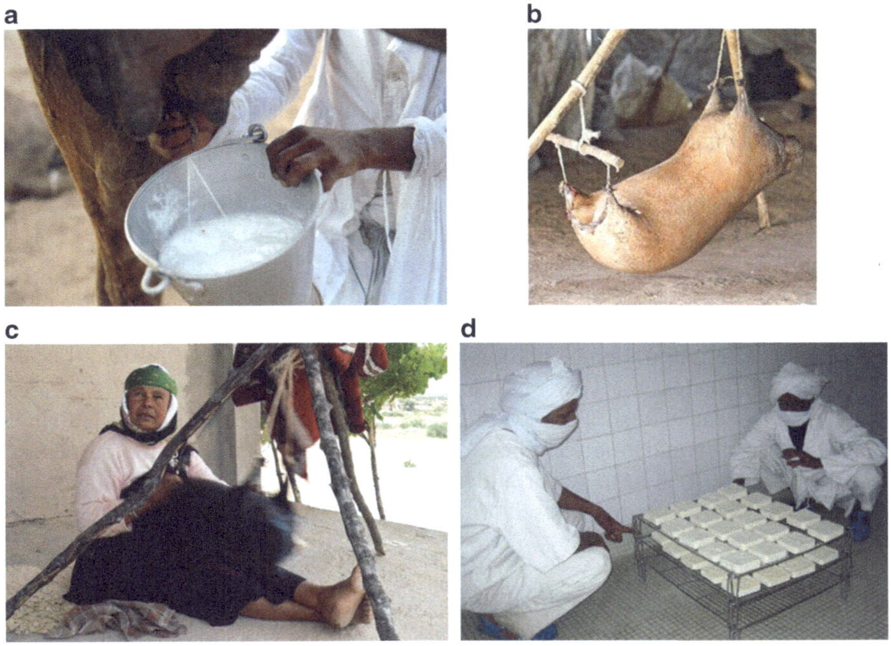

Fig. 4.24 Camel milking (**a**) and goatskin used for the fermentation (**b**), churning (**c**) and ripening stage of camel cheese (**d**)

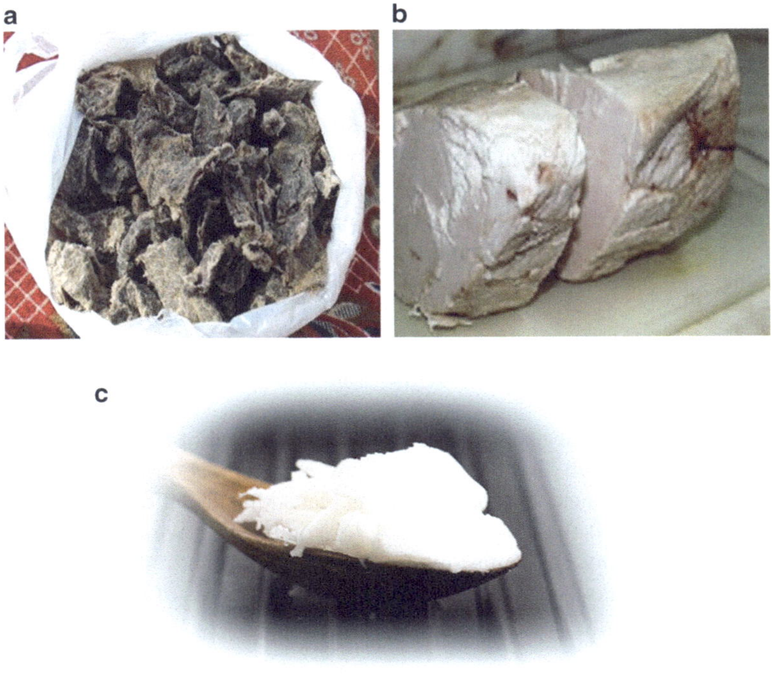

Fig. 4.25 Tichtar (**a**), camel hump (**b**) and Loudeg (**c**)

Tichtar can accompany culinary preparations. Among the main meals that could be prepared are rice with cooked or dried meat (Tchitar), which is a dish that is easy to prepare and nutritious at the same time. The second traditional product using camel meat is the fat of the camel hump (Fig. 4.25b) called Loudeg (or Loudek) (Fig. 4.25c), and characterized by its viscosity, white colour and typical taste. The production area of Loudek concerns the entire Saharan ecosystem of the Oued Eddahab Lagouira region of Morocco. It is an ancestral product whose secrets of use are preserved by the local population for a long time. The Loudek's preparation consists of cutting the fat from the camel hump into small pieces that will undergo traditional melting, filtration, conditioning and cooling processes.

The use of loudek is a very old Saharawi tradition of "fattening" girls before marriage. It is eaten fresh alone or in combination with aromatic and medicinal plants. The local population recognizes that loudek presents therapeutic properties when used in the food or to perform massages. If "Läalk" (Acacia radiana gum) is added to loudek, the product is used for therapeutic (Amina, 2017).

5.1.1 Nutritional Value of Camel Meat

Water content in fresh dromedary camel meat is in the order of 75% (Kadim, Mahgoub, & Mbaga, 2014), which is comparable to meats from other species. After drying the dromedary meat, the value of protein increases from 14 to 16%, while the fat content that varies between 1.4 and 7% in fresh meat increases to more than 20% in dry meat. Minerals varied between 1.1 and 1.5% of the meat. The calcium content (~6.5 mg/100 g) appears to be higher than in beef meat, which is due to the tighter structure of camel meat (Table 4.3) (Ismaili et al., 2016).

Concerning the vitamin contents, it is ~0.12 mg/100 g for thiamine (B1), 0.18 mg/100 g for riboflavin (B2), 0.25 mg/100 g for pyridoxine (B6) and 0.61mg/100 g for α-tocopherol (vitamin E), which represents a commercial argument to affirm the healthy character of this product. The hump, which is counted with fat, represents 8.6% of the weight of the carcass, and it is composed entirely of fats (Table 4.3).

5.1.2 Processing Technology of Tichtar and Loudek

The production of Tichtar needs in a first step the removal of fat. Then the meat is dried (Fig.4.26a) through direct exposure to solar origin heat during 3 weeks. Once completely dried, the meat called "Tichtar" is placed in special bags and stored for future uses. The meat is then cooked (Fig. 4.26b) on low heat with a little water, and it is moulded until a dough form is obtained. A small amount of the liquid in which the Tichtar has been cooked is added thereafter. Special types of fats are sometimes added to the Tichtar.

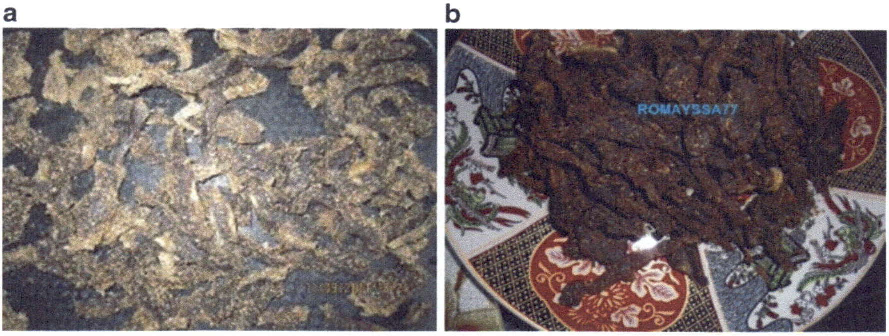

Fig. 4.26 Drying of meat (**a**) and cooked Tichtar (**b**)

Fig. 4.27 Cutting of fat hump (**a**) melting of fat hump (**b**), filtration (**c**) and conditioning of Loudek (**d**)

Regarding the production of loudek, the fat of the dromedary hump is cut (Fig.4.27a) into small pieces and then melted by heating over low heat in a deep bowl with frequent mixing. When all the fat is dissolved and the color turns golden (Fig. 4.27b), the liquid fat is filtered (Fig. 4.27c) and then packaged in boxes and left to cool (Fig. 4.27d).

5.2 *Kaddid and Khaliî*

Kaddid or Gueddid is a dried meat usually prepared from mutton meat, salted and dried in the open air. The traditional method for their preparation requires patience and attention (Fig. 4.28a). Once ready, the Gueddid is incorporated in many winter dishes especially in couscous to replace meat. In Algeria, Morocco and Tunisia, the main dishes prepared with Gueddid are couscous of wheat or barley, and also with the typical couscous called Berkoukes in Algeria and Berkoukech in Morocco.

In Tunisia, kaddid is exclusively chop meat and tripe, usually from the sheep scarified on the occasion of Eid Al-Adha. Couscous with kaddid and vegetables is prepared to celebrate New Year's Day in the Muslim calendar. In Morocco, Gueddid is consumed with couscous on the tenth day of the first month of the year in the Muslim calendar; this day is called Achouraâ. For centuries, Gueddid has been the only way to preserve meat. Today, despite the arrival of freezers, this type of product is still produced due to its typical characteristics.

"Khaliî" (Fig. 4.28b) is a recipe from Fes (Morocco). It is prepared during the summer and winter periods. Many mothers still respect this practice today and seize the opportunity of Eid Al-Adha, a time when meat is available in abundance to perpetuate this ancient culinary practice (Fig. 4.28c, d).

Fig. 4.28 Kaddid (**a**), Khaliî (**b**), couscous with Kaddid in Tunisia (**c**) and Khaliî with eggs in Morocco (**d**)

5.2.1 Nutritional Value of Kaddid and Khaliî

Lamb meat contains a high proportion of saturated fatty acids. In fact, they account for about 50% of the total amount of fat found in these meats, which is comparable to the proportions found in beef meat (Cabrera & Saadoun, 2014). The fat content of lamb meat contains ~ 44% of monounsaturated fatty acids, for which more than 80% constitute oleic acid. In general, the consumption of monounsaturated fatty acids would allow the reduction of total cholesterol without modifying the concentration of HDL cholesterol in the blood. The lamb chops used in the preparation of Kaddid and Khaliî are excellent sources of iron, zinc, phosphorus, vitamin B2 and vitamin B12 (Table 4.3).

5.2.2 Processing Technology of Kaddid and Khaliî

Process technology of Kaddid: The meat is cut into long strips whose width should not exceed 3 cm. Then crushed garlic and coriander, cumin and salt are mixed in a large container (Fig. 4.29a). The cut meat is added to the spice mixture and left to rest for 2–3 days in the refrigerator. The spiced meat strips are then spread out in the sun to dry (Fig. 4.29b). Late in the afternoon, the meat is covered with a clean cloth

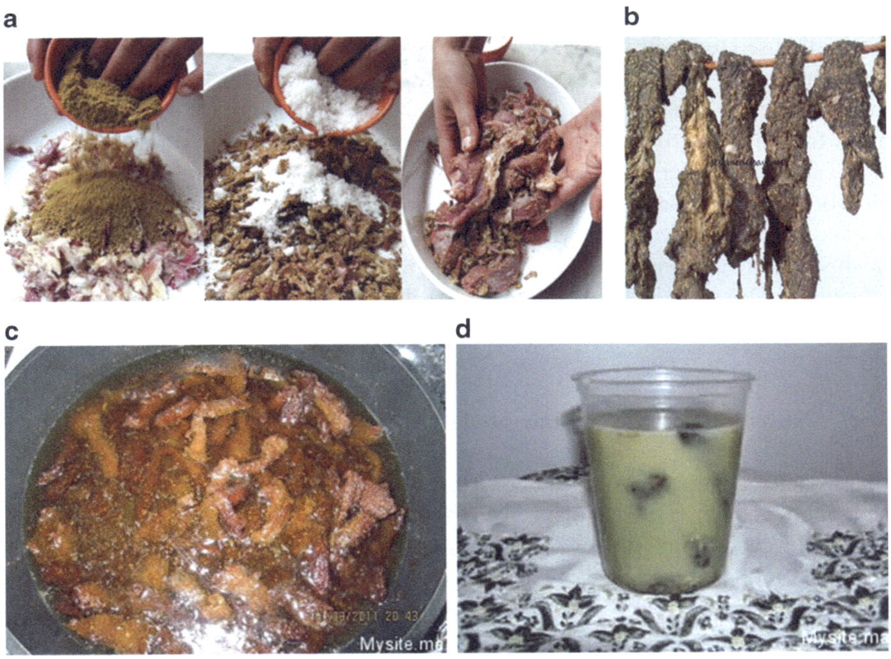

Fig. 4.29 Preparation of the spices and their mixture with meat (**a**), drying of meat to obtain Kaddid (**b**), cooking (**c**) and packaging (**d**) of Khaliî

so that it does not become impregnated with the humidity of the evening. The next day the linen is removed and the meat is allowed to further dry at the sun. In all, the drying period takes 4–5 days until the meat becomes hard and breaks between the fingers. When the meat strips are dry (turning into Gueddid) they are kept in a cloth bag protected from moisture.

Process technology of Khaliî: The first step consists in the preparation of the fat and marinade. For this, 4 kg of beef fat is cut into small pieces. Eight liters of water are placed in a pot, heated until boiling and fat is added and covered. The mixture is simmered 10 min and then 7 L of vegetable oil, 1 L of olive oil and the rest of the marinade in the form of a powder are added. The dried meat is then incorporated in the pot is covered and heated until boiling; the lid is then removed and the mixture is allowed first to cook (Fig. 4.29c) at high temperature for 30 min, then on a moderate heat treatment until the water is completely evaporated (Fig. 4.29d). Once the Khaliî is cooked, the fat is placed into a jar; the pieces of meat are removed with a pin and placed in a deep container. Once cooled, the meat is placed in the jars, covered with the liquid fat and left uncovered for a few hours and then the jars are closed tightly.

5.3 Charmoula

Charmoula (Fig. 4.30a) is a marinade and relish used in Tunisian cooking. It is traditionally used to flavor fish or seafood, but it can be used on other meats or vegetables. Originally, there were two versions known as the current Charmoula and the raw Charmoula. The latter is a mixture of very sweet raisin juice of the variety "zbib asli", from a vine grown in Sfax and Djerba, sprinkled with vinegar and seasoned with caraway powder and olive oil. The raw charmoula is still consumed by some elderly people, still attached to this method of preparation. On the other hand, the cooked Charmoula, currently more widespread, is a kind of marmalade based on raisins, onions seasoned with ingredients and aromas such as clove, rose bud, and cinnamon. This dish is served particularly with salted fish (Fig. 4.30b) and bread (Fig. 4.30c) in Sfax the day of Eid.

5.3.1 Nutritional Value of Charmoula

Charmoula sauce provides a source of many nutrients including potassium, fiber and vitamins A, E, and C. Dietary fiber from vegetables helps to reduce blood cholesterol levels and may lower risk of heart disease. In addition, fish is among the healthiest foods on the planet. Fish is also the world's best source of omega-3 fatty acids, which are incredibly important for the body and brain. It is loaded with other important nutrients, such as protein and vitamin D.

Fig. 4.30 Charmoula (**a**) served with salted fish (**b**) and bread (**c**)

5.3.2 Processing Technology of Charmoula

The following diagram represents the different steps for producing salted fish. To prepare the accompanying sauce, the following steps should be realized: (1) Heat the oil and add the finely chopped onions. Cook over low heat the mixture for 2 h (Fig. 4.31a). (2) Wash the black raisins well and mix them with 1/2 L of water. Then let the resulting mixture rest 40 min (Fig. 4.31b). (3) Filter the mixture on a fine sieve to extract the juice of black raisins (Fig. 4.31c). (4) Mix several times the mixture with the extracted juice to extract the maximum of aroma, and until obtaining dough (Fig. 4.31d). (5) Pour the raisin juice (only raisin juice and nothing else) on cooked onions (Fig. 4.31e). (6) Add the ground spices (Fig. 4.31f) and cook the mixture another hour on low heat (Fig. 4.31g).

5.4 Tagines in Maghreb

A tagine is a culinary preparation from Northwest Africa, cooked in specific utensil, a kind of stew stewed, which may be composed of a mixture of meat, poultry or fish, vegetables or fruit and spices, and olive oil (Fig. 4.32). The preparation of tagine, the ingredients and the spices used differ according to the countries of the Maghreb and even in the same country from region to region.

Some sources say that the origin of tagine is Berber, while there are sources that tell that tagine dates back to the Abbasid Caliphate in Iraq. The earliest writings about the concept of cooking of tagine appear in the famous "One Thousand and One Nights", an Arabic-language story collection from the ninth century. The dish would have been already famous amongst the nomadic Bedouin people of the Arabian Peninsula, who added dried fruits like dates, apricots and plums to give it its unique taste.

Fig. 4.31 Different steps for producing charmoula (**a–g**)

5.4.1 Moroccan Tagine

The preparation recipient of this dish is also called tagine. It is a cooking and serving recipient, wide and shallow, surmounted by a conical lid, terracotta sometimes glazed, resulting from the Berber cuisine, and also widespread in the traditional cuisine of the Maghreb (Fig. 4.32). The tagine is a symbol of generosity, so it must be rich in flavours and aromas. Thus, usually meat, vegetables and spices were added: turmeric, curry, nails clove, cinnamon, etc. Today, there are a different kinds of tagines, which contain meat, vegetables and fruits (there are many sweet recipes salty in Moroccan cuisine: tagine with olives, quince, date figs or apricots), all this with beef, lamb or chicken. The utensil used in the preparation of tagine is resistant to very high temperatures and can be used for cooking over a wood fire (called kanoun), or on gas or an electric plate, or even in the oven for a more modern use.

Its manufacture meets a local need to cook food without water (and without fat) to retain maximum natural flavour. The food vapour condenses inside the conical walls of the lid and falls to moisten the assembly. Food becomes tender, fragrant and melting. Traditionally made in terracotta, it is solid, slightly porous and hardens even more with use. In addition, it is resistant to thermal shock. Before its first use, it must be soaked for at least one night in water. Decorated tagines are also used for decoration. A real traditional tagine is in the ground sometimes glazed.

Fig. 4.32 Different kinds of tagine in Maghreb

5.4.2 Tunisian Tagine

What Tunisians refer to as a "tagine" is very different from the Moroccan dish. Tunisian tagine is more like an Italian frittata. It is scrambled eggs, enriched with additional ingredients such as meats, cheeses and vegetables (Fig. 4.32).

5.4.3 Nutritional Values of Tagine

It is very difficult to define the amount of calories brought by a tagine. Indeed, there is no one single recipe for tagine. This typical dish of West Maghreb can be declined in a multitude of recipes. What brings these recipes together is their method of preparation and the fact that they all revolve around a meat or a fish. We can find a tagine of beef, lamb, chicken but also cod or even shrimp.

The average total calories of tagines are about 165 kcal for a lamb tagine with vegetables. However, this nutritive value can vary greatly depending on the fat content of the meat used and its accompaniment. Although most tagines are cooked with vegetables, some recipes incorporate legumes or cereals. However, it should be noted that recipes rarely require adding more fat than a little oil for cooking because the dish is stewed. The nutritional value of lamb Tagine with Vegetables is shown in Table 4.4.

5.5 Mloukhiya

Mulukhiyah, mloukhiya, molokhia, molohiya, mulukhiyya, malukhiyah, or moroheiya is the leaves of Corchorus species (Jute leaves) (Fig. 4.33a) used as a vegetable in Middle Eastern and North African cuisine. Mloukhiya is generally eaten cooked, not raw, and is most frequently turned into a kind of soup or sauce (Fig. 4.33b). In Tunisia, the dish is generally prepared quite differently from the Egyptian method. The leaves, already separated from the stems are dried then ground to produce a very fine powder (Fig. 4.34a) and stored in jars or other tightly closed containers.

In Tunisian cooking, Mulukhya, or Mloukhiya, takes 5–7 h to prepare. The powder is prepared with olive oil and some tomato paste. Pieces of beef are often

Table 4.4 Nutritional values of tagine (Cabrera & Saadoun, 2014; Slavin & Lloyd, 2012) Pastilla (Charlton et al., 2008; De Oliveira Sousa et al., 2011)

Nutrient value	For 100 g of Tagine	For 100 g of Pastilla
Energy	165 kcal	389 kcal
Carbohydrates	14.9 g	27 g
Protein	7.3 g	25 g
Fats	8.4 g	20 g
Fibers	–	3 g

Fig. 4.33 Leaves of Corchorus (a) and Mloukhiya dish with piece of beef (b)

Fig. 4.34 Mloukhiya in the form of powder (a) and sauce (b)

added to the sauce halfway through cooking. The sauce is served in small deep plates with a piece of beef (Fig. 4.34b). Mloukhiya is rich in vitamins A and B, minerals (such as sodium, potassium, magnesium, and iron), fiber and carbohydrates. In addition, it stimulates the stomach, strengthens immunity and protects the mucous membranes, the digestive system and brain cells.

6 Sweet Products

6.1 Samet and Raisins

The precious Samet (Fig. 4.35a), fermented braised grape juice, is one of the outstanding agricultural and artisanal products of the region of Jbala. This product is a witness of a past wine activity. The Samet, a Moroccan traditional product; originated in the middle Ages (sixteenth century), where it knew a very great use in all Morocco. In addition, the cultivation of the vine was very extensive and based on

Fig. 4.35 Samet (a) and raisins (b)

the use of different local varieties. This culture was so important that some ancient texts name these mountainous regions of the Rif: "Jebel Zebib". Production was too important for fresh consumption, hence the need to transform the surplus. Part of the production is therefore transformed into Samet and the other into raisins (Fig. 4.35b).

6.1.1 Nutritional Value of Raisins

Like all dried fruits, raisins are excellent energy sources, available all year round, easy to carry and easy to store. As grapes, raisins contain many phenolic substances with formidable antioxidant power; and especially a high amount of Resveratrol, a phenol that protects the heart and would help prevent the risk of cancer associated with a healthy lifestyle. Raisins also contain many isoflavones, natural antioxidants that improve the function of the blood system. They are four times richer in trace elements than fresh grapes: calcium, iron, magnesium, potassium; carbohydrates; as well as vitamins, especially of group B (Table 4.5).

6.1.2 Processing Technology of Samet and Raisins

Process technology of samet: All vine varieties can be used for the production of Samet. When the vines reach maturity, the grapes are harvested, rinsed, and then crushed in the manner of the harvest, with the feet, in large barrels. The resulting juice is filtered before being placed in terracotta containers for cooking in a wood stove for more than 8 h. After cooling, the samite is preserved for a long time in jars or, currently, in bottles.

Process technology of raisins: In the case of grape drying and the production of raisin (locally called zbib), a particular process is used: with rinsing water for washing fruits, lentisque ashes, and inula leaves (*Dittrichia viscosa*) are added to aromatize the mixture. Then, the grapes already harvested and dried in the sun during 2–3 weeks are immersed in this mixture.

Table 4.5 Nutritional values of raisins (USDA Branded Food Products Database Release v0.0/April 2018), dried fig (USDA Branded Food Products Database Release v0.0/April 2018) and Assidat Zgougou with layer of cream and dry fruits

Nutritional value	For 100 g of raisins	For 100 g of dried fig	For 100 g of Assidat zgougou
Energy	296 kcal	249 kcal	523 kcal
Water	27 g	30 g	3 g
Carbohydrates	78 g	63 g	65 g
Proteins	4.16 g	3.3 g	13 g
Fats	0.54 g	1 g	19 g
Fibers	6.8 g	9.8 g	3 g
Vitamins			
Vitamin-B1	0.11 mg	–	–
Vitamin-B2	0.14 mg	–	–
Vitamin-B6	0.22 mg	–	–
Vitamin-B9	9 μg	–	–
Vitamin-C	4 mg	1.2 mg	–
Vitamin-K	–	15.6 μg	–
Minerals			
Sodium	23 mg	–	–
Potassium	783 mg	680 mg	–
Phosphorus	85 mg	67 mg	379 mg
Calcium	40 mg	162 mg	–
Magnesium	31 mg	68 mg	–
Iron	2.4 mg	2.03 mg	–

6.2 *Dried Fig*

Like the olive tree, the fig tree is closely associated with the ancient Mediterranean society. Beyond civilizations and religions (monotheistic polytheists) the fig tree and its fruit remains at the heart of beliefs, customs and practices that evokes its protective, fertilizing and regenerating virtues.

Each territory presents specific figs. In Morocco, the most famous are the Taounate figs, especially those from the lands of the Mtiwa and Rghiwa tribes.

The shelf life of the fruit at is 24 h, and 1 week for figs stored at 25 and 4–5 °C, respectively. Black and purple skinned varieties are eaten fresh, while green skinned varieties are mostly dried (Fig. 4.36).

6.2.1 Nutritional Values of Dried Fig

Figs contain different antioxidants, especially phenolic compounds and carotenoids. Dark coloured figs contain more antioxidants than pale varieties. In addition, fresh figs have a higher antioxidant power. Fresh and dried figs contain 30% soluble fibre

Fig. 4.36 Fresh and dried figs

and 70% insoluble fibre. The fresh one helps to normalize blood levels of cholesterol, glucose and insulin, which helps treat cardiovascular diseases and diabetes. The dried figs maintain adequate intestinal function and are distinguished by its rich content of potassium, calcium and iron (Table 4.5).

6.2.2 Processing Technology Related to the Extraction of Dried Fig

The fig can partially dry out on the tree without losing its quality. The skin is wrinkled, and the water content of the fruit is ~70% without the fruit rotting. The fruit with a sugar content of more than 20% is much better preserved in a cold room.

Harvesting stage and Transportation: The figs intended for drying must be very clean, intact and not crushed; their delivery to the drying site must be immediate to avoid the beginning of any fermentation. The transport must be done in good conditions to avoid crushing. It would be even better to carry out the drying on the places of harvest.

Sorting and calibraton: The objective of sorting (Fig. 4.37a) is to eliminate crushed or unripe products. The drying time is proportional to the size of the fruit and without the calibration, the drying would be irregular.

Washing: The product being dusty, it is necessary to carry out a washing by soaking several times in water (Fig. 4.37b). Treatment in an appropriate solution (figs: boiling solution of sodium chloride and potassium metabisulphite) in addition to the destruction of parasites adhering to the product, to increase the pace of drying.

Drying: To avoid the alterations of the product during drying, it must be protected against: insect attacks, dust, condensations and nocturnal or morning dew, which cause a partial rewetting of the product. The drying could be performed by using traditional or modern solar drying (Fig. 4.37c). Modern solar dryers can avoid the disadvantages of traditional solar drying, and improve the drying conditions.

Fig. 4.37 Sorting (a), washing (b) and modern drying of figs (c)

6.3 Pastilla

Pastilla, also called bastilla or Bstilla (Fig. 4.38a), is one of the most favoured dishes in Morocco. This original Berber dish (Moroccan and Algerian) perfectly combines the flavours of meat with a crisp pastry (warka) and sweet spices. Pastilla is a dish that can be enjoyed in North Africa, as in Europe or even in the United States. In Morocco, it is a dish made with pigeon or chicken. But what makes this dish even more unique is the meat with all the other ingredients that make the stuffing of the dough.

The dish combining sugar and salt is one of those dishes inseparable from any traditional event in Morocco. Thus, whether it is a wedding, a baptism of a new birth or simply a circumcision, the Pastilla is always present on the table. There are multiple pastilla forms that differ as much in the ingredients that compose them as in the tastes they offer. The pigeon Pastilla is made, as its name suggests, with pigeon meat instead of chicken meat, and contain beaten eggs, grilled almonds, onions and spices. On the other hand, the fish Pastilla is prepared with shrimp, squid and whiting, this one can vary according to the choice. It does not contain eggs or sweet spices. The nutritional values of Pastilla are given in Table 4.4.

Fig. 4.38 Pastilla (**a**) and cooking chicken (**b**)

6.3.1 Processing Technology of Pastilla: Ingredients

Twelve leaves of briks (thin pastry), 1 chicken of 1.5 kg, 200 g butter, 1 bunch of parsley, 4 onions, peeled and chopped, 1 dose of saffron, 1 tsp. cinnamon powder, 1 glass of powdered sugar, 8 eggs, 250 g of blanched almonds, 1 glass of oil, 2 tbsp. Tablespoon icing sugar for decoration, salt and pepper.

6.3.2 Processing Technology of Pastilla: Preparation

Cooking chicken: In a large Dutch oven, melt the butter over medium heat. Drop the cockerels and brown all sides for a few minutes. Add the onions, parsley, ginger, turmeric, saffron, salt, pepper. Pour over 150 ml cold water, mix well and cook over medium heat for 35 min. Remove the cockerels from the sauce, boning and cutting into small pieces (Fig. 4.38b).

Preparation of the egg filling: Put the sauce on the fire to reduce until the liquid is completely evaporated, stirring constantly. Add cinnamon and sugar to caramelize the onions.

Add the eggs with caramelized onions, stirring constantly and vigorously. The device must have a texture of a thick cream. Evaporate all traces of water (Fig. 4.39a).

Preparation of almonds: Place a non-stick pan (very important) over medium/high heat. Without waiting for the oil to heat, immerse the blanched almonds and fry (without burning) for a few minutes until a golden colour is obtained. Drain the almonds and crush them with the cinnamon and icing sugar. Preheat the oven to 185 °C.

Mounting: In the round pan (such as deep pie dish) each sheet of brik is placed. It is necessary to note that the buttered side of the leaves should always be upwards. At the bottom of the dish arrange four sheets of brick by overlapping one over the other and letting out about half of these four leaves outside the mold. Add a sheet in the centre to consolidate the bottom of the pastilla. Arrange a layer of eggs on the leaves, distributing them homogeneously to cover the entire surface. Cover with a sheet of brik. Add the cockerels in the same way and cover with another sheet of brik. Finally, add the crushed almonds. Fold the edges of the overflowing leaves inward, folding them in the centre and pressing lightly so that they marry the layer of almonds. Coat egg white to solder the leaves. Cover with two leaves to close the pastilla, and return the edges to the inside of the mould.

Brush the edges of the last sheet of egg white so that it adheres well and does not come off during cooking. Brush the surface of the pastilla with melted butter. Bake 30 min or until golden brown. Arrange the pastilla on its serving dish. Sprinkle with sifted icing sugar and cinnamon and serve hot (Fig. 4.39b).

a

b

Fig. 4.39 Preparation of eggs (**a**) and mounting step of pastilla (**b**)

6.4 *Assidat Zgougou*

Assidat zgougou (Fig. 4.40a) is the name of this sweet made from Aleppo pine seeds unique of its kind in the whole Mediterranean area. The preparation of assidat zgougou is a Tunisian custom. This ritual is not found in neighboring countries. According to historians, zgougou is used in Tunisia since a major famine occurred in Tunisia in 1864, linked to political and climatic events. In front of the lack of cereals to make the flour, the inhabitants then turned to the Aleppo Pine. This tree (Fig. 4.40b) grows very well in Tunisia; it is often encountered during hikes. It is the mid-mountain regions of Siliana, Kef, Makhtar, Kasserine (but also Bizerte, Cap Bon) that produce most of the zgougou.

Aleppo pine was quickly appreciated by Tunisians for its good taste and its many food virtues. These have multiplied the recipes that kept always the principle of assida, sweet cream topped with dried fruits. Traditionally, the Tunisians exchange bowls of assida between neighbors and family members, making the decoration of assidat zgougou bowls as important as the taste of the recipe (Fig. 4.40a). Many use all kinds of dried fruit to vary the decorative shapes and colors (pistachios, walnuts, hazelnuts). The assida can be kept at 4 °C until 7 days.

6.4.1 Nutritional Values of Assidat Zgougou

According to the National Institute of Consumption (NIC), 100 g of assida, covered with a layer of cream and dried fruit, provides approximately 595 calories, corresponding to 25% of the daily energy needed by adults. According to the specialists, the consumption of several days of more than 100 g per person/per day of assida, makes it possible to increase the ratio of energy in the body. The proximate composition of Assidat Zgougou is given in Table 4.5.

a b

Fig. 4.40 Assidat zgougou (**a**) and Aleppo pine tree (**b**)

6.4.2 Processing Technology of Assidat Zgougou

Harvesting of Zgougou: The harvesting process is a daunting one, families spending half a year collecting, heating, breaking and sifting pine cones to extract zgougou. Indeed the harvest takes place from December to April. Climbers climb the pine forests and make the pine cones fall down. The cones are then collected and put on fire. Thus, pine cones open and release the seeds. A big work of departure, to separate seeds, fins, small pieces of herbs and dirt should be done and the yield is modest. It takes 45,000 seeds to produce just 1 kg. Each hectare of forest produces only 4 kg of zgougou.

Preparation steps of the Assida are: (1) To prepare assida, different steps should be realized: Sort zgougou carefully to remove all the stones, herbs, and dirt. (2) The zgougou is placed in a large container and washed with water carefully to rid them of any more dirt (Fig. 4.41a). (3) Dry zgougou on low heat. (4) Grind the zgougou in the blender. (5) Mix the ground zgougou with water in another container. The ground zgougou is poured and mixed with water into a sieve placed over another empty container (Fig. 4.41b). The paste that forms on top of the sieve is press hard to release the maximum amount of black juice from the ground zgougou leaving behind a dry paste that is discarded. (6) Mix the sifted flour with the black juice of the zgougou. Place the container with the mixture on the stove and turn the burner on at low heat while continuing to mix well to prevent the formation of lumps (Fig. 4.41c). (7) Keep mixing constantly with a woody spatula and the assida will thicken gradually. (8) As soon as the assida thickens, add sugar while continuing to stir the mixture so that it maintains consistency. Then, pour the mixture into small bowls (Fig. 4.41d).

The preparation of pastry cream are as follows: (1) Whisk the egg yolks, sugar and vanilla extract in a medium bowl until thoroughly combined. Add the cornstarch and whisk until the mixture is pale yellow and thick. (2) Heat the milk in a medium heavy-bottomed saucepan over medium heat until simmering. (3) Add the yolk mixture to the saucepan out-off the flame. Return to a simmer over medium heat, whisking constantly, until a few bubbles burst on the surface and the mixture is thickened and glossy. For the filling and topping, the following steps are used: (1) Spread the pastry cream over the cooled assida (Fig. 4.41e). (2) Top the assida with garnish, such as almonds, hazelnuts, walnuts, and pistachios (Fig. 4.41f).

6.5 Baklawa

Just like there is difference of opinion over baklawa origins, the source of the word baklawa is also disputed. The word baklawa entered the English language in 1650, a borrowing from Ottoman Turkish. Turkish etymologists claim of its Turkish origin (baklağı or baklağu); whereas some say "baklawa" may come from the Mongolian root bayla- 'to tie, wrap up, pile up'. Bayla- itself in Mongolian is a Turkic loanword. Though the suffix –va suggests its Persian origins; however the word 'baqla'

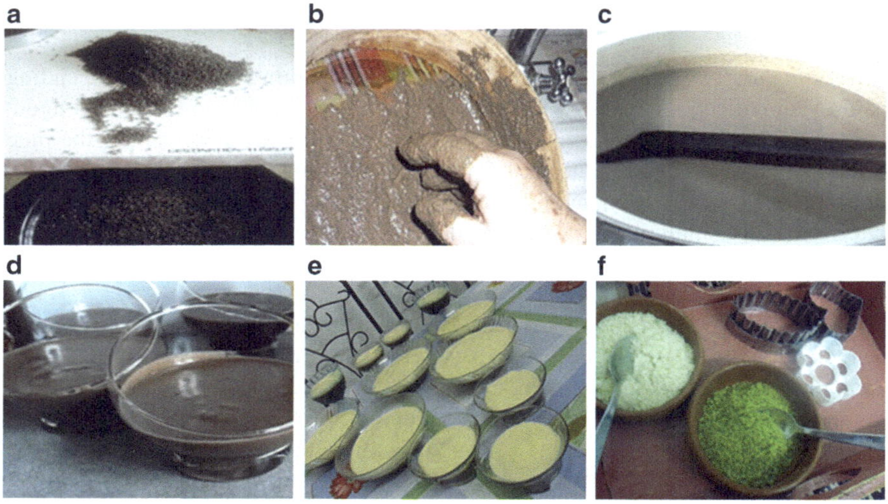

Fig. 4.41 Different steps for preparation (a–d) and decoration (e, f) of Assidat zgougou

is not Persian rather it is of Arabic origin meaning bean, but its Arabic name baqlāwa is doubtless a borrowing from Turkish. According to another source "The Armenians even insist that the word itself baklawa reveals its Armenian origins as the word appears to be related to the Armenian word for bakh (Lent) and halvah (sweet)." The name baklawa is used in many languages with minor phonetic and spelling variations. In the Arab world Baqlawa/baklawa is used; whereas, the Greeks call it baklava.

6.5.1 Nutritional Values of Baklawa

Baklawa provides with the consumer essential nutrients to help maintain his overall health. A 100 g provides 10.4 g of protein, which meets 20.8 percent of the daily needed value. Baklawa micronutrient composition is rich in: (1) unsaturated fats, (2) monounsaturated fats, (3) phosphorus, (4) magnesium, (5) copper, (6) manganese, (7) vitamin E and vitamin B2.

6.5.2 Processing Technology of Baklawa

To prepare baklawa, different steps should be realized. In a large bowl, the flour, salt, oil and butter were mixed manually (Fig. 4.42a). The water was added slowly until obtaining spherical dough. All the mixture was kneaded (Fig. 4.42b). The dough is divided into 20 different small balls and left for 30 min. The balls are then

Fig. 4.42 Different steps for the preparation of baklawa

flattened very well on a worktop sprinkled with starch (Fig. 4.42c). The dough is covered with a plastic and a kitchen towel to keep from drying out as work progress (Fig. 4.42d).

In a pan, very lightly toast the nuts until obtaining a very little browning and that some spots should retain their raw color/texture) (Fig. 4.42e). The product is removed from the heat and processed. Almonds are processed in a food processor into coarse consistency (like thick couscous) (Fig. 4.42f). Then, the same procedure is applied for pistachios that should be coarser than the almonds; for best results, it is recommended to chop them by hand with a knife (Fig. 4.42g). The chopped nuts are then placed in a mixing bowl and sugar and vanilla sugar are added. The different components are mixed with clean hands or a fork and the orange blossom water is added; the mixture is stirred well and set aside (Fig. 4.42h).

Then the following operations are performed: (1) Brush the bottom of a pan with melted butter. Lay down two sheets of dough onto the tray. One on each side to fully cover the tray. Brush the sheets with melted butter (Fig. 4.42i). (2) Continue to layer 2–3 sheets of dough at a time making sure to brush each layer with more melted butter until 15 sheets of dough are laid down finishing the last layer with melted butter (Fig. 4.42j). (3) Empty the dry fruit mixture onto the dough and gently spread it evenly with a fork to fully cover the dough (Fig. 4.42k). (4) Begin the layering process again making sure to brush each layer with melted butter until reaching another 15 sheets layered on top. Make sure to brush the last sheet evenly with the melted butter. (5) Using a sharp knife carefully cut the baklawa into diamond shapes (Fig. 4.42l) and bakes in the oven until golden brown. (6) To make the syrup, in a medium pot, stir together water, sugar, lemon juice, and orange blossom water (Fig. 4.42m). Then cook the syrup on medium low for 30 min and set aside. (7) Take out the baklawa of the oven and pour the warm simple syrup evenly over the entire tray (Fig. 4.42n) (*Note: The simple syrup can also be mixed with 2 cups of real honey. This gives the baklawa even more sweetness and a thicker, stickier consistency you commonly find in Tunisia and Algeria).* (8) Then run a knife through the original cut marks to help remove the baklawa pieces (Fig. 4.42o).

6.6 Makroudh

Makroudh is a sweet pastry filled with dates and nuts or almond paste, which has a diamond shape – the name derives from this characteristic shape. The dough is made with a combination of semolina and flour, which gives the pastry a very specific texture and flavor. There are two types of preparation: Makroudh can be fried in oil or oven-baked.

Makroudh is prepared according to several recipes, especially with dates or more traditionally with figs. This multitude of varieties is explained by the rich influence that Tunisia has known for centuries with the Vandals, Byzantines and Phoenicians. The makroudh is the reputation of the Tunisian city of Kairouan, ancient capital of the Aghlabid caliphate, where the first national festival of makroudh was organized on 20 May 2008.

It is also a popular product in neighboring countries such as Libya and Algeria where many varieties of Makroudh are present, some of which are pastries that do not share much in common with the traditional Makroudh except the shape. Among these varieties we find, Makroudh tmar, Makroudh lassel, Makroudh el louz, Makroudh msaker, Makroudh el koucha, Makroudh wahrani, Makroudh el mdawer, Makroudh el dgig and makroudh bel qarmouss. Makroudh with dates and honey is also popular in Morocco in the holy month of Ramadan. In this version, the semolina dough is lined with date paste, rolled into a log and then sliced, fried and dipped in honey.

6.6.1 Nutritional Values of Makroudh

From Tunisia to Morocco, makroudh finds its way in the cuisine of many Maghreb countries. The Makroudh with dates is a very popular pastry that provides around 72.2 calories/piece. It is made from good semolina, high quality olive oil and best season dates.

6.6.2 Processing Technology of Makroudh

Preparation of the dough: The semolina, flour and salt are blended in a large bowl. The melted butter (or oil) is added and hands are used to toss and massage the mixture for several minutes to ensure that each grain of semolina is individually coated with the butter (Fig. 4.43a). If time allows, cover and set the oiled semolina aside for an hour or longer before proceeding. Again using hands, gradually work the orange flower water into the semolina mixture until the formation of moist ball of dough (Fig. 4.43b). *If necessary, a little more water can be added.* Then, the dough is coved and left for at least 1 h.

Preparation of date paste: The pits are removed from the dates, and the dates are placed in a steamer basket or metal colander, which has been set over a pot of simmering water. The dates are steamed, uncovered, for 20–30 min, or until soft enough to easily mash. Then, the dates are transferred to a food processor. At this moment, butter, orange flower water, cinnamon and nutmeg are added, and processed until a smooth paste is formed. The paste is put aside to cool and firm (Fig. 4.43c).

Shaping and filling the cookies: When the date paste has cooled and firmed a bit, the paste is divided into four portions, and each portion is shaped into a thin paper about the diameter of the fingers. After make a deep indentation that runs the length of the dough (Fig. 4.44a), the date is then filled. The dough is gently pinched around the filling to enclose it (Fig. 4.44b). Special presses/molds are available for flattening and decorating the top of makrouth dough (Fig. 4.44c). The log is sliced with diagonal cuts into wide pieces. The cookies are transferred into a baking sheet or tray, and the process is repeated with the remaining date paste and dough. The cookies are left uncovered, for 30 min or longer before cooking (Fig. 4.44d).

Cooking: The vegetable oil is added in the frying pot and placed over medium heat. At the same time, 2 cups of honey are heated in smaller pot and tablespoon of orange flower water is added. When the oil is hot, the cookies are fried in batches until golden. The fried cookies are then immediately removed with a slotted spoon and transferred to the hot honey; the cookies are allowed to soak for a few minutes (Fig. 4.45a). The honey-drenched cookies are transferred to a strainer and after a few minutes to a rack or tray to finish cooling. The *makroudh samples* (Fig. 4.45b) *are then cooled* for several hours before storing in an airtight container.

Fig. 4.43 Preparation of dough (**a**, **b**) and date paste (**c**)

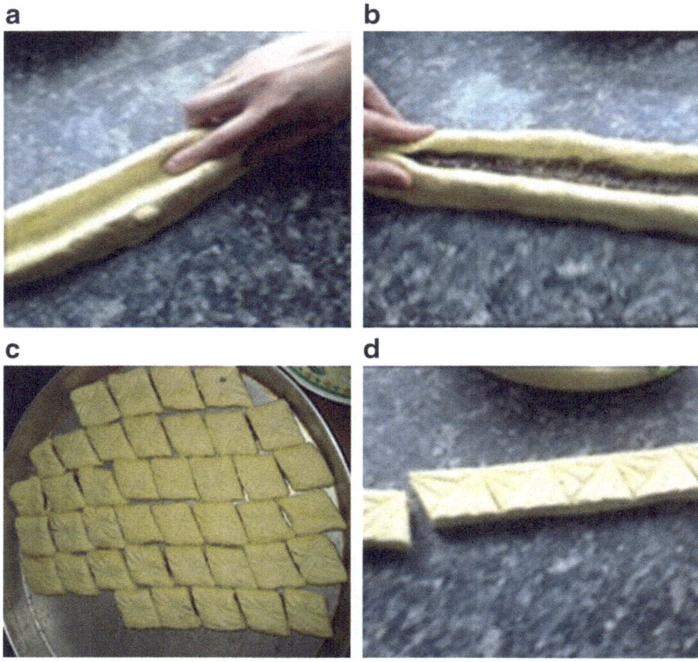

Fig. 4.44 Different steps for shaping and filling the cookies

Fig. 4.45 Cooking process of paste (**a**) and makroudh (**b**)

6.7 Zrir

Zrir is a tasty and fragrant dessert that is eaten in eastern Algeria and Tunisia. The Tunisian zrir is a mixture of hazelnuts and garnish with some toasted pine nuts or almonds, which is prepared exclusively for the new mother who has just given birth. A teaspoon of this delight provides strength because it is very caloric. For this reason, it is served in small glasses because it is very consistent.

6.7.1 Nutritional Value of Zrir

The Tunisian zrir is a kind of mixture based on sesame and dried fruits. From a nutritional point of view, this traditional sweet is very nutritious. In fact, zrir is rich in vitamins A and B, minerals (such as sodium, potassium, magnesium, and iron), fibres and carbohydrates. Moreover, zrir is considered among famous energizing dessert with 500 kcal/100 g. These are the reasons why this recipe is recommended for any new mom to recover all its forces after the pain and blood loss.

6.7.2 Processing Technology of Zrir

The different steps for producing zrir are indicated below. The sesame seeds are carefully sorted to remove all the stones, herbs, and dirt. Then, sesame seeds are placed in a large container and washed with water (Fig. 4.46a). The sesame seeds are dried and grilled on low heat, then grinded in the blender until obtaining smooth

Fig. 4.46 Different steps for the preparation of zrir

paste (Fig. 4.46b). The hazelnuts are roasted, peeled and grinded also until obtaining smooth paste (Fig. 4.46c). The two smooth pastes are mixed (Fig. 4.46d). Butter, honey and syrup are placed over low heat (Fig. 4.46e). The sesame seed and hazelnut paste are added and turned slowly with a woody spatula and put off when it starts to boil (Fig. 4.46f). The mixture is poured into a serving dish and served cold (Fig. 4.46g). The zrir could be conserved a few months at 4 °C.

7 Other Typical Products

7.1 Canned Capers

Capparis spinosa is an evergreen perennial shrub known from ancient times and used in the preparation of Moroccan and international culinary dishes (Fig. 4.47a). Collection and exportation of capers to France dates back to the Protectorate era. At that time, the region of Safi (Marrakech-Safi region) was the main capers producing region. The valorization and the exploitation of this crop started in the region only in the 1980s. This local product was labeled in 2014 as Meknes International Agricultural Exhibition. Currently, the area planted in the Safi region has reached almost 7000 ha instead of 20 ha twenty years ago and caper production in Morocco reached 10,000 tons for the season of 2016–2017 representing 45% nationwide. As part of the Green Morocco Plan, the extension of caper areas is planned by planting an additional 5000 ha.

In addition to the Safi region, the two Taounate and Taroudant regions also contribute the bulk of domestic caper production, and thanks to the production of these three regions, Morocco is considered the world's leading exporter of capers with two-thirds of world exportation. The exportation oscillates between 4000 and 7000 tons a year, and the turnover of the export of capers represents 10% of Morocco's plant exports. Capers are an important source of income for Moroccan farmers. Capers have a sharp piquant flavor and add pungency, a peculiar aroma and saltiness to comestibles such as pasta sauces, pizza, fish, meats and salads. The flavor of caper may be described as being similar to that of mustard and black pepper.

7.1.1 Nutritional Value of Canned Capers

Table 4.6 indicates the nutritional value of capers. Canned capers are composed on 84% water, 5% carbohydrates, 2% protein, and 1% fat. As for natural antioxidants, canned capers contain polyphenols, including the flavonoids quercetin (173 mg per 100 g) and kaempferol (131 mg per 100 g), as well as anthocyanins.

Fig. 4.47 Fermented (**a**), harvesting (**b**), softening, crimping and capping (**d**) of capers

Table 4.6 Nutritional values of canned capers in brine (USDA Branded Food Products Database, Release v0.0/July 2018) and Harissa (USDA Branded Food Products Database Release v0.0/Feb 2018)

Nutritional value	For 100 g of canned capers in brine	For 100 g of Harissa
Energy	36 kcal	62 kcal
Water	84 g	82.9 g
Carbohydrates	3.57–5 g	12.5 g
Proteins	0–2 g	–
Fats	0–1 g	–
Vitamins		
Vitamin-A	357 µg	–
Vitamin-C	4.3 µg	–
Minerals		
Sodium	3107 mg	1125 mg
Potassium	–	438 mg
Calcium	71 g	–
Iron	1.29 mg	–
Others		
Polyphenols	304 mg	29.7 mg
β-Carotene	–	3.35 mg

7.1.2 Processing Technology of Canning Capers

Harvesting: The harvesting stage of the caper (Fig. 4.47b) is the most expensive operation since it can absorb about 50% of the cost price because it is a long manual operation. The difficulties of this operation is related not only to the daily harvest period (early, before sunrise and opening of flower buds), but also in the difficulty of gathering (presence of thorns), squatting or curving, the picking, usually done by female and family labor, starts from April and lasts until September depending on the clones used. The harvesting period is located between April and June, while the late clones continue their production until November. The workers make passes, every 2–3 days, to pick the flower buds called capers before they flourish. As soon as the white petals begin to appear, the capers are unfit for marketing. In the production areas of the province of Safi, yields remain average and do not exceed 1.5 q/ha, which is 10 to 20 times less than those achieved in Spain (15–30 q/ha).

Softening: After sorting and removing the leaves, the caper is placed in fools and submerged with water. The water is changed every 24 h for 3 days until the color and odor of the capers change (Fig. 4.47c).

Juicing: The juicing consists in the filling of the containers (boxes or pots) with oil (soy, sunflower, olive), tomato sauce, or brine. This operation is performed either mechanically or manually, depending on the nature of the finished product, the availability of machinery and labor. The purpose of this step is to avoid a number of chemical or physical alterations (oxidation, corrosion, bending of boxes) in addition to the organoleptic improvement of the quality of the capers. The oil is added cold to avoid the presence of acidic taste of the hot oil in the finished product.

Crimping and capping: Once the boxes are filled with juice, they are sealed mechanically or manually by crimping or capping depending on the availability of the machine and the nature of the box (Fig. 4.47d).

7.2 *Harissa*

Harissa is a puree of red peppers originating from Tunisia, but also found in Morocco in a dry form. The origin of the word comes from the Arabic verb Harassa that literally means “crush”.

Harissa was originally a very beautiful woman and not a hot sauce. This woman, who lived in the late fifteenth century in North Africa, was excited about peppers from New World. So she decided to not let it down and realized that the quality of the pepper in North Africa was higher than the others originating from different countries.

“National sauce” in Tunisia is an important element of local cuisine, especially in Djerba, in the Sahel region, Cap Bon and in the eastern Algeria. The rural population makes their own Harissa, which can also be found in other kitchens in North Africa as well as in Palestine, where Tunisian and Moroccan Jews introduced it.

Fig. 4.48 Dishes prepared using Harissa (a) and Fresh (b) and dried (c) peppers

Tunisia produces about 20,000 tons of Harissa per year, of which about 6000 tons are for export. Production is growing by around 6% per year while exports are up 12–15% per year. This wonderful product is known in the four corners of the globe. Today, the challenge is to register this sauce made of red peppers and spices blend to the gastronomic heritage of UNESCO.

Harissa is usually used as a condiment or as an ingredient (Fig. 4.48a). It is often utilized to season dishes, such as couscous or kefteji, and also to prepare sandwiches such as "bocadios" in Morocco which are tuna sandwiches, lettuce and Harissa. However, harissa must not be heated or cooked together with the dish in order to preserve its original taste properties.

7.2.1 Nutritional Value of Harissa

Harissa is a source of many nutrients. Its ingredients include mashed peppers, coriander, and garlic. This mixture gives it a great richness in vitamins B_2, B_6, C, and E; trace elements such as sodium, copper and manganese and capsaicin (Table 4.6). An active condiment acts both as a natural anti-inflammatory and provides antioxidants that protect cells from premature aging due to free radicals. On average, products in the salted sauces and condiments category provide an energy value characterized by an average energy density. Harissa contains low levels of fats and proteins and high contents of carbohydrates and potassium (Table 4.6).

7.2.2 Processing Technology of Harissa

The traditional process consists in drying peppers (Fig. 4.48b, c) in the sun and then grinding it with spices, garlic and dried tomatoes. Harissa can also be prepared with fresh peppers. There are different processes depending on the type of peppers, the taste and the preparation. For one kilo of red peppers, 300–400 g of garlic, 250 g of whole blond caraway, salt and dry mint are added (Fig. 4.49a).

To have a delicious Harissa with a beautiful red color, it is necessary to use a red pepper dried naturally by the sun, not in the oven. This could be explained by the fact that the former process gives the true taste of dried pepper, without any after-taste, unlike oven-dried pepper, whose color is duller than dried pepper by the sun. Traditionally prepared Harissa can be kept perfectly in its tube or pot in a cool place for 2–3 weeks. The commercial process consists of following steps:

Reception of red peppers: The peppers are checked at the reception to avoid defects related to the raw material.

Sorting and washing: By manually removing damaged stems and peppers before washing peppers of good quality (Fig. 4.49b).

Fig. 4.49 Ingredients used for the production of traditional Harissa: preparation (**a**), sorting (**b**) and drying (**c**) of peppers

Drying and crushing: These two operations are performed by using an industrial size dryer machine and a crusher (Fig. 4.49c).

Adding ingredients and cooking: The operation consists in adding fresh garlic, herbs (caraway and Cumin), grass (Coriander) and artificial flavour to enhance the taste and salt. The percentages of a base preparation are as follows: Red pepper 86%, Garlic: 4%, Caraway: 2%, Coriander: 4%, Salt (NaCl): Max 0.5% of product weight. The recipe submitted a heat treatment at 60 °C for 20 min. Then, the product is filled in metal containers (cans or tube) and submitted to pasteurization at 90 °C during 45 min. The Harissa prepared according to the commercial process present a shelf life of 3 years. Once the tube is opened, it is recommended to keep it in the refrigerator and consume it in the following month.

8 Conclusion

Different traditional products are produced in the Maghreb since several decades that are unfortunately not labeled in Europe as protected designation of origin (PDO), protected geographical indication (PGI). These large traditional products are due to different culture since traditional products represent culture, heritage and identity factors that contribute to the development and sustainability of rural areas from depopulation. Taking into account the evolution of regulation regarding food products, changes in consumer habitudes and the development of process technology, traditional products in the Maghreb face the challenge to ameliorate their quality and safety in order to be: i) maintained in the local markets; and ii) extended on the international markets. Several different technological processes are an important part of the cultural heritage of Maghreb foods, but the development is still not enough and much research remained to be performed by applying innovation strategy.

Acknowledgments We would like to acknowledge Professor Néjib Guizani from the Department of Food Science and Nutrition, College of Agricultural and Marine Sciences, Sultan Qaboos University, for the reviewing of the chapter and for his valuable remarks and comments. Mr. Zaroual is grateful to Erasmus+ MIC for its financial support of his Ph.D. during his stay at Artois University.

References

Amina, F. (2017). Évaluation de l'activité anti-inflammatoire de l'extrait brut de la graisse de la bosse de Camelus dromedarius sur un modèle murin d'arthrite expérimentale - A thesis. Université des Frères Mentouri Constantine1. Retrieved from http://fac.umc.edu.dz/snv/bibliotheque/biblio/mmf/2018/Evaluation de l'activité anti-inflammatoire de l'extrait brut de la graisse de la bosse de Camelus dromedarius sur un modèle murin d'inflammation aiguë.pdf

Bajoub, A., Medina-Rodríguez, S., Gómez-Romero, M., Ajal, E. A., Bagur-González, M. G., Fernández-Gutiérrez, A., & Carrasco-Pancorbo, A. (2017). Assessing the varietal origin of extra-virgin olive oil using liquid chromatography fingerprints of phenolic compound, data fusion and chemometrics. *Food Chemistry, 215*, 245–255. https://doi.org/10.1016/j.foodchem.2016.07.140

Boggia, R., Borgogni, C., Hysenaj, V., Leardi, R., & Zunin, P. (2014). Direct GC-(EI)MS determination of fatty acid alkyl esters in olive oils. *Talanta, 119*, 60–67. https://doi.org/10.1016/j.talanta.2013.10.055

Brahmi, F., Dabbou, S., Flamini, G., Edziri, H., Mastouri, M., & Hammami, M. (2011). Fatty acid composition and biological activities of volatiles from fruits of two Tunisian olive cultivars. *International Journal of Food Science and Technology, 46*(6), 1316–1322. https://doi.org/10.1111/j.1365-2621.2011.02616.x

Cabrera, M. C., & Saadoun, A. (2014). An overview of the nutritional value of beef and lamb meat from South America. *Meat Science, 98*(3), 435–444. https://doi.org/10.1016/j.meatsci.2014.06.033

Charlton, K. E., Epid, M., Probst, Y. C., Tapsell, L. C., & Blackall, P. J. (2008). Food, health and nutrition: Where does chicken fit? *Journal of the Home Economics Institute of Australia, 15*(2), 5–17. Retrieved from http://www.chicken.org.au/files/_system/Document/Nutrition/PRINTVERSION REPORT Low res0808.pdf

Chimi, H. (2006). Technologies d'extraction de l'huile d'olive et gestion de sa qualité. Bulletin mensuel d'information et de Liaison du PNTAA 037, 77–80. Retrieved from https://www.agrimaroc.net/bulletins/btta_141.pdf

De Oliveira Sousa, A. G., Fernandes, D. C., Alves, A. M., de Freitas, J. B., & Naves, M. M. V. (2011). Nutritional quality and protein value of exotic almonds and nut from the Brazilian savanna compared to peanut. *Food Research International, 44*(7), 2319–2325. https://doi.org/10.1016/j.foodres.2011.02.013

Dheeb, B. I., Al-mudallal, N. H., Salman, Z. A., Ali, M., Nouri, M. A., Hussain, H. T., & Abdulredha, S. S. (2015). The inhibitory effects of human, camel and cow's milk against some pathogenic fungi in Iraq. *Jordan Journal of Biological Sciences, 8*(2), 89–93. https://doi.org/10.12816/0027553

Gharby, S., Harhar, H., Farssi, M., Ait Taleb, A., Guillaume, D., & Laknifli, A. (2018). Influence of roasting olive fruit on the chemical composition and polycyclic aromatic hydrocarbon content of olive oil. *25*(3). Retrieved from https://www.ocl-journal.org/articles/ocl/pdf/2018/03/ocl170044.pdf

Goodson, A. (2018). Is Couscous Healthy? Top 5 Health and Nutrition Benefits. Retrieved from https://www.healthline.com/nutrition/couscous-benefits#section7

Guillaume, D., Gharby, S., Harhar, H., & Charrouf, Z. (2018). Alwana oil: The traditional Amazigh olive oil of the Pre Rif (Northern Morocco). *International Journal of Gastronomy and Food Science, 13*, 90–94. https://doi.org/10.1016/j.ijgfs.2018.07.004

Ismaili, M. A., Saidi, B., Zahar, M., Hamama, A., & Ezzaier, R. (2016). Composition and microbial quality of raw camel milk produced in Morocco. *Journal of the Saudi Society of Agricultural Sciences, 18*, 17–21. https://doi.org/10.1016/j.jssas.2016.12.001

Jukanti, A. K., Gaur, P. M., Gowda, C. L. L., Chibbar, R. N., Jukanti, A. K., Gaur, P. M., & Chibbar, R. N. (2012). Nutritional quality and health benefits of chickpea (Cicer arietinum L.): a review. *British Journal of Nutrition, 108*, S11–S26. https://doi.org/10.1017/S0007114512000797

Kadim, I. T., Mahgoub, O., & Mbaga, M. (2014). Potential of camel meat as a non-traditional high quality source of protein for human consumption. *Animal Frontiers, 4*(4), 13–17. https://doi.org/10.2527/af.2014-0028

Khallouki, F., Voggel, J., Breuer, A., Klika, K. D., Ulrich, C. M., & Owen, R. W. (2017). Comparison of the major polyphenols in mature Argan fruits from two regions of Morocco. *Food Chemistry, 221*, 1034–1040. https://doi.org/10.1016/j.foodchem.2016.11.058

Khallouki, F., Younos, C., Soulimani, R., Oster, T., Charrouf, Z., Spiegelhalder, B., & Owen, R. W. (2003). Consumption of argan oil (Morocco) with its unique profile of fatty acids, tocopherols, squalene, sterols and phenolic compounds should confer valuable cancer chemopreven-

tive effects. *European Journal of Cancer Prevention, 12*(1), 67–75. https://doi.org/10.1097/01.cej.0000051106.40692.d3
Manai-Djebali, H., Krichène, D., Ouni, Y., Gallardo, L., Sánchez, J., Osorio, E., & Zarrouk, M. (2012). Chemical profiles of five minor olive oil varieties grown in central Tunisia. *Journal of Food Composition and Analysis, 27*(2), 109–119. https://doi.org/10.1016/j.jfca.2012.04.010
Olmo-García, L., Bajoub, A., Monasterio, R. P., Fernández-Gutiérrez, A., & Carrasco-Pancorbo, A. (2017). Metabolic profiling approach to determine phenolic compounds of virgin olive oil by direct injection and liquid chromatography coupled to mass spectrometry. *Food Chemistry, 231*, 374–385. https://doi.org/10.1016/j.foodchem.2017.03.139
Rueda, A., Seiquer, I., Olalla, M., Giménez, R., Lara, L., & Cabrera-Vique, C. (2014). Characterization of fatty acid profile of argan oil and other edible vegetable oils by gas chromatography and discriminant analysis. *Journal of Chemistry, 2014*. https://doi.org/10.1155/2014/843908
Seiquer, I., Rueda, A., Olalla, M., & Cabrera-Vique, C. (2015). Assessing the bioavailability of polyphenols and antioxidant properties of extra virgin argan oil by simulated digestion and Caco-2 cell assays. Comparative study with extra virgin olive oil. *Food Chemistry, 188*, 496–503. https://doi.org/10.1016/j.foodchem.2015.05.006
Slavin, J., & Lloyd, B. (2012). Health benefits of fruits and vegetables. *Advances in Nutrition, 3*(4), 506–516. https://doi.org/10.3945/an.112.002154.506
Vossen, P. (2007). Olive oil: History, production, and characteristics of the world's classic oils. *HortScience, 42*(5), 1093–1100.

Chapter 5
Chinese Traditional Foods: Preparation and Processing

Zhang Yu, Md Ramim Tanver Rahman, and Zaixiang Lou

1 Introduction

China has the most popular culinary heritage in the world. The history of their cuisine dates back to about 1000 years with varied cooking styles, techniques and ingredients that have evolved over time. Chinese traditional food has a rich and unique cultural connotation. It is the integration of long-term experience and wisdom. It has good flavor, nutrition, health and safety. Historically, the prosperity of many commercial business districts were developed under the auspices of traditional foods. For example, Dazhalan in Beijing, Nanjing's Confucius Temple, and traditional snacks on Shanghai's Nanjing Road. Those traditional foods make an indelible contribution to the prosperity of the national industry and the development of the national economy, which has become an important part of Chinese traditional culture.

Chinese people have a keen taste for foods. Sophisticated flavor and unique texture are most emphasized for food processing and preparation. Unlike Western food, creaminess is not preferred texture; traditional foods generally are less greasy, relatively low in caloric value, and require longer preparation times to achieve the desired sensory characteristics. Many traditional Chinese foods, such as sesame oil chicken and clam soup, are associated with a therapeutic or health-promoting effect besides their pleasant taste and basic nutritional values, while others are linked to a historical or folkloric story, with *zongzi* and Beggar's chicken as examples. Traditional Chinese Food is integrated in dietary menu of Sinosphere Asia. Having

Z. Yu
Institute of Biology, Qilu University of Technology (Shandong Academy of Sciences), Jinan, China

M. R. T. Rahman · Z. Lou (✉)
State Key Laboratory of Food Science & Technology, School of Food Science, Jiangnan University, Wuxi, China

M. Al-Khusaibi et al. (eds.), *Traditional Foods*, Food Engineering Series,
https://doi.org/10.1007/978-3-030-24620-4_5

hot (yang) and cool (yin) food are therapeutic strategy for chronic diseases (Chu, Yao, & Tan, 2018). Drying, salting and pickling are oldest and most widely used methods of food preservation.

Traditional Chinese foods rely greatly on these basic processing and preservation methods. These foods encompass a wide range of products, such as pickled vegetables, salted fish and jellyfish, winterized (semidry) meats, and salt-cured eggs. Fermentation has also been used in the preservation of foods for centuries. Almost all food from both animal and plant sources are usually fermented. There are numerous kinds of traditional fermentation products including, to just name a few, sauces, vinegar, rice spirits, fermented vegetables, stinky tofu, etc., and the processes used to make them vary extensively based on the products. Traditional Chinese foods can be characterized by their materials, ingredients, processing methods, functions or styles. A brief description of commonly found, selected traditional Chinese foods based on the food categories (Tables 5.1 and 5.2) is given in this chapter.

Table 5.1 Few chinese traditional foods are catagorized by area, food groups and processing ingredients (methods)

Area	Food catagory	Item	Common processing used
Inner Mongolia	Meat	Roast Lamb	Seasonings by shallot, ginger, salt, etc., and baked
	Tea	Milk tea	Boiled brick tea with cow's or sheep's milk, along with a little salt added
Guizhou, Hunan, Yunnan	Fish	Fish soup	Fish is boiled in the soup base with garlic, ginger, salt, chili, wine and special fragrant spices
Xinjiang, Gansu, and Ningxia	Noodles	Noodles and beef	Twisting, stretching and folding the dough into strands, using the weight of the dough
	Tea	Three-course Tea	The first course of tea is called bitter tea, the second course called sweet tea and the third course called aftertaste tea with sweet, bitter, sour, spicy flavors.
Chengdu	Meat, soup, vegetables	Variety of food items	Boiling the water then put the items

Table 5.2 Selected tarditional chinese food catagorized by common food group, food item, processing methods

Food group	Item	Methods
Vegetables	Cured vegetables	Sun/solar drying, brine
Soy foods	Soy sauce	Fermentation, pasteurization, filtration
	Soy milk, beverage	Grinding, boiling, pasteurization
	Tofu/Tofu products	Coagulation, pasteurization, molding, pressing
Sea food	Fish ball	Different layer of salting

2 Rice and Rice Bound Traditional Foods

Rice is the most important staple food in China. As much as 80–90% of the daily caloric intake of people in China is derived from rice. China contributes 38% of the world's rice production using 24% of the world's growing area. There are many forms of rice products produced as alternative food sources due to their nutritive values and special sensory attraction. In China, rice cereals are consumed as cooked rice and are served simultaneously with prepared vegetable dishes, pork, poultry, beef, seafood, and others. In addition to its standard form, rice is consumed in the form of noodles, puffed rice, breakfast cereals, rice cakes, fermented sweet rice and snack foods made by extrusion cooking. Rice is also used in making beer, wine, and vinegar. There are many kinds of rice snack foods, prepared for attractive taste, texture and aroma. These are served in some cases for special occasions, and in others for their special taste or for convenience. Many rice snack foods are made from glutinous rice (sweet or waxy rice) containing largely amylopectin (98% of total starch), but little amylose (less than 2% of total starch), while others are made from both types. A typical glutinous rice flour contains 11–13.5% moisture, 1% ash, 75–80% total starch, 5.5–6.5% protein and 0.5% total fat. Glutinous rice flour is often used in making snack foods because the sticky characteristics resulting from its high amylopectin content are necessary in many specialty rice foods. Another reason for the application of glutinous rice in baked and popped snack foods is that glutinous rice flour expands readily and produces a more porous texture.

Zongzi is another very popular rice snack food in which glutinous rice is wrapped in bamboo leaves. The bamboo leaves give a distinctive, pleasant flavor to the rice after cooking and serve as a packaging material for the product. *Zongzi* was made and consumed by people in memory of a famous patriotic poet, Qu Yuan (340–278 BC), who committed suicide by drowning himself in a river to protest the corruption of the ruler during the Warring State. When people from his village heard this sad news, men tried to rescue him from the river by rowing dragon boat as fast as possible but it was too late. Thus women in the village wrapped seasoned rice and meat in bamboo leaves, cooked and dropped them into the river so that the fish in the river would not disturb their beloved poet. Since then, every year on the day of Qu Yuan's death, people make and eat *zongzi* and hold dragon boat races to remember his death and patriotism. This has become the origin of the Dragon Boat Festival, one of the three major festivals in China (Li & Hsieh, 2004).

3 Fish Products

Salted jellyfish have been eaten in China for more than a thousand years and China is the first country to process edible jellyfish (*Rhopilema esculentum*) for human consumption. Jellyfish is more than a gourmet delicacy in China; it is a tradition. A Chinese wedding or formal banquet is rarely completed without a jellyfish salad.

Today, Semi-dried salted jellyfish represent a multimillion dollar seafood business in Asia.

Suan yu is also a Chinese low-salt fermented whole fish snack with a characteristic flavor. It is stable during storage and free of fishy odor and taste and retains all the nutritional advantages of fish compared with other traditional fermented fish products. It is manufactured using traditional technologies without the addition of starter cultures in small-scale processing units. Fish used are mainly freshwater fish mixed with cooked carbohydrate, salt, and spice. Long-term spontaneous fermentation is conducted to develop flavor in anaerobic conditions.

4 Poultry Products

Among the many traditional Chinese foods, poultry products are recognized not only as a delicacy, but also for their nutritional and health promoting values. Chicken soup cooked with a variety of herbs such as ginseng, ginger, black mushrooms, wolfberry fruits, dried dates and/or special ingredients such as wine and sesame oil has been commonly used for patients for a speedy recovery. For example, chicken stewed with sesame oil, rice wine and fresh ginger root is a typical food for women to eat daily during the month after delivering their babies in South China and Taiwan. These food items are often prepared at home and served to healthy people as well as invalids to enjoy the flavor and promote good health in general.

Beggar's chicken, which is cooked in leaves and clay, is a famous baked dish, specific to the southern Chinese provinces such as Zhejiang, Changshu and Guangdong. As the name suggests, beggar's chicken was originally eaten by homeless beggars, who had to invent a cooking method for a stolen chicken without using any cooking utensils. The chicken was cleaned and wrapped in lotus leaves unseasoned, with a thick layer of clay that enveloped the chicken.

5 Eggs Products

Century egg or Pidan has been an alkaline-fermented ethnic food for many generations in China. The dark greenish color and sophisticate flavor of this product has resulted in it being known as the "thousand-year-old egg" in Western society. Mixing crushed or cut-up *pidan* and tofu and seasoning with soy sauce and sesame oil makes a delicious, easy-to-prepare family side dish often served at table for any meal when the housewife is busy. Today, this side dish often appears on restaurant menus due to its increased popularity. Depending on the processing methods, several types of *pidan* are available, such as the pine-floral *pidan*, softyolk *pidan* and hard-yolk *pidan*. In traditional *pidan* processing, many variations of methods can be found from different sources. Generally the traditional processing of *pidan* can be classified into three types: the rolling powder method, the coating method and the immersion method.

6 Chinese Traditional Fermented Foods

Many kinds of fermented foods have appeared and enjoyed great popularity in China since ancient times. A range of raw materials including grains, vegetables, milk, meat, tea, and beans has been used for fermentation, resulting in a myriad of fermented products like Chinese liquor, traditional aged vinegar, Chinese suancai, and soy sauce. Microorganisms, the soul of fermented foods, play a crucial role in determining the overall quality attributes of any fermented food. Selected strains isolated from traditional fermented foods can be used to standardize their production while maintaining their traditional traits. Extensive research has been conducted to uncover the microbial composition of all kinds of Chinese traditional fermented foods and a large variety of microorganisms including archaea, bacteria, and fungi has been revealed.

7 Chinese Liquors

In Chinese history, alcohol has been termed the "water of history" because of its role in folklore. The use of alcoholic fermented products has great significance in the dietary culture of China. Among these products, Chinese liquor enjoys unique status in the cultural life of Chinese people. There are many varieties of Chinese liquor based on different production techniques and distinct liquor starters. It is usually produced with the use of aunique starter culture called Daqu, a saccharifying and fermenting agent. Daqu is usually prepared by solid state fermentation from wheat, barley, and/orpeas, usually in the form of cakes or bricks. Bricks are subjected to a long period of incubation during which large amounts of enzymes are produced and high microbial loads, including bacteria, yeasts, and molds, develop. Daqu preparation is manipulated under various specific time-temperature control schemes to obtain liquors with distinctive flavors (Fig. 5.1).

7.1 Shaoxing Rice Wine

Shaoxing rice wine, named after the city Shaoxing, has a unique flavor and high nutritional value. It is quite popular among Chinese people and was first recorded more than two thousand years ago. It is an excellent representative of Chinese traditional wines and is internationally well known. Shaoxing rice wine is an amber-colored clear liquid. It has a mixed aroma and taste. For those who are unfamiliar with it, Shaoxing rice wine does not smell much like alcohol. Some people think it has a very particular taste: vinegary, spicy and caramel-like. Shaoxing rice wine is fermented from gluten free rice with yeast and spring water. However, it is worth noting that there is also a small amount of wheat in Shaoxing rice wine. The alcohol percentage is around 14.

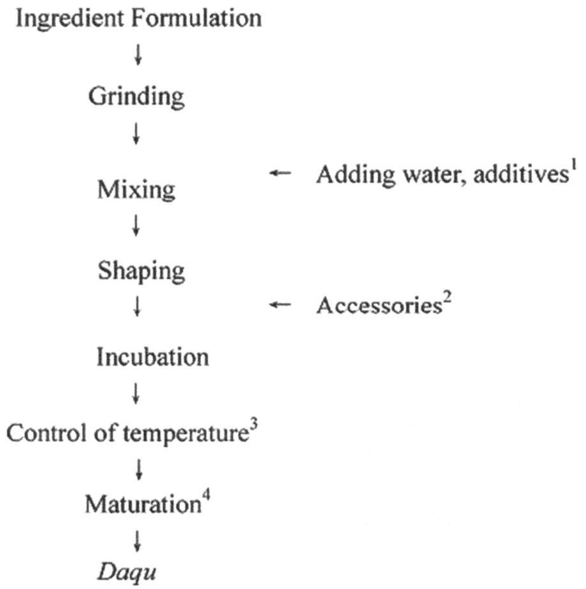

Fig. 5.1 The production process of *Daqu*.[1]Some *Daqu*such as *Moutai-Daqu*use previous Qu as additives. [2]Accessories such asrush mats, bamboo, etc. are used as covers to control the temperature. [3]Involves low temperature incubation phase, high temperature converting phaseand aroma creating phase. [4]Maturation aims at achieving spatial equilibrium of moisture, acidity, enzyme activities and microbiota

8 Chinese Sourdough-Steamed Bread

Sourdough is a mixture of flour and water that is fermented by yeast and LAB is employed as an ideal inoculum in making Chinese steamed bread (Zhang et al., 2016). Zhang and He (2013) investigated the major bacteria in five sourdough samples collected from different places using the Illumina HiSeq 2000 system and found that *Lactobacillus*, *Leuconostoc*, and *Weissella* were the predominant genera among the five samples. Zhang et al. (2011) investigated the diversity of LAB and yeasts in traditional sourdoughs collected from the western region of Inner Mongolia Autonomous Region in China by culturing and PCR-DGGE at the species level (Fig. 5.2).

9 Pu-Erh Tea

Pu-erh tea or Pu'er is a fermented dark tea, black to brown in color, with a moderate taste, and is produced only in Yunnan Province, China. The role of Pu-erh tea as an anti-oxidative agent for reducing cholesterol levels and aiding digestion is well acknowledged. There are two main categories of Pu-erh tea, namely raw and ripened. The processing of raw Pu-erh tea includes the fixation, rolling, and drying of leaves before the tea is compressed into bricks, cakes, or other shapes.

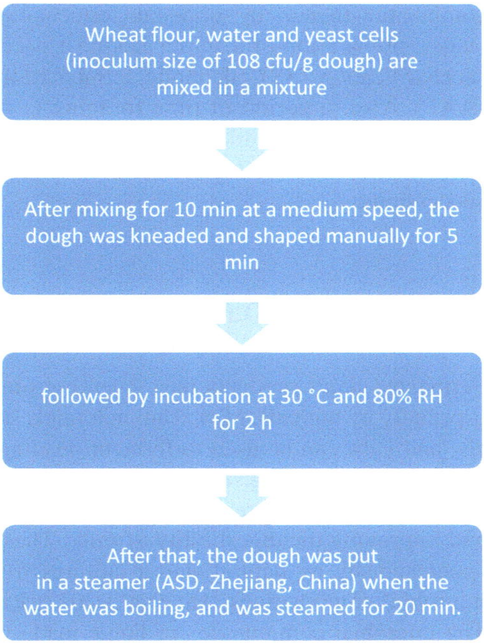

Fig. 5.2 Chinese steamed bread making

The processing of ripened tea involves an extra step called wet piling which is similar to composting and is performed before drying to facilitate the aging process. This step involves fermentation of the leaves in a warm and humid environment by bacterial and fungal cultures (https://www.youtube.com/watch?v=CrVVglIWtC8).

The processing of raw tea includes sun fixation, rolling, and sun drying before the tea is compressed into tea bricks. The ripened type is produced in the same way as the raw but with the additional step of wet piling, a process similar to composting, after sun drying and before compression. After the tea is compressed into bricks, both types undergo a natural post fermentation when they are exposed to the environment during storage before consumption (Tian et al., 2013). People believe that the quality of Pu'er tea increases with storage (aging) time.

10 Chinese Vinegar

Chinese traditional vinegar (CTV) or Cu, also known as cereal vinegar, is not only an important seasoning for Chinese dishes but also a significant element in Chinese daily diet. The production of vinegars in China can date back 3000 years. Chinese vinegars are sweeter than western white or cider vinegars. Their average acidity is about 5%–6% with some residues of sugar remaining after fermentation. Most CTVs are brewed with starchy cereals such as rice, wheat, millet, sorghum, or a combination thereof, by solid-state fermentation that consists mainly of the following stages: preparation of a starter Daqu (if needed, e.g. Shanxi aged vinegar), starch saccharification (SS), alcoholic fermentation (AF), and acetic acid fermentation (AAF). Also, a period of time foraging is often indispensable. Unlike the pure-culture liquid fermentation process of vinegar production in European countries, the fermentation of CTV involves various microbes and complicated but regular successions.

11 Fermented Dairy Products

Fermented dairy products have been consumed in China for thousands of years, especially by nomads. Milk from mammals such as cows, horses, ewes, goats, camels, and yaks can be used to ferment, leading to various fermented dairy products throughout China, e.g. *Tibetan kefir*, *tarag*, yoghurt, *kumiss*, and cheese. The fermentation process increases the shelf-life of the products, while enhancing the taste and improving the digestibility of milk. These home-made products are fermented spontaneously by natural starter cultures from previous batches having complex microbial compositions (He et al., 2017; Liu & Tong, 2017; Peng et al., 2018). The microbial communities of home-made yoghurts vary with their geographical origin and manufacturing process (Zhang et al., 2004).

12 Traditional Fermented Vegetables

Fermented vegetables, like other types of fermented products, are popular and traditional edible foods in China,and are comprised of a variety of vegetables such as mustard, cucumber, cabbage, radish, pepper and bamboo, developed using different manufacturing processes. Presently, numerous anti-microbial substances, sugar polymers, sweeteners, aromatic compounds, essential fatty acids, vitamins, useful enzymes and other probiotic properties have been found to be closely associated with microorganism populations in fermented foods. Result indicated a high level of diversity in the bacterial community structure in traditional fermented vegetables and demonstrated the possible influences of manufacturing location and conditions, and/or the type of raw material, on bacterial diversity.

13 Fermented Tofu

When fresh tofu is fermented with a strain of certain fungi, such as *Mucor hiemalis* or *Actinomucor elegans*, it becomes a new product known as *sufu* or Chinese cheese. The product, known as *doufuru* or *furu* in mandarin Chinese, and *toufuju* or *fuju* in Cantonese, consists of tofu cubes covered with white or yellowish-white fungous mycelia, having firm texture, salty taste, and characteristic flavor. Although relatively unknown in some adjacent countries, such as Japan and Korea, sufu has been produced in China long before the *Ching* Dynasty and consumed mainly as an appetizer or relish by all segments of the Chinese people, including those living oversea.

13.1 Preparation Method

Preparation methods vary with type of sufu and region but all involve three basic steps; preparing tofu, molding (first fermentation), and brining (second fermentation). Firm tofu is prepared and then cut into uniform dices with a size ranging from 2.5 to 7.5 cm (1 to 3 in.). In natural fermentation, the dices are arranged on woven bamboo trays or on rice straw in some places. The trays are placed in direct sunlight for at least several hours to let solar radiation naturally kill many unwanted microorganisms in tofu dices, and then they are stacked on shelves in an incubation room at 20–35 °C. The molds already inhabiting the trays or rice straw begin to inoculate the tofu naturally. In pure culture fermentation, freshly cut tofu cubes are first immersed in an acid-saline solution (6% NaCl plus 2.5% citric acid) for 1 h and then subjected to sterilization at 100 °C for 15 min. The cubes are separated from one another in a tray with small openings in the bottom and top to facilitate the circulation of air. This helps mycelium development on all sides of the cubes. After cooling, the cubes are then inoculated over their surfaces with a suspension or dried powder of spores from a selected, pure cultured microorganism.

The inoculated tofu dices, either naturally or with a pure culture, are now transferred to an incubation room. The recommended temperature and relative humidity are 25–30 °C and 97%, respectively. After several days of incubation, depending on the temperature and the type of culture, each cube is covered with a fragrant cotton-like mycelium. This intermediate product is known as *pehtze*. To ensure good formation of dense and thick texture of a mycelial mat, the mycelium on the *pehtze* is normally rubbed flat before *pehtze* are salted, seasoned, and aged in earth ware crocks with brine. The crocks are tightly sealed, and put in a cool, dark place for several weeks or even several months.

The brine may contain different types of flavorings and colorings. This results in different types of *sufu* products. The typical brine is one containing 12% NaCl and rice wine (~10% ethanol). The end product has a characteristic flavor and color, and very salty taste. Some varieties may have putrid flavor and be objectionable to some individuals. In commercial production, the product is finally bottled with brine, sterilized, and marketed.

13.2 Processing Principles

Making *sufu* requires two stages of fermentation. The first stage allows molds to grow as much mycelia as possible. At the same time, various enzymes are produced, including lipases and proteases. The second stage of fermentation, known as brine aging, promotes major biological changes since many enzymes produced during the first fermentation are now released into the brine and become active. This process results in increased total soluble nitrogen, decreased total insoluble nitrogen, and increased free fatty acids, although total lipids and nitrogen remain unchanged. Many volatile compounds are produced. There are also changes in total amino acid and free amino acid profiles.

14 Conclusion

With the continuous development of marketization and modernization, China's traditional foods have achieved certain development. Although the mechanisation and moderization is coming to Chinese traditional food, food safety should be ensured.For generations (beer, wine, ham, etc.), domestic traditional foods should be appropriately referenced. Due to The difference in the external dietary culture, the modern development of domestic traditional food depends on independent innovation carry out. Over the past two decades, traditional foods in China have continued to innovate on the basis of inheritance.

References

Chu, H. S., Yao, C. K., & Tan, V. P. Y. (2018). Food therapy in sinosphere Asia. *Journal of Clinical Gastroenterology, 52*, 105–113.

He, G.-q., Liu, T.-j., Sadiq, F. A., Gu, J.-s., & Zhang, G.-h. (2017). Review: Insights into the microbial diversity and community dynamics of Chinese traditional fermented foods from using high-throughput sequencing approaches. *Journal of Zhejiang University-Science B (Biomedicine & Biotechnology), 18*(4), 289–302.

Li, J.-r., & Hsieh, Y.-H. (2004). Traditional Chinese food technology and cuisine. *Asia Pacific Journal of Clinical Nutrition, 13*(2), 147–155.

Liu, D., & Tong, C. (2017). Bacterial community diversity of traditional fermented vegetables in China. *LWT Food, 86*, 40–48.

Liu, T., Sadiq, F. A., Yang, H., Gu, J., Yuan, L., Lee, Y. K., & He, G. (2018). Predominant yeasts in Chinese traditional sourdough and their influence on aroma formation in Chinese steamed bread. *Food Chemistry, 242*, 404–411.

Peng, Q., Jiang, S., Chen, J., Ma, C., Huo, D., Shao, Y., & Zhang, J. (2018). Unique microbial diversity and metabolic pathway features of fermented vegetables from Hainan, China. *Frontiers in Microbiology, 9*, 399.

Tian, J., Zhu, Z., Wu, B., Wang, L., & Liu, X. (2013). Bacterial and fungal communities in Pu'er tea samples of different ages. *Journal of Food Science, 78*(8), M1249–M1256.

Zhang, G., & He, G. (2013). Predominant bacteria diversity in Chinese traditional sour dough. *Journal of Food Science, 78*(8), 1218–1223.

Zhang, G. H., Wu, T., Sadiq, F. A., Yang, H. Y., Liu, T. J., Ruan, H., & He, G. Q. (2016). A study revealing the key aroma compounds of steamed bread made by Chinese traditional sourdough. *Journal of Zhejiang University-Science B (Biomedicine & Biotechnology), 17*(10), 787–797.

Zhang, H., Chen, X., Dan, T., & Dong, J. (2004). Traditional Chinese fermented dairy foods. In H. Zhang & Y. Cai (Eds.), *Lactic acid bacteria*. Dordrecht: Springer.

Zhang, J. C., Liu, W. J., Sun, Z. H., Bao, Q., Wang, F., Yu, J., … Chang, H. (2011). Diversity of lactic acid bacteria and yeasts in traditional sour doughs collected from western region in Inner Mongolia of China. *Food Control, 22*(5), 767–774.

Chapter 6
Indian Traditional Foods: Preparation, Processing and Nutrition

Laxmi Ananthanarayan, Kriti Kumari Dubey, Abhijeet B. Muley, and Rekha S. Singhal

1 Introduction

India is a country embracing various cultures, languages, climates, religions and communities. By virtue of such diversity, India is blessed with many cuisines employing a variety of ingredients and preparation methods. Thus, Indian traditional foods, which are usually prepared in households or institutions like hotels/restaurants exhibit tremendous diversity and variation. Traditional foods are prepared according to the customary method(s), which has been practiced for many years and has been handed down from generation to generation. Although Indians predominantly follow a vegetarian diet, some regions, particularly in the coastal areas, are popular for their non-vegetarian preparations. It may be noted that in the Indian context, the term 'vegetarian diet' includes foods of plant origin as well as dairy based products, while the term 'non-vegetarian' also includes animal based products such as all types of meats, eggs as well as fishes. There are many vegetarians who consume eggs but abstain from other meat and fish based products. This unique trait is also recognized by the regulatory system in India wherein any food containing any animal based product is to be indicated as a dark brown circle enclosed in a brown square. Vegetarian foods are marked with a dark green circle enclosed in a green square box [e.g. ▣▣].

Each region of India has a unique style of cooking with distinct flavours and can proudly boast about a variety of traditional foods typical for that region. Punjab region is known for its *lassi* and *makai ki roti* with *sarson ka saag* while southern India is known for *idli*, *dosa*, *vada* and *sambhar*. West Bengal in the east is known for its milk-based confections such as *rosogolla* and *sondesh* and Gujarat in the west

L. Ananthanarayan · K. K. Dubey · A. B. Muley · R. S. Singhal (✉)
Food Engineering and Technology Department, Institute of Chemical Technology, Mumbai, Maharashtra, India

M. Khalfan Al-Khusaibi et al. (eds.), *Traditional Foods*, Food Engineering Series, https://doi.org/10.1007/978-3-030-24620-4_6

is known for *undhiyo* and *thepala* among other food items. Maharashtra is known amongst other things for *bhakri, thalipeeth,* and *chakli.*

A variety of ingredients are used in the preparation of traditional foods. Staple foods are usually based on rice or wheat. Occasionally other less known grains such as sorghum (*jowar*), pearl millet (*bajra*) or finger millet (*ragi/nachani*) are used in locally producing areas. Many ingredients are used in the preparation of meal accompaniments including whole legumes/sprouts, *daals* (pulses), vegetables, *dahi* (yoghurt), *paneer* (cottage cheese), acidulants such as tamarind and *kokum*, and a wide variety of spices. Desserts and confections may be grain-based such as *laddu* and *chikki* or milk-based such as *gulab jamun* and *rosogolla* and may use sugar or jaggery as a bulk sweetener. Beverages may incorporate milk, buttermilk and/or fruit ingredient and sugar. A variety of snacks having short to long shelf life are made using variety of ingredients including cereals, legumes, vegetables, spices and oil. Chicken and mutton tend to be the most commonly consumed meats.

Beef and fish are widely consumed in some parts of India such as coastal areas as well as the north-east part of India but they are not widely consumed across India. The most important and frequently used spices and flavourings in Indian cuisine are asafoetida (*hing*), black mustard seed (*sarso*), cardamom (*elaichi*), coriander (*dhania*), cumin (*jeera*), garlic (*lasun*), ginger (*adrak*), pepper (*kali miri*), red chilli (whole or powdered), and turmeric (*haldi*). *Garam masala* is a popular condiment used in the northern region of India that typically includes five or more dried spices, especially cardamom, cinnamon (*dalchini*), clove (*laung*), cumin seeds, peppercorns, coriander seeds, caraway seeds, nutmeg (*jaifal*), and dried ginger (*sunth* powder). Each culinary region has a distinctive *garam masala* blend with some variations. Some leaves commonly used for flavouring include bay leaves (*tejpatta*), *dhaniya patta* (coriander leaves), *methi* (fenugreek leaves), and *pudina* (mint leaves). People from Gujarat and South India generously use curry leaves and roots for flavouring their cuisine. Numerous other condiments tailored for specific foods are made at home scale and available commercially. Some of these include *goda masala, chana masala*, and *panch phoran*.

2 Classification of Indian Traditional Foods

2.1 Region Wise Foods of India

2.1.1 North Indian Food

In North India, flatbreads made from whole wheat flour (*atta*) are eaten as a staple and almost with all major meals of the day. *Roti* and *paratha* are common types of flatbreads. Lentils (*daal* items) and preparations of vegetables are also mostly eaten with rice and flatbread. There is liberal use of *ghee* or clarified butter in North Indian cuisine. Food in North India is also more oily or buttery as compared to South Indian dishes. Traditional preparation of North Indian food involves the use of ground spices like *garam masala* (five spice mix also known as a mixture of hot spices).

2.1.2 East Indian food

Mustard seeds and paste, chillies (both green and red), *paanch phoran* (a mix of five spices—mustard seeds, cumin seeds, onion seeds, fennel seeds and fenugreek seeds) are common spices used in cooking. *Dahi* (yoghurt), *nariyal* (coconut), maize and *besan* (gram flour) are commonly used ingredients. Milk and dairy products play a major role in the preparation of sweets in Eastern India.

2.1.3 South Indian Food

South Indian food mostly involves the use of *rava* (wheat semolina) and whole rice. In South India, a lot of rice is consumed in the form of *dosa* (Indian savoury crepe), *idli* (steamed savoury rice cake), *idiyappam* (rice flour pressed into noodle form and then steamed), *appam* (a kind of pancake made with coconut milk and fermented rice batter). *Puliogre* (tamarind rice), coconut rice, *bisibele bhath* (lentils and rice dish), *parotta* (layered flatbread made with white flour), and *puttu* (rice portioned in cylindrical form) are also popular dishes. Rice is combined with *sambaar* (a soup-like lentil dish tempered with whole spices and chillies) and *rasam* (a hot-sour soup like lentil dish), dry and curried vegetables and meat dishes and a host of coconut-based *chutneys* and *papadum* (deep-fried crispy lentil pancakes). South Indian meals are mostly prepared in coconut oil or refined oil. Gingelly (sesame) oil is also common in the south since it imparts a fragrant and nutty aroma. In South Indian cuisine, there is a generous use of curry leaves and dried red chilli for tempering. Mustard seeds, asafoetida, pepper, tamarind, chillies and fenugreek seeds are generally used as spices for flavour and aroma generation.

2.1.4 West Indian Food

The western region of India has the most assorted varieties of food dishes. Food is spicy and largely vegetarian in Rajasthan, while Gujarati cuisine is acknowledged for its slight sweet touch and is traditionally vegetarian. In Maharashtra, *Malvani* cuisine (fresh coconut-based hot and sour curries with fish and seafood) is famous in coastal regions, while the interiors have the more prudent, *Vidharba* cuisine, which uses lot of dry coconuts. Goan food is rich, piquant and strongly flavored by coconut, red chillies, and vinegar. In Gujarat and Rajasthan, corn, lentils and gram flour, dry red chillies, buttermilk, yoghurt, sugar, and nuts; in Maharashtra, fish, rice, coconut and peanuts and, in Goa, fish, pork and rice are the common ingredients for cooking. Vegetable oils like sunflower, peanut oil, and *ghee* are commonly used for cooking. Spices mainly include dry red chillies, sesame seeds, coconut, nuts, and vinegar.

2.2 Breakfast Foods, Staples, Savouries, Confections and Beverages

Indians traditionally eat four meals daily: breakfast, lunch, tea-time snack, and dinner. These are discussed in the following sections.

2.2.1 Breakfast Foods

Breakfast is usually a freshly cooked meal which includes snack foods such as *idli*, *dhokla*, *pohe*, *poori bhaji*, and *paranthtas* and a beverage usually tea/coffee. Tables 6.1, 6.2, 6.3 and 6.4 give a broad classification of Indian traditional breakfast and snack items of all four regions of India (north, east, south and west) and include some examples of each class.

2.2.2 Staples

Lunch is a freshly home cooked full meal and consists of staples, meal accompaniments with desserts/confections and deep fried snack items served on special occasions. Dinner is like lunch but a smaller version with variety and quantity. Staples in the Indian diet vary by region, but usually include rice, *roti* (flatbread), and lentils/*dal* along with some vegetable based accompaniments. *Roti* is the general name used for different types of flatbreads. They are generally unleavened and prepared from whole wheat flour.

Naan is cooked in tandoor oven, while various types of *parathas* like plain *parathas*; stuffed *parathas* (*aloo/paneer/mooli*, and others) are shallow fried on *tava* (Table 6.5). *Parathas* and *rotis* are often eaten with *subzi* or curry dishes. In Eastern India, *mughlai parota* and *petai parota* are very popular dishes (Table 6.6) while, *Akki roti, jolada* and *obattu* are common flatbread items in southern India (Table 6.7). *Bhakri* and *jowar ni rotlo* are common millet flatbread items in western Indian households (Table 6.8). Rice is a staple food consumed all over the Indian subcontinent. It is used more as a staple in southern and eastern regions of India as it is the most cultivated crop in these regions. Most meals use rice as their base, with other foods accompanying it.

In Eastern India, rice is generally the plain white variety, which is popularly accompanied with *daal*/fish curries. *Bisibele bhath*, *pongal*, *puliyodharai*, *mosarrana and hyderabadi biryani* are popular rice dishes of South India. Lentils and beans are staples of North Indian region, which serve as major protein sources. *Rajma*, *chole*, *kadhi*, *daal (masoor daal/moong daal/toor daal.*) are vegetarian

Table 6.1 Breakfast foods/snacks of North India [(V) = vegetarian and (NV) = non-vegetarian]

***Chole bhature* (V)** Major Ingredients: *Chole* [white chick peas, tomatoes, onion, spices (chilli powder, turmeric, coriander powder, cumin powder, dried mango powder, *garam masala*), salt], *bhature* (refined flour, salt, yeast, oil) Major steps in preparation: *Chole* [soaking (chickpeas soaked overnight), gravy preparation (sautéing (garlic and ginger paste, finely chopped onions, tomato paste and spices), cooking (addition of soaked chickpeas to gravy and pressure cooking), *Bhature* preparation (kneading, dough fermentation, rolling, deep frying) Common accompaniments: Pickles/cut onions/cut lemon
***Kulcha* (V)** Major Ingredients: Dough (refined wheat flour, curd, salt, sugar, oil, baking soda), Filling [Potatoes, green chillies, ginger, coriander leaves, spices (coriander powder, dried mango powder, red chilli powder, *garam masala*), salt] Major steps in preparation: Dough preparation (mixing salt, sugar, curd, baking powder and oil to refined flour and fermentation for 30 min), stuffing (boiling, peeling and mashing the potatoes, addition of salt, spices and coriander leaves), filling and baking (filling of stuffing to the dough balls and rolling), dusting with carom seeds and baking in oven/tandoor/*tava* (griddle) Common accompaniments: *Chole*
***Daal puri* (V)** Major Ingredients: Dough (wheat flour, salt, oil), stuffing [split Bengal gram, spices (turmeric powder, *garam masala*), salt, oil] Major steps in preparation: Dough preparation (kneading), stuffing (boiling split Bengal gram with turmeric and salt, mashing cooked *daal* and mixing spices, filling in dough balls and rolling, followed by shallow frying in a pan) Common accompaniments: Curd/*curry*/*chutney*s
***Daal Phara* (V)** Major Ingredients: Dough (rice flour/whole wheat flour, oil, salt), stuffing [skinless split Black Bengal gram (*urad daal*), green chillies, ginger, fennel seeds, black pepper], *Frying* (ginger, garlic onions, tomatoes, spices (cumin powder, pepper), salt, oil) Major steps in preparation: Dough preparation (kneading, preparing dough balls) s*tuffing*: (soaking, draining and grinding *daal* mix, mixing spices and salt to it), boiling pharas (stuffing filling to dough balls followed by cooking in boiling water for 10–15 min), frying (sautéing ginger garlic paste, chopped onions and spices. Boiled pharas cut cross-wise into 4–5 discs and cooked for few minutes) Common accompaniments: Tomato- coriander green *chutney*/cucumber *raita*
***Aloo tikki* (V)** Major Ingredients: Potatoes, green peas, chopped coriander leaves, spices (chilli powder, *chaat masala*), lemon juice, salt, oil Major steps in preparation: Boiling (potatoes and green peas boiled separately), mashing (boiled and peeled potatoes and peas together, adding salt, spices lemon juice to mashed potatoes, mixture divided into small balls and rolled into thick round discs), shallow frying (discs are shallow fried in a pan) Common accompaniments: Tomato- coriander green *chutney*/cucumber *raita*

preparations, which are served with flatbread and rice items. In southern India, *sambhar* and *rasam* are soup like preparations using lentils, which are served with breakfast items and lunch items. An Indian meal is incomplete without one or more *subzi* (vegetable preparation), which play an important role in complementing the meals on nutritional front. Recipes of subzis vary according to local produce, spices used and method of cooking. In Punjab, *makke ki roti* is accompanied with *sarson*

Table 6.2 Breakfast foods/snacks of East India [(V) = vegetarian and (NV) = non-vegetarian]

***Cheera Doi* (V)** Major Ingredients: Beaten rice, curd and jaggery Major steps in preparation: washing and soaking beaten rice for few minutes, mixing curd and jaggery to soaked rice
***Pitha(Sunga Pitha, Tekeli Pitha, Uhuwa pitha)* (V)** Major ingredients: Dough (rice flour, water), stuffing [*Sunga Pitha* (jaggery), *Tekeli Pitha* (coconut, sugar, milk, cardamom, dried orange rind), *Uhuwa pitha* (rice flour, jaggery/salt)] Major steps in preparation: Dough preparartion (kneading ingredients, stuffing filling into dough dumplings), cooking [*Sunga Pitha* (roasting in a bamboo tube), *Tekeli Pitha* (baking in earthern ware), *Uhuwa pitha* (boiling)]
***Luchi cholar daal* (V)** Major ingredients: Dough (refined wheat flour, water), frying (oil), *cholar daal* [split Bengal gram, spices (chilli powder, turmeric powder), salt, sugar], tempering (bayleaf, cumin seeds, cinnamon, dried coconut, oil) Major steps in preparation: *Luchi* (kneading the dough and cutting small balls, rolling *luchis* into a small circle shape, deep frying puris in oil), *Cholar daal* (pressure cooking soaked and drained Bengal gram *daal*, tempering cooked *daal* with bayleaf, cumin seeds, cinnamon, red chillies and sliced coconut in oil)
***Ghoogni* (V)** Major ingredients: Black small chickpea, tomatoes, onion, bayleaf, cumin seeds, turmeric powder, coriander powder, chilli powder, salt, oil Major steps in preparation: *Soaking* (chickpeas soaked overnight), *gravy preparation* (sautéing garlic and ginger paste, finely chopped onions, tomato paste and spices and salt) Major steps in preparation: Addition of soaked chickpeas to gravy and pressure cooking, consistency is semi-dry. Common accompaniments: *Muri* (puffed rice)
***Alu Pitika* (V)** Major ingredients: Potatoes, onion, spices, salt Major steps in preparation: Boiling, peeling and mashing of potatoes, mixing cut onions and spices to it. Common accompaniments: *Chutneys*/dips
***Mangsher singhara* (NV)** Major ingredients: Dough (refined wheat flour, salt, water), stuffing (minced meat, peas, lemon, spices, oil), frying (Oil) Major steps in preparation: Dough (kneading, cutting round small balls and rolling into small circles), stuffing (filling stuffing inside the rolled dough and giving triangle shape to it), cooking (deep frying) Common accompaniments: Mint *chutney*, tea
***Momos* (V)** Major ingredients: Dough (refined wheat flour, salt, oil), stuffing [(meat/chopped vegetables/ cheese), chilli powder, salt] Major steps in preparation: Dough preparation (kneading refined flour), filling (stuffing of choice is filled inside the dough balls), cooking [steaming in moktu (bamboo container for steaming *momos*)] Common accompaniments: Chutneys/dips

Table 6.3 Breakfast foods/snacks of South India [(V) = vegetarian and (NV) = non-vegetarian]

***Dosa* (V)** Major ingredients: *Batter:* Rice, skinless black *urad daal, stuffing:* potatoes, onion, tomatoes, spices (turmeric powder, coriander powder, chilli powder), salt, *curry* leaves, oil Major steps in preparation: Batter preparation [soaking rice and *urad daal* (3:1), grinding, fermentation], stuffing (boiling, peeling and mashing potatoes, tempering with mustard seeds, *curry* leaves and oil), *Dosa* preparation (cooking batter as thin pancakes on *dosa tava*, layering it with stuffings, cut onions, tomatoes folding it in half or rolled like a roll, shallow frying till it becomes crisp golden brown) Common accompaniments: *Sambhar*, coconut *chutney*
***Iddiyappam/Nool puttu* (V)** Major ingredients: Rice flour, salt, grated coconut, water Major steps in preparation: Dough preparation (kneading rice flour with hot water), *Iddiyappam* making (requires *idiyappam* maker which helps to make rice noodles), dough filling (in *Idiyappam* maker) and pressing on heated cavities in concentric circles, *cooking* (by steaming for few minutes), cooked *idiyappam* is garnished with grated coconut. Common accompaniments: Coconut *chutney*
***Idli* (V)** Major ingredients: Rice, skinless black *urad daal* Major steps in preparation: Batter preparation [soaking [rice and *urad daal* (3:1), grinding, fermentation], cooking (by steaming in *idli* vessel) Common accompaniments: *Sambhar*, coconut *chutney*
***Nandu omelette* (NV)** Major ingredients: Whole eggs, onion, chillies, tomatoes, crab (nandu), *curry* leaves, oil, pepper, salt Major steps in preparation: Steaming deshelled crab flesh and shredding, sautéing onion, green chillies and *curry* leaves, adding sautéed onion and cracked eggs to the cooked crab, cooking (cooking the mix like a pancake in a pan)
***Kuzi Paniyaram/Paddu/Yeriyappa/Gundponglu* (V)** Major ingredients: Beaten rice, rice, *urad daal*, fenugreek seeds, onion, chillies, salt, oil Major steps in preparation: Batter preparation (soaking rice, *urad daal* and fenugreek seeds separately for approx 6 h, beaten rice blended in the batter), *Paniyaram* making (It needs special paniyaram vessels with fissures, batter is placed in a greased vessel and cooked both sides) Common accompaniments: *Curry/chutney*/dips
***Puttu* (V)** Major ingredients: Rice flour and coconut Major steps in preparation: Rice flour kneading, scraping coconut, upper portion of *Puttu kutti* vessel is layered with alternate layers of coconut and rice mixture, lower portion of vessel filled with water for steaming, steamed cylinders of *puttu* are removed after cooking. Common accompaniments: Gravies (fish *curry*/chicken *curry*/chickpea *curry*)
***Upma* (V)** Major ingredients: Semolina (*sooji*), onion, carrot, peas, mustard seeds, *curry* leaves, oil Major steps in preparation: Sautéing [*curry* leaves, mustard seeds, groundnut peas, finely chopped onions and vegetables (optional) in oil], adding roasted semolina and water to it and cooking. Common accompaniments: Coconut *chutney*

(continued)

Table 6.3 (continued)

***Uttapam* (V)** Major ingredients : Batter [rice, skinless black gram *daal* (*urad daal*)], toppings [vegetables (onion, tomatoes, green chillies), chilli powder, salt, oil] Major steps in preparation: Batter preparation: soaking [rice and *urad daal* (3:1)] grinding, fermentation, pan frying like a thick pancake, topped with cut vegetables. Common accompaniments: Veg korma, sweetened coconut milk
***Medu wada* (V)** Major ingredients: Skinless *urad daal*, salt, oil Major steps in preparation: Batter preparation [soaking *urad daal* (6–8 h), grinding, fermentation], cooking (deep frying) Common accompaniments: *Sambhar*, coconut *chutney*
***Mysore bonda*(V)** Major ingredients: Potatoes, *curry* leaves, mustard seeds, turmeric powder, salt, oil spices Major steps in preparation: Boiling (potatoes and green peas separately), mashing (boiled and peeled potatoes, adding salt, spices to mashed potatoes), tempering of mashed potatoes with mustard seeds, *curry* leaves in oil, mixture divided into small balls, coated with gram flour batter and deep frying (balls/*bondas*).

Table 6.4 Breakfast foods/snacks of West India [(V) = vegetarian and (NV) = non-vegetarian]

***Pohe* (V)** Major ingredients: Beaten rice, onion, potato, *curry* leaves, spices (turmeric powder, chili powder), mustard seeds, salt, oil Major steps in preparation: Tempering (mustard seeds, *curry* leaves in oil), sautéing (onion, tomatoes and spices), adding washed and soaked beaten rice to it, covered cooking with occasional stirring. Common accompaniments: *Chutney*s
***Dhokla* (V)** Major ingredients: Batter [Bengal gram flour (besan), oil, salt, spices (turmeric powder, chili powder), salt, baking powder], tempering (mustard seeds, *curry* leaves, grated coconut, oil) Major steps in preparation: Batter preparation (mixing *besan* and other ingredients in water and ferment for 10 min), steaming (filling the container with batter and steam cooking), tempering (steamed dhokla is cooled and cut into square pieces and temepered with mustards seeds and other ingredients). Common accompaniments: *Chutney*s
***Khandvi* (V)** Major ingredients: Batter [Bengal gram flour, buttermilk, *curry* leaves, lemon juice, spices (turmeric powder, chili powder), salt], tempering (mustard seeds, salt, sugar, oil) Major steps in preparation: Batter preparation [mixing gram flour, buttermilk (3:1) and spices to get a smooth consistency], cooking (in a pan by continuously stirring and spreading on a greased plate, thin layers are rolled and cut into equal sized strips), tempering (with *curry* leaves, mustard seeds in oil) Common accompaniments: Coconut/mint/tamarind *chutney*
***Thalipeeth* (V)** Major ingredients: Whole wheat flour, millet flour, rice flour, gram flour, onions, chillies, spices (carom seeds, cumin powder, turmeric powder, coriander powder), finely chopped ginger, salt, *ghee*/oil Major steps in preparation: Dough preparation (mixing all the ingredients and kneading), cooking (rolling and flattening dough on muslin cloth to get a flatbread and transferring to the hot greased griddle) until it turns brown and crisp Common accompaniments: Pickles, *chutney*s/yoghurt

(continued)

Table 6.4 (continued)

***Pav Bhaji* (V)** Major ingredients: *Pav* (refined flour, yeast, salt, oil), *Bhaji* [cabbage, onion, capsicum, peas, cauliflower, potatoes, spices (turmeric powder, chili powder, coriander powder), salt, oil] Major steps in preparation: *Pav* [kneading and fermentation, baking (in tray/mould at 270°C)], *Bhaji (*tempering (mustard seeds, *curry* leaves in oil), sautéing (cut vegetables, spices and cooking *curry* style) Common accompaniments: Cut onion, cut lemon
***Vada pav* (V)** Major ingredients: *Vada* [gram flour (for the batter), potatoes, *curry* leaves, spices (chilli powder, turmeric powder), salt, oil], *Pav* (as discussed in *Pav Bhaji*) Major steps in preparation: *Vada* [mashing (boiled and peeled potatoes, addition of salt, spices), tempering (with mustard seeds, *curry* leaves and oil), mixture divided into small balls and dipped in gram flour batter, deep frying (*vada* deep frying in oil); *Pav* (as discussed in *Pav Bhaji)* Common accompaniments: Mint *chutney*, tamarind *chutney*, *lasun chutney*
***Kombdi vade*(NV)** Major ingredients: Rice, black gram *daal*, salt, spices (cumin seeds, coriander seeds, black peppercorns, fenugreek seeds turmeric powder, chilli powder), oil Major steps in preparation: Batter preparation (soaking and grinding rice and black gram *daal*), mixing salt and spices in the batter, cooking (deep frying in oil) Common accompaniments: Chicken *curry*

Table 6.5 Staple Foods of North India [(V) = vegetarian and (NV) = non-vegetarian]

***Chapati/Roti* (V)** Major ingredients: Whole wheat flour, oil, salt Major steps in preparation: Dough preparation (mixing flour, salt, oil and water, kneading), rolling *chapatis* and roasting on hot *tava* or griddle Common accompaniments: *Curry*/cooked vegetables
***Naan Roti* (V)** Major ingredients: Refined wheat flour, curd, oil Major steps in preparation: Dough preparation (mixing flour, salt and water and kneading and fermentation for 30 min), cooking (baking in a hot *tandoor*) Common accompaniments: *Curry*/cooked vegetables
***Paratha* (V)** Major ingredients: Wheat flour, salt, oil Major steps in preparation: Dough preparation (mixing flour, salt, water and kneading), cooking [rolling and layering (with oil) *parathas* in a triangle shape, shallow frying with oil] Common accompaniments: *Curry*/curd/pickles/onion/chillies
***Rumali roti* (V)** Major ingredients: Refined wheat flour Major steps in preparation: Kneading, rolling, roasting Common accompaniments: *Curry*/curd/pickles/onion/chillies
***Stuffed paratha* (V)** Major ingredients: Dough (wheat flour, salt, oil, wheat flour), stuffing (potato mixture/*paneer*/ radish/vegetables, spices, oil) Major steps in preparation: Dough preparation (mixing flour, salt, water and kneading and preparing dough balls), stuffing (filling dough balls with the desired filling), rolling stuffed dough balls and shallow frying in a pan Common accompaniments: *Curry*/curd/pickles/onion/chillies

(continued)

Table 6.5 (continued)

Makke di roti **(V)** Major ingredients: Corn flour Major steps in preparation: Kneading, rolling, baking Common accompaniments: *Sarson ka saag*, jaggery/*ghee*
Plain rice **(V)** Major ingredients: Rice Boiling in water and straining Common accompaniments: Indian *curries* and/*daal* items
Vegetable pulaav **(V)** Major ingredients: Rice, vegetables (onion, pea, carrot), spices (turmeric powder, chili powder, peppercorns, bayleaf, cardamom, cloves, saffron strands, salt, *ghee*/oil Major steps in preparation: Tempering (cumin seeds, peppercorns, bayleaf, cardamom and cloves in *ghee*/oil), sautéing (vegetables and spices), addition of washed and strained rice to it and pressure cooking Common accompaniments: *Curries* and/*daal* items
Khichdi **(V)** Major ingredients: Rice, split pigeon pea, cumin seeds, bayleaf, turmeric powder, chili powder, salt, oil Major steps in preparation: tempering cumin seeds, bay leaf in *ghee*/oil, addition of washed and drained rice and lentil, pressure cooking Common accompaniments: Curd, *papad*, pickles
Aloo curry (Aloo baingan, Aloo gobhi, Aloo methi, Aloo shimla) **(V)** Major ingredients: Potatoes, onion, tomatoes, garlic, ginger, vegetable of choice (brinjal/cauliflower/pea/fenugreek leaves/capsicum), spices (turmeric powder, chili powder, coriander powder, cumin seeds), salt, *ghee*/oil Major steps in preparation: Sautéing ginger garlic paste in cumin tempered oil, addition of potatoes and vegetable of choice, cooking and simmering with spices Common accompaniments: Indian bread items/rice items
Baingan bharta **(V)** Major ingredients: Brinjal, chili, coriander leaves, spices Major steps in preparation: Roasting and mashing brinjal, mixing cut chillies, mustard oil, salt Common accompaniments: Indian bread items
Stuffed vegetable **(V)** Major ingredients: vegetables (capsicum/potatoes/tomatoes/gourds), filling (mashed potatoes/peas/other vegetables/spices/dry mango powder), salt, oil Major steps in preparation: vegetable preparation (peeling, scraping, and removal of pulp and/or seeds from vegetables), stuffing (mashed potatoes/peas/spices/dry mango powder), cooking (shallow pan frying) Common accompaniments: Indian bread items/rice items
Saag **(V)** Major ingredients: Leafy vegetables (fenugreek/mustard, spinach/gram and others), garlic, oil and spices Major steps in preparation: Sautéing cumin seeds, green chillies in oil, adding green vegetables, turmeric powder and salt, cooking in a covered pan Common accompaniments: Indian bread items/rice items

(continued)

Table 6.5 (continued)

***Murgh musallam* (NV)** Major ingredients: Whole chicken, medium-sized onions, garlic, ginger, green chillies, raw papaya paste, *ghee*, spices (coriander seeds, cumin seeds, red chili powder, turmeric powder), curd, salt Major steps in preparation: *Masala* preparation [(dry roasting spices on the griddle and then grinding to a fine paste, frying and grinding of ginger, garlic and onion), adding these two pastes with curd in a bowl and blending], marination (chicken cleaned and washed thoroughly with water and pricked with a fork all over, marination with papaya paste and salt for a minimum of 2 h), cooking [sautéing (onions, turmeric powder, red chili powder, ginger and garlic paste, and green chillies) in heated oil in a pan, addition of marinated chicken and boiling till *ghee* surfaces on the chicken and turns tender] Common accompaniments: Indian bread items/rice items
***Paneer tikka masala* (V)** Major ingredients: Cottage cheese, tomatoes, onion, ginger, garlic, lemon, *kasoori methi*, spices (cinnamon, cardamom, cloves, turmeric powder, coriander powder, chili powder), corn flour, cream, salt, curd, oil Major steps in preparation: Marination*:* [*Paneer* (Cottage cheese) capsicum, tomato and onion marination with ginger garlic paste, spices, lemon juice, oil and curd for 20 min], making paneer tikka (grilling or shallow frying of marinated *paneer* cubes and diced vegetables), gravy preparation (heating oil in a pan and roasting cinnamon, cardamom and cloves), adding tomato puree, *kasoori methi*, roasted cumin powder, dry ginger powder, salt, diluted corn flour and cooking on medium flame, addition of *paneer tikka* and cream/*malai* to the gravy and cooking for few minutes Common accompaniments: Indian bread items/rice items
***Tandoori Fish Tikka* (NV)** Major ingredients: Boneless fish, ginger, garlic paste, green chilli, lemon, salt, mustard oil, carom seeds Major steps in preparation: Marination (same as *paneer tikka*), cooking in open fire Common accompaniments: Indian bread items/rice items
***Daal fry with tadkha* (V)** Major ingredients: Lentils, onion, tomatoes, spices (cumin seeds, turmeric powder, chili powder), salt, oil/*ghee* Major steps in preparation: Boiling (washed and drained lentils with turmeric powder and salt), sautéing (cumin seeds, onion, tomatoes and chillies in oil/*ghee),* tempering cooked *daal* with sautéed onion and tomatoes Common accompaniments: Indian bread items/rice items
***Kadhi Pakode* (V)** Major ingredients: *Kadhi* [gram flour, yogurt, spices (cumin seeds, turmeric powder, chili powder), salt], *pakode:* Diced onions, gram flour, spices, salt Major steps in preparation: *Kadhi* [thin watery consistency cooking (gram flour, curd, turmeric powder, chilli powder and salt), tempering with mustard seeds, *curry* leaves], *Pakode:* [batter preparartion (onion, gram flour and spices), frying crispy friiters (*pakode*)], dipping *pakode* in *kadhi* Common accompaniments: Cooked rice
***Rajma* (V)** Major ingredients: Rajma (red kidney beans, tomatoes, onion, spices(chilli powder, turmeric, coriander powder, cumin powder, *aamchur* powder (dried mango powder), *garam masala*, salt Major steps in preparation: Soaking (kidney beans soaked overnight), gravy preparation, sautéing (garlic and ginger paste, finely chopped onions, tomato paste, spices and salt), cooking (addition of soaked *rajma* to gravy and pressure cooking) Common accompaniments: Cooked rice

Table 6.6 Staple foods of East India [(V) = vegetarian and (NV) = non-vegetarian]

Bread items
***Luchi* (V)** Major ingredients: Major ingredients: Refined wheat flour, hot water, oil Major steps in preparation: Kneading, rolling, deep frying Common accompaniments: *Curry*/curd/*chutneys*
***Moghlai porota* (NV)** Major ingredients: Refined flour, egg, vegetables (onions, tomatoes, ginger, garlic, potatoes, chillies, coriander leaves), spices (cumin powder, turmeric powder, coriander powder, five spice powder (*garam masala*), salt, oil Major steps in preparation: Preparation of filling: whipped egg, cut, boiled and mashed vegetables, spices and salt, kneading (kneading the dough with salt and oil and rolling into square shape), stuffing (filling the rolled bread with filling mixture and further rolled to bigger square shape bread), cooking (deep frying on the big curved griddle) Common accompaniments: Onion, salad
***Petai porota* (V)** Major ingredients: Refined flour, oil, salt, sugar Major steps in preparation: Dough preparation (kneading of dough with salt, oil and keeping it for 30 min), rolling into a large circle, cooking [roasting on big upside down heated curved griddle *(kadahi)*], followed by shallow frying of roasted *parota* in refined oil, beating and tearing into fragments Common accompaniments: Curd/cream/jaggery
Rice items
***Komol chawal* (V)** Major ingredients: *Komol chawal* (Assamese soft rice) Major steps in preparation: soaking and mixing Common accompaniments: *Ghee*/jaggery/cream
***Ghee Bhat* (V)** Major ingredients: Rice, salt, sugar, spice (cardamom, cloves, bayleaf), dryfruits (cashewnut, raisins), *ghee* (clarified butter) Major steps in preparation: Cooking (Steaming rice to make it very soft), sautéing (spices and dryfruits in *ghee),* addition of cooked rice, salt, sugar mixed properly and topped with *ghee* Common accompaniments: *Dum aloo*
Vegetables (subzi)/*Daal* items/*Curry* items (vegetarian/non-vegetarian)
***Khalo Daal* (V)** Major ingredients: Black gram *daal*, onion , tomatoes, chillies, mustard oil, garlic cloves, spices (turmeric powder, coriander powder), salt, oil Major steps in preparation: Cooking (pressure cooking black gram *daal* along with salt and spices), tempering (with cumin seeds, garlic cloves in oil) Common accompaniments: Rice
***Til ko alu* (V)** Major ingredients: Potato, onion, sesame seeds (*til*), spices Major steps in preparation: tempering sesame seeds in oil. Sautéing onion, potatoes, spices, salt in oil. Common accompaniments: *Sidra ko achar*, cooked rice
***Bengena Pitika* (V)** Major ingredients: Brinjal, onion, spices, salt, mustard oil Major steps in preparation: Roasting (black brinjal on the flame), skin removal and mashing of roasted brinjal, addition of salt and mustard oil Common accompaniments: Rice

(continued)

Table 6.6 (continued)

Bread items
***Misa Mach poora* (NV)** Major ingredients: Shrimps (shelled and de-veined), onion, lemon, spices, mustard oil Major steps in preparation: Sautéing, grilling/roasting
***Begun bhaja* (V)** Major ingredients: Brinjal (egg plant/aubergine), spices, mustard oil Major steps in preparation: Brinjal cutting (into circular discs), marination (with turmeric powder, chilli powder and salt), deep frying in mustard oil Common accompaniments: Plain cooked rice and *daal*
***Macher Paturi* (NV)** Major ingredients: Fish, poppy seeds, mustard seed, coconut, mustard oil, banana leaf, salt, green chilli Major steps in preparation: boiling fish and removal of bones, filling of paste (developing paste with crushed poppy seed, mustard seed and crumbled coconut), roasting (greased banana leaf to develop burnt flavour). filling of paste in greased roasted banana leaf cut in square shape, steaming of the filled paste under cover for 10 min until the smell of mustard seed evolves, topping with mustard oil before serving Common accompaniments: Plain cooked rice
***Muri ghonto* (NV)** Major ingredients: Fish head, rice, vegetable, mustard oil, cumin, coriander, spices, salt Major steps in preparation: Frying (vegetable semi fried, fish head broken and fried in mustard oil/*ghee*), boiling (rice semi boiled and starch removed), mixing of spices and sautéing until spices cooked, addition of semi cooked rice and fish and addition of water, preparation kept in steam to get fully cooked, topping with *ghee*/five spice powder
***Mochar ghonto* (V)** Major ingredients: Banana flower, coconut, spices, peanut, refined oil Major steps in preparation: Cutting, washing, soaking of banana flowers, sautéing, adding spices, peanut and water, cooking under cover till flowers are soft and flavor of spices released
***Macher jhal* (NV)** Major ingredients: Fish, mustard paste, mustard oil, spices, salt Major steps in preparation: Fishes semi fried, mustard paste and other spices cooked in mustard oil for short times and water added till oil released, fishes added in the end, topping with mustard oil and green chillies before serving

Table 6.7 Staple foods of South India [(V) = vegetarian and (NV) = non-vegetarian]

Bread items
***Akki roti* (V)** Major ingredients: Rice flour, oil, salt Major steps in preparation: Dough preparation (mixing rice flour, salt, oil and kneading with hot water), cooking (rolling *chapatis* with hands and roasting on hot *tava* or griddle) Common accompaniments: *Curry*/vegetable
***Jolada roti* (V)** Major ingredients: *Sorghum* flour Major steps in preparation: Dough preparation (mixing flour, salt, oil, hot water and kneading), cooking (rolling *chapatis* and roasting on hot *tava* or griddle) Common accompaniments: *Curry*/vegetable

(continued)

Table 6.7 (continued)

Bread items
***Obbattu* (V)** Major ingredients: Wheat flour Filling/stuffing: (*moong* gram *daal* and jaggery or coconut) Major steps in preparation: Dough preparation (mixing flour, salt, oil, hot water and kneading), stuffing (boiling and mashing green gram *daal*) adding jaggery and coconut to stuffing, cooking (stuffing dough balls with filling mixture and rolling *chapatis*), shallow frying in a pan Common accompaniments: *Curry*/vegetable/pickles/curd
***Parotta* (V)** Major ingredients: Refined wheat flour, oil Major steps in preparation: Dough preparation (mixing flour, salt, oil, hot water and kneading), rolling chapattis and layering into multiple layers using oil, cooking (shallow pan frying) Common accompaniments: *Curry*/vegetable/pickles/curd
Rice items
***Bisibele bhath* (V)** Major ingredients: Rice, lentil, vegetables, spices Major steps in preparation: Cooking rice and lentils along with spices and tempering Common accompaniments: *Chutney*, *boondi*, salad or *Papadum*
***Hyderbadi Murg Dum Biryani* (NV)** Major ingredients: Marination [Yogurt (whisked), *Kashmiri* red chili powder, turmeric powder, coriander powder, shahi jeera, ginger paste, garlic paste, golden brown onion, cinnamon, cardamom, cloves, bay leaf, *garam masala* powder, freshly ground black peppercorn, coriander leaves, finely chopped mint leaves, finely chopped, green chillies, lemon juice, salt], cooking rice [long grain *Basmati* rice, lemon juice, black cumin (shahi jeera), star anise, salt], layering: [*ghee*, milk, saffron strands, mint leaves finely chopped, coriander leaves, green chillies, kewra water, raisins, fried cashew nuts, cardamom] Major steps in preparation: frying onions (salt is applied on thinly sliced onions and kept aside for 10 min, water is squeezed and deep fried in oil), marination (chicken with the ingredients mentioned for 2 h.), cooking the rice and layering the biryani (rice is cleaned, washed and soaked for 30 min, rice is halfcooked and strained), layering (half cooked rice and marinated chicken layered along with spices, *curry* leaves, dryfruits and other ingredients in a *handi* and tightly sealed), kept on a high flame for 15 min (placing hot griddle underneath the handi and keep it on "Dum') Common accompaniments: *Boondi* and *Khajur Raita*
***Pongal* (V)** Major ingredients: Rice, green gram *daal*, spices (black pepper, green chilli, ginger, cumin seeds), *curry* leaves, cashew nuts, *ghee*, salt Major steps in preparation: Sautéing (rice and green gram *daal* for 2–3 min), adding water and salt and pressure cooking, tempering (adding *ghee*, cashew nuts, black pepper, cumin seeds, green chilli, *curry* leaves and ginger in a pan and frying for 1 min and pouring over the dish) Common accompaniments: *Chutney* and *sambhar*
***Puliyodharai* (V)** Major ingredients: Rice, Bengal gram *daal*, black gram *daal*, salt, peanuts, tamarind, grated jaggery, red chillies, spices (turmeric powder, chilli powder, asafoetida), oil Major steps in preparation: Heating the oil in a pan on medium heat, adding mustard seeds, dry red chillies, *curry* leaves and peanuts, then adding tamarind water, freshly ground spice powder, turmeric powder, jaggery and salt, cooking (till it thickens and becomes a paste like consistency), adding cooked rice Common accompaniments: *Curry*, curd, *chutney*s

(continued)

Table 6.7 (continued)

Bread items
Daddojanam/Thayir sadam/Mosaranna/curd rice Major ingredients: Rice, milk, curd, spices, overcooked white rice, yogurt, chillies, *curry* leaves, chopped coriander leaves, chopped ginger, salt, mustard seeds, black gram *daal*, Bengal gram *daal*, asafoetida, oil Major steps in preparation: Overcooking rice in water to make it soft, mixing (yoghurt, coriander leaves, ginger, green chillies, rice and salt), tempering (heating oil in a pan and adding mustard seeds, adding *daals*) and sautéing until golden, tempering (with *curry* leaves, red chillies and asafoetida) Common accompaniments: *Chutney*, salad, pickle
Vegetables (*subzi*)/*Daal* items/*Curry* items)
***Avial* (V)** Major ingredients: Yam, white and red pumpkin, snake gourd, carrot, beans, drumstick, raw bananas, raw mango, coconut oil, spices Major steps in preparation: *Masala* [(grinding coconut, chillies and cumin seeds into a fine paste), adding whipped yoghurt to the paste], Vegetable preparation (vegetables cut into small pieces and added to boiling water), cooking (*Masala* paste and *curry* leaves are added to half cooked vegatables boiled until the vegetables are completely cooked), coconut oil was drizzled before serving Common accompaniments: Rice
***Chettinad Chicken* (NV)** Major ingredients: Chicken, oil, onions, tomatoes, cinnamon sticks, cloves, cardamoms, red chilli, coconut, fennel seeds, cumin, *curry* leaves, turmeric powder, salt, oil Major steps in preparation: Grinding the ingredients (onions, ginger, garlic, fennel seeds, cumin seeds, peppercorns, red chillies and coconut for the paste), cutting and marinating chicken with paste, tempering (cinnamon, cardamom, cloves and cumin), addition of chopped tomatoes, onion and coriander leaves and sautéing, addition of marinated chicken and turmeric powder, salt and sautéing for 10 min, sprinkling water at intervals and covered cooking until done
***Haleem* (NV)** Major ingredients: Wheat, meat (goat/meat/beef/chicken), *ghee*, lime, vegetables, milk, lentils, dry fruits, spices, cloves, bay leaves, saffron, green chillies, turmeric powder, mace, cardamom, garlic paste, ginger, onion, lamb stock, salt, oil Major steps in preparation: Washing and soaking all the lentils and broken wheat together for 2 h, tempering of whole spices (cloves, bay leaf and green cardamom in *ghee* in a heavy bottom *handi*, adding grated ginger and garlic paste till the color turns brown), adding lamb along with all the powdered spices (saffron, green chilli paste, turmeric powder), cooking (until it is half done and then fried onions are added), addition of washed and drained lentils and lamb stock and salt to the lamb, cooking (until lamb gets tender and mixes with the lentils and becomes thick in consistency)
***Mor kuzambu* (V)** Major ingredients: Split chickpea, rice, butter milk, green chilli, ginger, coconut, red chilli, curry leaves, turmeric, salt, oil, mustard seeds, asafoetida Major steps in preparation: Soaking and grinding (chickpea lentil and rice), mixing (buttermilk, turmeric powder and salt), heating the *kuzhambu* until it raises and becomes frothy, tempering (with the *curry* leaves, asafoetida, mustard seeds, chillies, garlic, ginger and coconut) Common accompaniments: Rice
***Neimeen porichathu* (NV)** Major ingredients: Seer fish, ginger, garlic, rice, groundnut, chillies, salt, spices, oil Major steps in preparation: Marination, deep frying. Common accompaniments: Rice

(continued)

Table 6.7 (continued)

Bread items
***Rasam* (V)** Major ingredients: Tamarind, green chillies, turmeric powder, curry leaves, asafoetida powder, coriander leaves, salt, mustard seeds, oil, *rasam powder masala* (black peppercorn, cumin seeds, coriander seeds, *toor* daal, red chillies, garlic, curry leaves) Major steps in preparation: *masala* preparation (coarse grinding of ingredients), soaking tamarind in warm water and juice extraction, cooking [temepering (heating and popping mustard seeds and *curry* leaves in oil), adding *rasam masala*, crushed green chillies, chopped tomatoes and cooking for few minutes, adding tamarind juice, turmeric powder, asafoetida, salt, finely chopped coriander leaves and water, cooked till boiling] Common accompaniments: Rice
***Sambhar* (V)** Major ingredients: Lentils, vegetables, spices, oil, *sambhar* powder (roasted lentils, dried whole red chillies, fenugreek seeds, coriander seeds, asafoetida and curry leaves). Major steps in preparation: Cooking (vegetables, tamarind pulp, *sambhar* powder, turmeric, salt, and asafoetida are boiled together until the vegetables are half-cooked), cooked lentils added to half cooked vegetables and further cooked until the vegetables are done, tempering [popping mustard seeds and other ingredients (curry leaves, red chillies) in hot vegetable oil] Common accompaniments: Rice/*idli*/*dosa*/*pongal*/*upma*

Table 6.8 Staple Foods of West India [(V) = vegetarian and (NV) = non-vegetarian]

Bread items
***Bajri no rotlo* (V)** Major ingredients: *Bajri* millet flour *flatbread* Major steps in preparation: Dough preparation (mixing thick millet flour, salt, oil, and water and kneading), cooking (rolling and roasting flatbreads on hot *tava* or griddle) Common accompaniments: *Curry*/vegetable
***Bhakri* (V)** Major ingredients: millet flour Major steps in preparation: Dough preparation (mixing whole wheat flour, salt, oil, and water and kneading), cooking: (rolling and roasting flatbreads on hot *tava* or griddle) Common accompaniments: *Curry*/vegetable
***Jowar no rotlo* (V)** Major ingredients: *Sorghum* flour Major steps in preparation: Dough preparation (mixing flour, salt, oil, and water and kneading), cooking (rolling and roasting flatbreads on hot *tava* or griddle) Common accompaniments: *Curry*/vegetable
***Pooranpoli* (V)** Major ingredients: Wheat flour, soaked gram, oil, cardamom, nutmeg. Major steps in preparation: Dough preparation (mixing refined wheat flour and oil), preparation of flat bread by rolling, preparation of filling (mixing cooked gram flour with jaggery, cardamom), cooking of soaked split Bengal gram *daal*, sieving the mixture to form a smooth paste, followed by addition of cardamom and nutmeg, filling of smoothened paste in dough balls and rolling into circles, cooking (shallow frying both sides in oil on *tava*) Common accompaniments: Milk/*ghee*

(continued)

Table 6.8 (continued)

Vegetables (*subzi*)/*Daal* items/*Curry* items
***Ambot Tik* (V)** Major ingredients: Fish (shark or catfish), soaked red chillies, tamarind, grated coconut, spices Major steps in preparation: Preparation of paste (mixing and grinding grated coconut, ginger, garlic, soaked red chillies to form a smooth paste), sautéing the paste in oil, addition of spices (turmeric powder, and red chili powder), preparation of curry (addition of water to the above mixture), cooking of fish fillets (marinated with chili powder, turmeric and salt) in boiling *curry* Common accompaniments: Rice/bread items
***Amti* (V)** Major ingredients: Split lentil, tamarind/*kokum*, jaggery, oil, spices, salt Major steps in preparation: Boiling (washed and drained split lentils with turmeric powder, and salt), sautéing (mustard seeds, turmeric powder, chili powder in oil/*ghee*), addition of cooked lentils, tamarind/*kokum*, jaggery and salt, tempering (with mustard seeds, curry leaves and spices) Common accompaniments: Rice/bread items
***Bombil fry* (NV)** Major ingredients: *Bombay Duck* (fish), semolina, spices, oil Major steps in preparation: Marination (*Bombay duck* with turmeric, red chili powder, ginger-garlic paste, and salt), coating the fish with semolina, deep frying/shallow frying of the fish till it turns golden brown Common accompaniments: Rice/bread items
***Crab Xec Xec* (NV)** Major ingredients: Crab meat, grated coconut, coriander, spices Major steps in preparation: Roasting (grated coconut, onion, chillies and whole spices separately), grinding the ingredients to form a paste, sautéing (finely sliced onion on coconut oil in a pan), cooking crab meat in the same pan with some water, continued cooking by addition of ground paste on cooked crab meat Common accompaniments: Rice/bread items
***Daalithoy* (V)** Major ingredients: Pigeon pea *daal*, coriander, *ghee*, spices Major steps in preparation: boiling (washed and drained pigeon pea *daal* with green chillies), mashing (cooked *daal* with water for thinner consistency), tempering (with mustard seeds, red chillies, asafoetida and curry leaves) Common accompaniments: Rice
***Turiya Patra sabji* (V)** Major ingredients: Taro leaves, gram flour, spices Major steps in preparation: Preparation of *patra* [mixing (gram flour, coriander, cumin powder, red chili powder, *garam masala*, powdered sugar, and salt, a pinch of soda, tamarind juice and oil), addition of little water to make a medium thick paste, removing veins of taro leaves and cutting length wise into two halves, spreading the prepared batter over it and covering it with remaining part of taro leaf, followed by rolling, steaming and disc cutting]. Preparation of *Turiya patra sabji* [sautéing (cumin seeds, asafoetida, and ginger chili paste), cooking chopped taro leaves (*turiya*) discs with red chili powder, coriander cumin powder, *garam masala*, salt and sugar, simmering (for around 12–15 min or till the *patra* is cooked well) Common accompaniments: Bread/rice items

(continued)

Table 6.8 (continued)

***Undhiyu* (V)** Major ingredients: *muthiya* (Gram flour, wheat flour, salt, fenugreek leaves, ginger, green chillies, coriander, cumin, water and oil), *undhiyu* (green leaf garlic, brinjal, small potatoes, sweet potatoes, yam, broad beans, raw bananas, sugar, tamarind, oil, coconut, spices) Major steps in preparation: Muthiya [dough preparation (mixing all the *muthiyia* ingredients except oil), dividing into small portions and shaping each into one-inch long half-inch thick rolls, deep frying in hot oil Preparation of *undhiyu:* making (paste of garlic, green chillies and ginger and mix cut coriander), stuffing (mixture in potatoes and brinjal, string beans), layering (putting the rest of the vegetables in layers one on top of the other), mixing fried *muthiya*, salt and turmeric powder, cooking the layers of vegetables Common accompaniments: *Puri* or *roti*
***Veg Kolhapuri* (V)** Major ingredients: French bean, onion, tomatoes, mixed vegetables, spices, oil Major steps in preparation: Preparation of *kolhapuri* gravy (blending tomatoes, cashews, *kashmiri* red chillies to form a paste), sautéing the paste on oil, followed by addition of the cumin seeds, bayleaf, cloves, cinnamon, cardamom, ginger garlic paste, and sautéed onion), Preparation of veg *kolhapuri: s*autéing (cauliflower, green peas, frenches beans, capsicum, carrot, *paneer*, cashewnuts and salt), cooking the vegetables further by the addition of fresh cream and gravy Common accompaniments: Bread/rice items
***Vindaaloo* (NV)** Major ingredients: Goan pork, ground spices, vinegar, brown sugar, salt Major steps in preparation: Preparation of vindaaloo paste (sautéing finely sliced onion till golden brown, grinding of sautéed onion to form a paste), mixing (ground spices, brown sugar, onion paste and ginger garlic paste), cooking (pork cubes cooked untill light brown on all sides), sautéing (ginger garlic paste, turmeric, salt, cooked crab meat and *vindaal*oo paste) Common accompaniments: Rice items

ka saag. *Avial* is an integral part of a keralite feast. **Indian non-vegetarian, traditional dishes** include chicken, fish, mutton, pork and seafood **recipes**. Impact of the *mughlai* style of cooking is obvious in North Indian cuisines. *Murg masallam*, *tandoori fish*, and *tandoori chicken* are popular *muglai* cuisines of North India. *Tandoor* dishes are cooked using a special earthen oven known as *tandoor*. *Macher paturi*, *muri ghonto*, *mochar ghonto*, and *macher jhaal* are lip smacking fish preparations in the eastern region like West Bengal. Down south *Chettinad chicken*, and egg *omelette*s are non-vegetarian delicacies. Some of the delicious preparations in western regions are *Bombay Prawn* and *Pomfret*, while *vindaloo* is a sweet-sour non-vegetarian dish of Goa.

2.2.3 Savouries

Tea snacks are usually deep fried snacks either high moisture content and low shelf life or stored snacks (savouries) with longer shelf life, which are consumed in accompaniment with tea/coffee. Tables 6.9, 6.10, 6.11, and 6.12 give a broad classification of savoury items of all four regions of India (north, south, east and west) and include some examples of each class.

Table 6.9 Savouries of North India

Kachori
Major Ingredients: Dough (refined flour, salt, oil), stuffing: (lentils/peas/potatoes) turmeric powder, chili powder, salt, oil
Major steps in preparation: Dough: kneading all the ingredients and rolling into small circles, stuffing (filling of stuffing into dough discs), deep frying.
Common accompaniments: Mint or tamarind sauce
Samosa
Major Ingredients: Dough (refined flour, salt, baking soda, carom seeds), *stuffing* (potatoes, onions, peas, coriander seeds, turmeric powder, chili powder, salt, oil)
Major steps in preparation: *Filling* (Sautéing boiled and mashed potatoes and spices in oil), *dough* (kneading all the ingredients and rolling small circles), *stuffing*: filling into triangle shaped *samosa* and deep frying in oil
Common accompaniments: Mint or tamarind sauce
Mathri
Major Ingredients: *Maida* (refined wheat flour), oil/*ghee*, salt, carom seeds, warm water, oil
Major steps in preparation: *Dough* (mixing all the ingredients to make the soft and pliable dough), rolling to make small discs *mathri* and poking with a fork, deep frying
Common accompaniments: Tea/coffee
Besan sev
Major Ingredients: Gram flour, turmeric powder, chili powder, asafoetida, baking powder, salt, oil
Major steps in preparation: Thick batter preparation, pouring batter into *sevai* machine and pressing thin strands into heated oil, deep frying.
Common accompaniments: Tea/coffee
Gol-gappe/Pani puri
Major Ingredients: *Puri* (Semolina/wheat flour, salt, oil), stuffing (potatoes/green gram/*boondi*, spices), *meethi chutney* (tamarind, jaggery), *Teekha paani*: fresh mint leaves, chillies, salt, chilli powder
Major steps in preparation: Dough preparation (an equal amount of semolina and wheat flour is mixed and kneaded), *puris* (kneaded dough rolled into small puris and deep fried), filling [(boiled, peeled and mashed potatoes mixed with spices)/(boiling of whole green gram with salt and turmeric)/*boondis* (fried gram flour balls)], *Meethi chutney* (soaking of tamarind in water and separation of pulp, tamarind juice cooked with jaggery), *teekha paani* (mint leaves paste, chillies, salt and spices mixed in cold water)
Serving: round puris broken in the centre and filled with stuffing. The desired amount of *meethi chutney* and *teekha paani* is filled in stuffed *puris* according to personal choice

Table 6.10 Savouries of East India

Chanachur
Major Ingredients: Gram flour, nuts, spices, flattened rice
Major steps in preparation: Deep frying, mixing
Common accompaniments: Puffed rice, tea
Chaal bhaja
Major Ingredients : Rice grits, refined oil
Major steps in preparation: Deep frying of rice grits in oil
Common accompaniments: Tea
Daalmuth
Major Ingredients: Split Bengal gram, refined oil, salt
Major steps in preparation: Deep frying split of Bengal gram in heated oil
Common accompaniments: Tea

(continued)

Table 6.10 (continued)

Chira bhaja
Major Ingredients: Flattened rice, salt, refined oil
Major steps in preparation: Shallow frying of flattened rice
Common accompaniments: Tea
Kordoi
Major Ingredients: Refined wheat flour, salt, carom seeds
Major steps in preparation: Kneading and rolling into small circles, cutting stripes joined at tips and rolling into a star shape. Deep frying in oil
Common accompaniments: Tea
Nimki
Major Ingredients: Gram flour, refined flour, carom seeds, salt, refined oil
Major steps in preparation: Kneading dough, rolling and cutting stripes. Deep frying in heated oil
Common accompaniments: Tea
Gur gutta/gur tokti
Major Ingredients: Peanut, sugar syrup, jaggery
Major steps in preparation: Semi frying/sautéing of peanut and mixing in jaggery and putting them in thick sugar syrup and allowed to set in the form of thick round plates

Table 6.11 Savouries of South India

Murukku
Major Ingredients: Rice flour, black gram *daal* flour, sesame seeds, carom seeds, salt, chili powder, oil
Major steps in preparation: Dough preparation, shaping and deep frying
Common accompaniments: Tea/coffee
Kada boondi
Major Ingredients: Gram flour, rice flour, turmeric powder, chili powder, asafoetida, baking powder, salt, oil
Major steps in preparation: Thick batter preparation, pouring batter through perforated spoon in the heated oil, deep frying
Common accompaniments: Tea/coffee
Banana chips
Major Ingredients: Unripe bananas, turmeric powder, salt, coconut oil
Major steps in preparation: Slicing peeled banana in ice cold salted water and keeping for 10 min, deep frying drained banana slices
Common accompaniments: Tea/coffee
Tapioca chips
Major Ingredients: Tapioca/cassava, salt, oil
Major steps in preparation: Washing, peeling and slicing of tapioca, deep frying in heated oil
Common accompaniments: Tea/coffee
Groundnut masala
Major Ingredients: Groundnut, gram flour, rice flour, corn flour, ginger garlic paste, turmeric powder, chili powder, salt, oil
Major steps in preparation: Mixing all the ingredients to make the batter, coating groundnuts with batter and deep frying in oil
Common accompaniments: Tea/coffee

Table 6.12 Savouries of West India

Chakali Major Ingredients: Mixed grain flour, chili powder, salt, oil Major steps in preparation: Mixing ingredients to make the batter, filling in molds, extruding and deep frying in oil Common accompaniments: Tea/coffee
Chivdo Major Ingredients: Beaten rice, groundnut, split black gram , spices, salt, oil Major steps in preparation: Shallow frying of beaten rice, groundnuts, split black gram individually and mixing, sprinkling salt, and spices on top of it Common accompaniments: Tea/coffee
Chorafali Major Ingredients: Chickpea flour, black gram flour, dry mango powder, seasoning and spices, oil Major steps in preparation: Deep frying Common accompaniments: Tea/coffee
Khakhra Major Ingredients: Wheat flour, fenugreek (*methi)*leaves, spices, oil Major steps in preparation: Kneading, rolling and roasting Common accompaniments: Tea/coffee
Fafda Major Ingredients: Gram flour, carom seeds, oil Major steps in preparation: Deep frying Common accompaniments: *Jalebi, kadhi* and chillies
Farsi Puri Major Ingredients: Wheat flour, refined wheat flour, spices Major steps in preparation: Deep frying Common accompaniments: Sauces
Thepla Major Ingredients: Wheat flour, fenugreek leaves, spices, oil Major steps in preparation: Roasting with oil. Common accompaniments: Pickles, green chillies

2.2.4 Confections

India is well accepted across the world as a country of cultural and traditional festivals as it has a population belonging to many cultures and religions. The first thing that comes to mind related to festivals is sweets/confections. Indian traditional confections include *mithai*, *halwa*, *kheer*, *payasam*, *sevai*, etc. Popular ingredients like condensed milk, *khoya*, grain flours, lentils, semolina, chickpea flour and vegetables such as carrots and pumpkins are used to make popular sweets like *laddu*, *barfi* and *halwa*. Sweet dishes are often flavoured with cardamom, saffron, nutmeg, and rose petal essences. Tables 6.13, 6.14, 6.15, and 6.16 give a broad classification of Indian traditional confections/sweet items of all four regions of India (north, east, south and west) and include some examples of each class.

Table 6.13 Confections of North India

Name of the food/ dish	Major Ingredients	Major steps in preparation
Sooji halwa	Semolina, clarified butter, cashew nuts	• Roasting semolina in *ghee*, addition of sugar and dryfruits. Cooking by addition of water and keeping the consistency thick.
Petha	White gourd/pumpkin, milk, saffron, sugar, cardamom	• Peeling, cutting, deseeding and pricking of white pumpkin. Dipping in chemical lime solution for 2 h. Draining, washing and squeezing to remove lime traces. Boiling till it becomes soft. • Preparing sugar syrup of one thread consistency and dipping pumpkin pieces into it. Draining excess syrup and cooled.
Shahi tukra	White Bread, thickened milk, sugar, sliced almonds	• Removal of bread sides and cutting it diagonally into a triangular shape, frying in oil. • Thickening of milk to make *rabdi* by constant stirring on low flame. • Making sugar syrup: Addition of sugar and cardamom to water and boiled to make sugar syrup. • Dipping fried bread slices to sugar syrup for few minutes Topping with *rabdi* and dry fruits.
Phirni	Milk, rice flour, saffron , cardamom, dry fruits (almonds, cashew, pistachio), sugar	• Condensing fresh milk over medium heat. • Mixing ground rice and ground almonds together in the milk and boiling by frequent stirring. Adding condensed milk, sugar, dryfruits and cooking on low heat for few minutes.
Kheer	Rice, milk, sugar, dry fruits	• Cooking rice in milk by constant stirring on the flame. Addition of sugar, cardamom powder, saffron strands and dryfruits.
Jalebi	Rice flour, sugar, gram flour (besan), baking powder, baking soda, yoghurt, food colour, cardamom, oil	• Mixing refined wheat flour, gram flour (*besan*), baking powder, baking soda, yoghurt, food colour, cardamom powder and water to form a flowing consistency batter. Fermentation for 0–12 h. Whisking of fermented batter and frying spiral *jalebis* in hot oil. Dipping fried *jalebis* in cooked sugar syrup (sugar, cardamom powder, water).
Gajar ka halwa	Carrot, condensed milk, *ghee*, cardamom, cashew.	• Grating carrots and cooking it in *ghee* for few minutes. • Adding condensed milk, sugar, cardamom and cooking.
Panjeeri	Coriander seeds, sugar, *ghee*	• Dry roasting coriander seeds and grinding it to make powder. Adding powdered sugar and dry fruits to it.

Table 6.14 Confections of East India

Name of the food/dish	Major Ingredients	Major steps in preparation
Balooshahi	Refined wheat flour, oil, cardamom, sugar, dry fruits	• Kneading cottage cheese to a soft consistency, mixing sugar, dryfruits, coating with refined flour paste, deep frying and dipping in thick sugar syrup.
Chhena gaja	Cottage cheese, flour, sugar	• Kneading cottage cheese to a soft consistency, mixing sugar, flour for binding, forming desired shapes, frying and dipping in sugar syrup.
Chhena poda	Cottage cheese, flour, sugar	• Kneading cottage cheese to a soft consistency, baking in closed flame, dipping in sugar syrup.
Goja	Maida, sugar, refined oil	• Kneading with oil and salt. • Cutting in finger like shapes, deep frying in refined oil, putting in thick sugar syrup and left to crystallize.
Gaja	Refined wheat flour, oil, cardamom, sugar, dry fruits	• Kneading, shaping, deep frying and dipping in sugar syrup.
Gokul peethe	Refined wheat flour, semolina, sugar, cardamom, coconut, jaggery, oil	• Preparing filling with coconut and jaggery into tight consistency, forming cutlet with the filling, dipping cutlet in the mixture of flour and semolina, frying of the cutlet and dipping the cutlets in thin sugar syrup.
Malpua/ Malpoa	Rice flour, refined wheat flour, sugar, *khoya*, cardamom, fennel seed, coconut, dry fruits, oil	• Preparation of filling with cottage cheese, sugar, *khoya*, and spices, a coating of the filling in refined flour mixture and frying, and dipping in sugar syrup.
Mishti Doi	Curd, sugar syrup/jaggery	• Heating milk with sugar syrup/jaggery, adding inoculum and fermentation.
Pantua	Cottage cheese, starch, sugar	• Kneading cottage cheese to paste, adding starch, forming shapes, frying and dipping in a sugar solution.
Patisapta	Rice flour, refined wheat flour, sugar, cardamom, *khoya*, condensed milk, coconut, dry fruits, oil	• Preparation of filling with coconut, *khoya*, sugar, cardamom, dry fruits, jaggery. • Preparation of flat bread mixture using wheat flour, semolina. • Filling of the mixture in flat bread, rolling and shallow frying in oil.
Puli pithe Payesh	Rice flour, refined wheat flour, sugar, cardamom, condensed milk, coconut, dry fruits	• Kneading dough into finger like shapes, filling with coconut and jaggery mixture, dipping in condensed milk.
Peda	Milk, cardamom, sugar, refined wheat flour	• Kneading cottage cheese with sugar and spices, and moulding into forms.
Rosogulla	Milk, sugar, starch	• Kneading cottage cheese with sugar and arrowroot and forming balls, and dipping in a sugar solution.
Nolen gurer shondesh	Cottage cheese, date jaggery, cardamom	• Kneading cottage cheese with jaggery, mixing in flame to form paste and molding.
Thekua	Wheat flour, cardamom, dry fruits, jaggery/sugar	• Kneading wheat flour, fennel seeds, sugar/ jaggery, *ghee*. Dough balls are cut and given different shapes with mold. Deep frying in oil.

Table 6.15 Confections of South India

Name of the food/dish	Major Ingredients	Major steps in preparation
Payasam	Vermicelli, sugar, cardamom powder, cashewnuts, raisins, *ghee*/oil	• Sautéing vermicelli in oil/*ghee* for few minutes. Adding milk, sugar and cardamom powder, and cooking till consistency becomes thick. Addition of sautéed cashewnuts and raisins for garnishing.
Sakkara pongal	Rice, green gram *daal*, jaggery, cashew, cloves, raisins, cardamom powder, edible camphor	• Soaking and pressure cooking rice and green gram *daal*. Jaggery syrup prepared by stirring and melting in water. Filtration of jaggery syrup and addition to cooked rice and *daal*. • Cooking on medium flame till syrup blends well with rice. Addition of *ghee* fried coconut pieces, dryfruits, cardamom powder and edible camphor.
Jangiri	Whole black gram *daal*, rice flour, cornflour , red orange colour, sugar, milk, rose essence, special thick cloth with a hole in the centre for making *jangiri*	• Soaking and grinding *urad daal* to make a smooth and fluffy paste. Mixing rice flour, corn flour water mix and the red orange color to the batter. • Making sugar syrup with sugar, rose essence and water. Piping *jangiris* using special cloth in heated oil in a shallow flat bottomed plate (*tambalam*). Dipping *jangiris* in cooked sugar syrup.
Mysore pak	Gram flour, sugar, *ghee*/oil	• Boiling sugar and water in a pan until the sugar attains one string consistency. • Adding sieved gram flour in little batches with constant stirring making sure no lumps are formed. Adding ladleful of hot *ghee*/oil over the gram flour mixture and stirring continuously until oil is absorbed completely. Repeating 4–5 times until the mixture starts to separate from pan. • Transferring the mixture to a greased tray of enough depth and cut into the desired shape.
Unniyappam	Rice flour, bananas, jaggery, cardamom, *ghee*, raisins, cashewnuts, coconut, baking soda.	• Mixing rice flour, crushed cardamom, jaggery syrup, mashed banana, baking soda, finely chopped coconut pieces, raisins, roasted cashew nuts (optional) and water to form a batter. Resting batter for 30 min. • Deep frying in heated oil or cooking in *paniyaram* vessel on both sides.

2.2.5 Beverages

Various regions serve drinks prepared with a different variety of ingredients including local fruits, flowers and spices. Tea and coffee are readily available and popular hot drinks in all regions of India. South Indians prefer filter coffee as compared to other hot beverage, while tea is relished in other regions. Filter coffee is prepared

Table 6.16 Confections of West India

Name of the food/dish	Major Ingredients	Major steps in preparation
Laddu	Gram flour, *ghee*/ oil, sugar	• Roasting gram flour in *ghee*/oil. Kneading roasted gram flour with sugar and molding into forms.
Modak	Rice flour, coconut, dry fruits, cardamom, jiggery	• Preparation of dough by cooking rice flour in water. Kneading, rolling and flattening of dough to form flat bread. • Preparation of filling by roasting coconut, cardamom, dry fruits, and jaggery. • Filling of the mixture in flat bread and shaping it in the form of dumplings. Steaming of dumplings.
Mohan thaal	Gram flour, *ghee*, dry fruits, milk, sugar	• Roasting gram flour in *ghee* till it turns golden brown. Preparation of sugar syrup by heating water and sugar, followed by addition of milk. Simmer till one thread consistency. • Mixing of cool roasted gram flour and sugar syrup. Spreading of the mixture evenly on a greased surface. Garnish with dry fruits. Cut into squares after cooling.
Shankarpali	Refined wheat flour, sugar, oil	• Preparation of dough by kneading refined wheat flour and sugar. Rolling of dough to form flat bread. Cutting of flat bread into diamond shaped forms (side = 2 cm). • Deep frying of the diamond shaped forms in refined oil.
Shira	Semolina, milk, *ghee*, sugar, dry fruits	• Roasting of semolina in *ghee* till it turns golden brown. Cooking of roasted semolina in milk, followed by addition of sugar and dry fruits.
Shrikhand	Hung curd, cardamom, dry fruits, and saffron	• Hung curd, cardamom, dry fruits and saffron mixed together and served when chilled.
Soan papdi	Gram flour, refined wheat flour, *ghee*/ oil, sugar, pistachio	• Roasting gram flour and refined wheat flour in *ghee* till it turns golden brown in color. • Preparation of sugar syrup by heating water and sugar, followed by addition of milk. Simmer till 2.5 thread consistency. Mixing of cool roasted gram flour and sugar syrup. Beating it with a large fork till it forms thread like flakes. • Pouring the mixture on a greased surface. Garnishing with pistachio. Cutting into squares on cooling.
Sukhdi	Wheat flour, jaggery, *ghee*	• Roasting wheat flour in *ghee* till it turns golden brown in color. • Mixing the roasted flour with jaggery when hot. Pouring the cooled mixture on a greased surface and cutting into pieces on cooling.

with thick full cream milk and freshly brewed strong coffee decoction. Froth in the coffee is achieved by pouring it from one glass to another manually as against using a machine. The traditional method of preparing *masala chai* is through decoction, by boiling a mixture of milk and water with loose leaf tea, sugar and whole spices such as cardamom, ginger, clove, etc. As for cooler drinks, North Indians enjoy *lassi*

Table 6.17 Beverages of North India

Name of the food/dish	Major Ingredients	Major steps in preparation
Shikanji	Lemon, black salt, sugar, water	• Mixing lime juice, black salt, sugar in water.
Sugarcane juice	Sugarcane sticks	• Extraction of sugarcane juice from sugarcane sticks by machine pressing.
Jal jeera	Cumin, ginger, black pepper, mint, fruit powder (mango/citrus), chilli, hot pepper powder	• Mixing all the ingredients in water, topped with *besan boondis*.
Lassi	Curd, cream, sugar, cardamom, dry fruits Flavor: Optional (rose/*kesar*/mango)	• Mixing curd, sugar, cardamom and dryfruits in water keeping the consistency thick • Addition of flavorants for flavored variety.
Chai (tea)	Tea leaves, sugar, ginger, cardamom, milk	• Boiling tea leaves with water, addition of grated ginger and cardamom powder for flavor. Adding milk and boiling again for few minutes. Strained and served hot.

Table 6.18 Beverages of East India

Name of the food/dish	Major Ingredients	Major steps in preparation
Gondhoraj ghol	Kaffir lime, butter milk, salt, sugar	• Mixing of butter milk, salt, sugar and kaffir lime
Aam pora shorbat	Raw mango, black salt	• Roasting of raw mango, extracting pulp, the addition of water, salt and mixing
Bael shorbat	Wood apple (*bael*), water, black salt, sugar	• Extracting pulp from fruit, mixing with water, salt, sugar straining
Sattu sherbet	Roasted gram flour (*sattu*), black salt, roasted cumin seeds, lemon juice	• Mixing *sattu* and salt in water. Sprinkling roasted cumin seed powder and adding a dash of lime juice

while southern counterpart enjoys coconut water. *Jigarthanda* and *Nannari sherbet* are popular cool drinks in South India. In North India, *sherbets* are made from flowers/almonds/spices which are mixed in water or milk. *Kokum sherbet* and *chaas* are popular drinks in western India. In eastern region, *aam pora sherbet* and *bael sherbet* are consumed during summers. Tables 6.17, 6.18, 6.19, and 6.20 give a broad classification of Indian traditional beverages of all the four regions of India (north, east, south and west) and include some examples of each class.

2.3 Food and Religious Beliefs

In India, food has turned out to be a symbol of diverse religious and social characteristics. Majority of religions prescribe, what is appropriate to eat in order to demonstrate faith. Every religion is extant with varying believes and preferences, such as

Table 6.19 Beverages of South India

Name of the food/dish	Major Ingredients	Major steps in preparation
Nannari sherbet	*Nannari* roots, sugar, lime, water	• Boiling *Nannari* roots with sugar. Addition of hint of lime.
Panakam	Jaggery, cardamom, ginger, lemon, mint leaves	• Dissolving jaggery into water. Adding cardamoms, ginger and lime juice. Garnished with mint leaves
Buttermilk/ Sambaram	Curd, water, salt, *curry* leaves, mustard seeds, asafoetida	• Thin buttermilk is prepared by dissolving curd into the water • Tempering it with mustard, *curry* leaves and green chillies
Neera	Palm juice	• Extraction of palm juice and fermentation of sap
Filter Kaapi (Filter coffee)	Coffee beans medium ground (powder), water, milk, sugar	• Adding the coffee powder to the upper compartment (of filter coffee maker assembly) and pressing down the powder firmly. Pouring boiling hot water in the upper compartment, until water level nearly reaches the brim • Keeping this covered for 3–4 h until thick coffee decoction collects in the bottom compartment • To prepare coffee: Mixing 3–4 tablespoon coffee decoction with boiling milk. Mixing sugar if required

Table 6.20 Beverages of West India

Name of the food/ dish	Major Ingredients	Major steps in preparation
Chaas	Curd, water, salt, *curry* leaves	• Mixing curd, water, mint leaves and salt in a watery consistency.
Panna	Raw mango, cardamom, pepper, black salt	• Boiling raw mangoes in water, mashing pulp and straining through the filter. • Addition of salt, spices and mint leaves
Feni	Ripened cashew apples	• Cleaning and deseeding of ripened cashew apples. Stomping/pressing of cashew apples to release juice • Fermentation of extracted juice in an earthen pot. Double distillation of fermented *feni*
Kokum sherbat	*Kokum* fruits (chopped or dry *kokums*), water, sugar, cardamom powder/crushed roasted cumin powder, black salt	• Cleaning, deseeding and chopping of *kokum* fruits • Grinding fruit and pulp with little water. Straining and boiling of *kokum* mixture with sugar and water until the syrup becomes a little thick • Adding roasted cumin powder and powdered cardamom

Hindus do not eat beef, Muslims do not consume pork but eat beef, and Jain population do not consume roots or subterranean vegetables. These eating habits further demonstrate the fact that Indian cuisine has absorbed a substantial impact of religion. The magic of Indian cuisine lies not only in the taste, but in its variety which marks Indian cuisine as the assortment of several traditions, and as the potpourri of various religions. The food in India is diverse and deeply embedded in traditions, with a combination of distinct cultural effects, similar to the religious heritage of India. The influence of religion is further compounded by geographical diversity, environmental factors, food preferences, and availability of plant or animal sources. Religion not only impacts the type of food, but affects ingredients and rituals associated with preparation process adopted by diverse religious groups. Hinduism dominates the religious landscape of India. Hinduism promotes reverence for life forms and contributes to the significant prevalence of vegetarianism in India; and they heavily depend on lentils and beans as the primary sources of protein. Milk is considered auspicious and dairy products such as curd, *paneer* (cottage cheese) and sweets made of milk solids are a vital part of the food. Some sects of Hinduism prohibit the use of onions and garlic in food, and so substitute flavourings such as asafoetida, cumin seeds, ginger, and cashew paste have been incorporated into the cuisine.

In the most sacred book of Hinduism, the *Bhagavad Gita*, foods are segregated into three different types, *sattvika*, *raajasika*, and *taamasika*, based on the property, quality, and sanctity. The *sattvika* food is often believed to bring prosperity, longevity, intelligence, strength, health, and happiness. This food type includes fruits, vegetables, legumes, cereals, and sweets. The *raajasika* food is believed to bring activity, passion, and restlessness, which includes hot, sour, spicy, and salty foods. The *taamasika* food is intoxicating and unhealthy, which is generally considered to cause dullness and inertia (Dubey, 2010). Traditional foods have social importance for celebrations, especially during festivals and on social events. Every festival is associated with the particular type(s) of food preparation. Foods are offered in temples for worshipping Gods and called *prasadam*. For example, milk and milk products are offered to deities during worship.

The culinary landscape of India started to reflect Islamic influences after Islamic invasion. The primary impact was the inclusion of meat and fish in the food. The cooking techniques (such as grilling of meat into *kebabs*) and ingredients (such as the use of dates and nuts in rice dishes), were adopted. Various foods are prohibited such as swine meat, the flesh of carrion (dead animals), blood in any form, and alcohol as an intoxicant. During Ramadan, month-long fasting, family members, friends, and relatives share common meals after sunset. Islam recommends halal food laws, which have been decreed to maintain strict guidelines around Muslim foods for killing an animal for food in a particular way. Similar to Muslims, Christians eat meat and fish, but cooking methods evolved over a period with regional influences. The cooking method of Christians in Kerala and Goa is different. In Kerala, food integrates local ingredients and cooking methods having a less European influence. In Goa and Calcutta, Christianity was introduced by the British and Portuguese with rapid conversions; hence there is significant influence of European way of food preparation, e.g. preparation of rum-flavoured cake on occasion of Christmas. Christians do not have any restrictions on the type of meat that can be consumed, unlike Muslims where pork is strictly prohibited. In Christian food culture, all family members sit together at a table and eat together after family prayers.

Traditional Buddhists may avoid eating meat and fish out of respect for life. However, non-vegetarian foods are not strictly forbidden. According to the Buddhist religious dietary code, if the animal flesh is eaten, the animal should be killed by non-Buddhists. It is expected that fasting for the entire day be observed on the new moon and the full moon days of each month. In the case of practitioners of Jainisim, they are also influenced by the principle of non-violence. Due to their belief, they do not consume any form of non-vegetarian food. They do not consume any food that has been obtained by hurting any kind of life. They are so strict in their principles that they do not consume root vegetables, alcohol, honey, and meat. Primarily, their eating practices are based on eating after sunrise in the morning and eating before sunset in the evening. People following Jain religion eat only those vegetables and fruits, which ripen on the tree. Thus, religious influence on Indian food has been profound and has resulted in the creation of a diverse range of cuisines.

2.4 Classification of Foods Based on Shelf Life

Home made Indian traditional foods with some exceptions have limited shelf life. They are usually consumed within a few hours after their preparation. Occasionally they may be eaten the next day with/without refrigeration. Only low moisture, deep fried snack preparations are stored in air tight containers for longer periods of time. Table 6.21 gives a classification of home made Indian traditional vegetarian foods according to their shelf life.

Traditional foods prepared at home have no added chemical preservatives. They tend to spoil quickly when stored under ambient conditions prevalent in India due to mould or bacterial growth. Furthermore, the eating quality of some traditional foods is observed to deteriorate with progress in storage due to staling, a commonly observed phenomenon in cereal based traditional foods, which leads to a loss in texture and eating quality. Thus staling and microbial spoilage are two important determinants of the shelf life of many traditional foods.

Table 6.21 Observed shelf life of selected home made Indian traditional foods under ambient storage conditions

No.	Category	Shelf life period	Examples of traditional foods
1	Short shelf life	Less than 1 or 2 days	Cooked rice, *chapatis*, steamed snacks like *idli*, vegetable and *daal* preparations, *samosa*, *batata vada*
2	Medium shelf life	Less than a week	*Puran poli, methi thepala*
3	Long shelf life	Less than a month	*Chivda, chakli, sev, ribbon pakoda*
4	Very long shelf life	Less than a year	*Muramba*, pickles

3 Sensory Characteristics of Indian Traditional Foods

The desirable organoleptic characteristics of traditional foods are important as they determine consumer acceptability. Shape, size, appearance, color, texture, mouth-feel, aroma, flavor, and taste are the sensory characteristics of foods in general and Indian traditional foods have unique characteristics. Sensory characteristics of these foods can be influenced by the quality of raw materials used and method of preparation. A lot of variation is observed in traditional foods prepared in different households making it sometimes difficult to identify a common benchmark quality that will be equally appealing to all regions. This may make standardization for large scale preparation a challenging task. One has to identify important characteristics for each traditional food and design appropriate methods for their evaluation in order to implement quality assurance during large scale manufacture of these items. Tables 6.22, 6.23, 6.24, and 6.25 describe the sensory characteristics of Indian traditional food items of all four regions of India (north, south, east and west) and include some examples of each class.

4 Preparation and Steps Involved

Variety of methods, singly or in combination, may be employed in the preparation of these traditional foods, such as boiling, concentrating, open pan cooking, pressure cooking, fermentation, germination/malting, dry or wet grinding, shallow or deep frying, grilling, roasting, dry or wet blending, sheeting and shaping with hands, casting in trays and cutting, extruding, garnishing, sautéing, marinating, brining, syruping, peeling, trimming, chopping, mashing, curdling and churning, extracting and draining.

4.1 Mixing

Mixing is a unit operation in food processing which involves the proper distribution of all ingredients in a media, which may include raw materials, liquids, spices, and condiments or a combination thereof. The purpose of mixing is to bring together all components of the food formulation in a stable condition before further processing of the cuisine, which will help to improve overall product quality. Mixing is a basic step for all food preparation. In Indian household mixing is done manually with a ladle either of stainless steel or of wood; beater/masher is specially used for recipes, which require broken eggs, pulses. Some examples of foods that require mixing are: *cheera doi* (snack item, East India), and *daddojanam/thayir sadam/mosaranna/curd rice* (staple item, South India).

Table 6.22 Sensory characteristics of some selected North Indian traditional foods

Food item	Picture/image	Appearance, colour	Aroma	Taste	Texture/ mouthfeel
Breakfast foods and snacks					
Chole bhature		*Bhature:* Golden brown puffed circles *Chole:* Dark brown, thick curry	*Bhature:* Fried *Chole:* Spicy	*Bhature:* Salty and chewy *Chole: Salty, spicy*	*Bhature:* Soft and crispy *Chole:* Semi solid
Aloo tikki		Golden brown discs	Burnt	Spicy	Semi soft
Staples and accompaniments					
Naan Roti		Whitish, soft	Roasted	Salty and chewy	Soft
Paneer tikka masala		Grilled cubes in golden brown paste	Spicy	Salty	Soft paneer and semi-solid gravy
Savouries					
Kachori		Deep brown puffed balls	Fried	Satly and spicy	Crunchy outer crust, soft filling outside

(continued)

Table 6.22 (continued)

Food item	Picture/image	Appearance, colour	Aroma	Taste	Texture/ mouthfeel
Samosa		Tetrahedral, dark brown savouries	Fried,	Salty and spicy	Crunchy outer crust, soft inner filling
Confections					
Sooji halwa		Mushy paste	Nutty	Sweet	Soft, palpable
Jalebi		Golden coolured concentric circles	Fried	Sweet	Crunchy, syrupy, juicy
Beverages					
Sugarcane juice		Pale yellow coloured liquid	Fresh	Very sweet	Thin liquid
Lassi		Thick, white coured	Milk and cardamom aroma	Sweet	Thick liquid

Table 6.23 Sensory characteristics of some selected East Indian traditional foods

Food item	Picture/image	Appearance, colour	Aroma	Taste	Texture/ mouthfeel
Luchi cholar daal		*Luchi:* Crispy fried golden brown *Cholar daal:* Dark brown gravy	Fried	*Luchi:* Salty, chewy *Cholar daal:* Thick spicy gravy	*Luchi:* Crispy, crunchy, *Cholar:* Thick gravy
Ghoogni		Dark yellow, thick	Spicy	Salty	Thick
Staples and accompaniments					
Macher Paturi		Light to golden brown	Burnt, mustard oil aroma	Very spicy	Thick, rubbery
Savouries					
Chanachur		Yellow and golden brown soild mixture	Spicy, nutty	Spicy, sour, salty	Crunchy
Chira bhaja		Roasted, white/light yellow mixture	Fried	Salty, nutty	Crunchy

(continued)

Table 6.23 (continued)

Food item	Picture/image	Appearance, colour	Aroma	Taste	Texture/ mouthfeel
Confections					
Rosgulla		Soft white balls dipped in syrup	Milky	Very sweet	Soft, spongy, juicy,
Patishapta		Golden brown folded pancakes	Mixed aroma of fried, sweet Jaggery, milk and coconut	Coating: sweet, chewy Filling: Very sweet	Soft spongy
Beverages					
Gondhoraj ghol		White thick liquid	Lime flavored/spiced/milky/ fermented	Fermented, spicy, sweet	Semi thick liquid
Bael shorbat		Thick, yellow/orange liquid	Fruity, sour	Sour/sweet/salty	Thick, pulpy

Table 6.24 Sensory characteristics of some selected South Indian traditional foods

Food item	Picture/image	Appearance, colour	Aroma	Taste	Texture/mouthfeel
Breakfast foods and snacks					
Idli		White round, spongy discs	Fermented	Slightly salty	Soft porous, spongy
Puttu		White cylinders	Coconut aroma	Slightly salty	Soft and crumbling
Staples and accompaniments					
Hyderabadi biryani		Fluffy	Spicy with dominant saffron flavour	Spicy, salty	Soft
Sambhar		Liquid	Spicy	Spicy, tangy, salty	Liquid

(continued)

Table 6.24 (continued)

Food item	Picture/image	Appearance, colour	Aroma	Taste	Texture/mouthfeel
Savouries					
Banana chips		Golden yellow thin discs	Fried	Salty	Crunchy
Kada boondi		Golden yellow small balls	Fried	Salty	Crunchy
Confections					
Shakkara pongal		Soft, sticky, brown in colour	Cardamom aroma	Sweet and nutty	Soft, palpable
Unniyappam		Soft, dark brown in colour small flat balls	Burnt	Sweet	Porous, spongy

Beverages					
Nannari sherbet		Liquid, red in colour	Herbal	Sweet	Thin Liquid
Filter Kaapi		Liquid dark brown in colour	Coffee aroma	Bitter/burnt	Sweet, hot steaming liquid

Table 6.25 Sensory characteristics of some selected West Indian traditional foods

Food item	Picture/image	Appearance, colour	Aroma	Taste	Texture/mouthfeel
Breakfast items/snacks					
Pohe		Soft, yellow	Spicy and nutty	Salty, sour	Soft
Vada paav		*Pav*: Golden brown, fluffy *Vada*: Golden brown coloured soft balls	Spicy, fried	Spicy, sweet, salty	*Paav*: Soft, spongy *Vada*: Soft
Staples and accompaniments					
Puranpoli		Golden yellow, circular, soft flat circles	Nutty and cardamom flavour	Sweet, chewy	Soft and fluffy
Undhiyu		Green, semisolid thick gravy	Rustic	Spicy	Thick gravy

Savouries					
Khakra		Crispy brown thin flat circles	Fried	Spicy	Crunchy
Chakali		Goden brown/cream/brown coloured solid concentric rings	Fried	Salty, spicy	Crunchy
Confections					
Shrikhand		Soft, mushy paste	Curd like	Sweet	Soft, creamy
Modak		Katori/cone shaped, soft, white coulred dumplings	Sweet	Sweet	Outer shell: soft Inner filling: crumbling

(continued)

Table 6.25 (continued)

Food item	Picture/image	Appearance, colour	Aroma	Taste	Texture/mouthfeel
Beverages					
Panna		Yellow to brown coloured liquid	Fresh, tangy	Sweet-sour	Pulpy, juicy
Kokum sherbet		Pink to red coloured liquid	Fresh, sour	Sweet-sour	Thin consistency

4.2 Grinding/Pulverizing

Grinding is a method where particles/ingredients are crushed against hard surfaces to reduce their sizes as well as increase their surface area. Solid substances like spices are generally ground into powdered form to extract their flavour and increase their surface area for ease of application. Apart from spices, ingredients, which require a semi liquid consistency can also be ground to get the desired matrix. Mortar and pestle is the oldest known grinder and still in use in traditional Indian Kitchen. However, with the advent of technology, electric mixer-grinder is available which is used not only to crush and grind but also to mix the ingredients into a homogenous consistency. Few examples of cuisines which require grinding as a part of the preparation are as follows: *macher paturi* (staple item, East India), *adai* (breakfast item, South India), and *handwo* (breakfast item, West India)

4.3 Homogenization

Homogenization can be technically defined as a method where particles are mixed under a particular force to break all particles into uniform size and bring unity among its phases. It is a very important step in preparation of cuisines, where different phases like fat, water are to be mixed together and should be stable as well. Milk items mainly require homogenization. Homogenization can be achieved through any tools, which can be used to apply pressure. Mortar and pestle or traditional tools called *Seal Nora* which are two stones of uneven shapes are generally used for solid materials. Churning machines are mostly used for milk based products which require homogenization. Traditional wooden churners are now being replaced by electric churning machines with higher efficiency and shorter operation time. Examples of cuisines with homogenization as a part of their preparation are *chaas* (beverage, North India), and *gondhoraj ghol* (beverage, East India).

4.4 Roasting

Roasting is a method of placing a raw or a semi-prepared item into open flame with the intention to develop flavour as well as to semi cook the material before final processing. Sometimes roasting is the final step and is a healthier alternative than frying or sautéing as the use of oils is negligible or completely avoided in this method. Cereal based items like flatbreads are generally roasted, but some items with vegetables have roasting as a major step. Roasting decreases the bacterial count in a raw ingredient, gelatinizes the starch and denatures proteins so the food can be consumed immediately as well. Roasted foods have unique flavour characteristics developed due to the breakdown of carbohydrates and proteins under heat treatment, which is desirable for consumers. Roasting can be done on charcoal ovens, flat pans, and iron grids depending on the requirement of the dish. Examples of roasted cuisines are *sunga pitha* (breakfast item, East India), *akki roti* (breakfast item, South India), and *baingan bharta* (staple item, North India).

4.5 Baking

Baking is a process of heating a food mixture without exposing it to open flame. With respect to cooking, it is a method of converting any semi liquid mass mixed with suitable ingredients into soft, spongy texture by the use of leavening agents like baking powder or baking soda where in the semi liquid mass is dried and filled with air that gives the end product a spongy texture. Baking is a controlled heating process done generally in ovens either microwave oven, or normal baking oven with temperature and time controller. Some examples of baked foods are as follows: *naan roti* (Staple item, North India), *sana* (staple item, West India), *kulcha* (Breakfast item, North India), and *naan khatai* (confection item, West India).

4.6 Boiling

Boiling is a method of cooking or blanching any raw material in water to change its texture and semi cook or fully cook it before further preparation. Boiling is generally done to soften raw vegetables before making them into curry as it reduces the cooking time and gives a better taste. Most Indian staple cuisines require boiling where vegetables, cereals and pulses are involved. The step also eliminates enzymatic reactions within the ingredients, which may be responsible for certain defects in colour or flavour. However, in Indian households, people prefer boiling mostly to eliminate the raw flavour of the ingredients. Boiling can be done in pressure cooker or pans or suitable utensils with bigger surface area. Food items requiring boiling as an essential step are as follows: *khichuyu* (breakfast item, West India), *karela bharta* (staple item, North India), and *chettinad curry* (staple item, South India).

4.7 Steaming

Steaming involves partially cooking of ingredients under the influence of moist heat. Steaming induces softness in ingredients without the application of direct flame or other methods like frying or baking. It is a simple procedure and used as a preparative step similar to boiling. However, unlike boiling in steaming the item is not softened completely but enzymatic degradation or semi cooking of the raw materials is possible. Steamed products generally require further processing.

Steaming is mostly done in a pressure cooker without allowing pressure to buildup. Some cuisines which involve steaming are *aloo pitika* (breakfast item, East India), *momos* (breakfast item, East India), *macher paturi* (staple item, East India), and *idli* (breakfast item, South India).

4.8 Cooking

Cooking is a generalized term referring to all the procedures that involve the conversion of raw material into final product. However, to be specific, cooking is the final step in the formation of a product which involves placing the raw material (in a vessel) in the open flame so as to soften the ingredient, develop taste, colour, flavour and texture. Cooking can be done in steel, iron equipment, which can conduct heat easily. *Kadhai* is a typical container, which is used in most Indian households. Most Indian food requires cooking. However to name a few are *Begun bhaja* (staple item, East India), *til ko alu* (staple item, East India), and *upma* (breakfast item, North India).

4.9 Seasoning

Food is said to be seasoned when its each and every part is coated or submerged with a typical spice/herb mixture so that the taste is imparted uniformly across the food once it is cooked. In many Indian food preparations, a seasoning of hot oil with added whole spices and nuts is added over the food at the end of its preparation. Some cuisines which have seasoning as a preparatory step are as follows *khandvi* (breakfast item, West India), and *laksa* (breakfast item, East India).

4.10 Frying (Shallow/Deep)

Frying is the method of cooking any item in oil by fully dipping (deep frying) or by semi dipping where each surface is partially covered with oil (shallow frying). Gravies and curries are never deeply fried, in fact, are sautéed, which requires even less quantity of oil. Cutlet, *pakora*, or basically snack items rich in carbohydrates are deep fried. Deep fried items are tastier but a less healthy option as they are high in oil content. Indian consumers have a greater preference towards deep fried items. *Tawa* or flat plate cookwares are most suitable for shallow frying as it is surface cooking but *kadhai* or deep bottom cookwares are needed for deep frying which is meant for inner cooking. Some examples of deep and shallow fried cuisines are *daal puri* (shallow frying) (breakfast item, North India), *aloo tikki* (shallow frying) (breakfast item, North India), *mangsher singhara* (deep frying) (breakfast item, East India), and *medu wada* (deep frying) (breakfast item, South India).

4.11 Marination

Marination is a simple process of exposing raw materials to spice, salt, sugar to impart a taste and flavour into the raw materials before cooking. The scientific explanation of marination, however, defines it to be a process, which helps to eliminate germs from the raw materials before cooking due to the action of spices which definitely includes turmeric and salt, two inevitable ingredients in the Indian kitchen. Apart from this, marination with ingredients like citrus fruits/curd/papaya breaks down certain meat enzymes and tenderizes it making it softer and juicer, which is preferred by consumers. It is done by ladle or hands whichever is suitable in a broad mouthed container. Examples include *murgh musallam* (staple item, North India), *tandoori fish tikka* (staple item, North India), and *masor tenga* (staple item, East India).

4.12 Brining

Brining is a step that involves dipping or storing raw vegetables/fruits in a salt solution of certain concentration to impart them with taste and flavour. In Indian cuisine, brining is restricted to making different types of pickles with raw/undeveloped fruits like *amla*, mango, lime, lemon after marinating them with spice mixture and brining them in oil. Brining has two purposes, which are to impart a taste and flavour and to protect from unwanted microbial degradation during the fermentation process. Brining is most preferably done in glass jars or containers with their mouth being closed with flexible materials for air circulation. Examples include all pickles from different parts of India.

4.13 Syruping

Syruping is a term suitably coined to mean any semi processed or fully processed food item that is dipped in sugar syrup to impart sweetness into it. Generally, confectionery items are suited for this treatment, and many Indian cuisines involve syruping as an important step. The syrup can be of an optimal thickness and depending on the dish to be prepared syrup brix or consistency is fixed. Syrup preparation involves mixing of sugar and water in required proportions and boiling until required consistency is reached. Stainless steel pans or teflon coated, non-stick pans are mostly preferred for preparation of syrups and wooden ladles which help in mixing properly. Syruping is either done by dipping the semi-cooked/fully-cooked items into the syrup or pouring syrup from the top into the cooked item depending on the texture of the food item. Examples include *gokul peethe* (confection item, East India) and *malpua/malpoa* (confection item, East India).

4.14 Fermentation Practices

This involves allowing raw fruit, vegetable, pulses or fish to ferment under natural conditions or environment to develop certain texture, flavour or taste characteristics. This is mostly practised in North eastern part of India when fermentation is done with many vegetables like *kimchi*, *sauerkraut* but on pan India basis, Dahi or curd is the most well-known fermented product. Fermentation not only increases taste but also incorporates healthy bacteria into the gut system, which is beneficial for the human health. Examples include ***mishti Doi*** (confection item, East India), *pani-yaram* (breakfast item, South India) and *naan roti* (staple item, North India).

4.15 Sun Drying

Sun drying is an old practise, which is still followed in many Indian households where raw fruits/vegetable pieces or any other raw items are kept under natural sunlight for days to dry them to improve storability. Though it may not be controlled drying and may depend on weather conditions, still the method is easy, economical and the natural components are retained. Fruit candies are generally prepared using sun drying. A variety of grain based products is also sun dried to facilitate storage throughout the year. These items are usually deep fried in oil prior to consumption. However, there might be issues of contamination from the environment if proper hygiene is not maintained. Examples include *amla* candy, *papad*, and *kurdai.*

5 Processing of Indian Traditional Foods

In India, food manufacture occurs in both the organized sector as well as the unorganized sector. Large scale manufacture of foods calls for some degree of mechanization in food production, standardization of quality and preservation to improve the shelf life of foods, thereby facilitating storage, transport and distribution of these foods. The manufacture of many traditional food products occurs primarily in the unorganized sector, which has several limitations as described below.

5.1 Availability of Processed Indian Traditional Foods (RTE, RTC and Instant Premixes)

Traditional foods require both time and skill in their preparation. With changing lifestyles and affordability, consumers today are seeking all kinds of conveniences with regards to food preparation. Thus, a large variety of convenience foods in the form of premixes, pastes, ready-to-cook (RTC) as well as ready-to-eat (RTE) preparations are available today in the market. Further today's discerning consumer is always on the look out for improved products, new products, increased convenience, better quality and all this at an affordable price. There are different degrees of conveniences offered, and the Indian consumer so far has not been strongly inclined to a RTE product as much as he/she has liked a RTC type of product. There are many reasons for this- RTE products have a high cost, they show a high degree of preservation, they sometimes possess inadequate quality, and there is little scope for innovation by the consumer. On the other hand, RTC type of product is more popular because of reasonable cost, yielding good quality product on preparation and with sufficient scope for innovation. Premixes are a type of RTC products, which usually comprise of a blend of appropriate powdered ingredients. Since they are dry, with low moisture content, they are shelf-stable with moderate packaging requirements. Usually, the consumer has to add required amount of water to the premix and subject the resulting dispersion/batter/dough to some form of cooking such as boiling, steaming, shallow frying, and deep-frying. There are many varieties of premixes available for several kinds of Indian traditional foods. The art of preparation of traditional foods is sometimes restricted to a region or community. People from other regions who enjoy these foods will, therefore, buy these premixes.

5.1.1 Ready to Cook Items (RTC)

RTC products are items which are packed in such a way that they need minimal preparation steps and can be directly cooked. They are less time consuming and easy to prepare. There are several companies in the market for RTC foods providing a range of breakfast items, meals, snacks, and savoury items to choose from. Table 6.26 compiles several categories of RTC food products belonging to different brands that are readily available in Indian market.

5.1.2 Ready to Eat Food Items (RTE)

RTE food products are pre-cooked or semi-cooked to such a consistency that they can be consumed directly with simple processing steps like heating or warming. They have a huge demand with the new generation consumers because of time saving and low processing steps involved with them. Different brands have a range of RTE foods under several categories like snacks, breakfast foods, gravies, sweet dishes or confectionaries. Table 6.27 presents a list of RTE foods that are available under several categories.

5.2 Challenges in the Promotion of Indian Traditional Foods

Traditional foods of India are mostly made by the unorganized sector including many small scale industries. There are a few exceptions to this (Haldirams, Amul—organised sector). In the unorganized sector, which is ridden by many problems,

Table 6.26 Ready to cook food items in Indian market

Category	Name of product	Company
Breakfast	*Dosa, upma, handvo, rice idli, rava idli, sambhar, uttapam, dahi vada, rava dosa, medu vada*	Gits
	Carrot idli, beetroot idli, spinach idli, poha, vermicilli upma, oats upma, ragi dosa, khata meetha poha, oats idli, multigrain dosa, masala idli, masala upma, puttu	MTR
	Ragi idli, rawa idli, sabudana khichdi	Kohinoor Foods
	Paratha, vatata vada, thepla	Fingerlix
Staple/curry/gravy	*Dal tadka, aloo matar, methi matar, veg biriyani, aloo raswala, dal makhani, palak paneer, matar paneer, rajma masala, punjabi chole, paneer makhani, paneer tikka masala*	Gits
	Amritsari chole, prawn curry-kashmiri, prawn curry malabar, steamed basmati rice, subz-e- biryani	Fazlaani
	Chana masala, methi matar, palak paneer, paneer butter masala, panner makhanwala, veg kadhai	MTR
	Bombay egg curry, Delhi butter chicken, Mumbai kolhapuri masala, Lucknow adhraki bhuna, Old Delhi nihari, punjabi haryali methi, rajasthani mirch masala	Kohinoor Food
Snacks/savouries	*Khatta dhokla, khamman dhokla, moong dal vada, sandwitch dhokla, dahi vada*	Gits
	Gobi Manchurian, bajji & bonda (pakoda), muruku	MTR
	Instant mix—methi thepla, moong bhajiya	Zappy
	Bhajni—thalipith	Bedekar
	Mung pakoda	Emar Foods
	Pav bhaji	Rasoi Magic
	Tomato omelette	Emar Foods
Confectionaries/sweet/dessert	*Kulfi, jalebi, rabdi, phirni, basundi, rasmalai, kalakand, rice kheer, gulab jamun, bombay halwa, vermicelli kheer, sahi gulab jamun, badam & pista, falooda, rose kheer, vermicelli*	Gits
	Mango kulfi, gulab jamun, vermicelli payasam/seviyan kheer, kulfi kesar badam, basundi/rabdi, kulfi kesar pista, besan laddoo, jalebi, rabdi	MTR
	Vermicelli payasam	Mothers Recipe
	Malai kulfi	Blue Bird
Beverages	*Badam drink*	MTR

Table 6.27 Ready to eat food items in the Indian market

Category	Product	Company
Breakfast	*Khatta meetha poha, magic masala upma, vegetable upma*	MTR
	Puliogare rice	Mother's recipe
Staple	*Aloo matar, punjabi chole, methi matar, veg biryani, paneer tikka masala, chana masala, palak paneer, dhansak, dal makhani, dal tadka, panjabi kadhi, dal palak, jeera aloo veg pulao, masala rice, paneer makhani, aloo raswala, rajma masala, matar paneer, veg biryani (no onion/no garlic), paneer tikka masala (no onion/no garlic)*	Gits
Snacks	*Pav bhaji*	Gits
	Khaman dhokla	Mother's recipe
Confectionaries	*Payasam*	Mother's recipe
	Kesari halwa	MTR

there is a lack of scientific culture. Food processing is attempted based on the knowledge of cooking with scant respect for parameters, which appear and are important for any scale up operation. There is a lack of technical know-how in this sector, and since their turnover is low, they are not willing to pay the requisite price for this know-how. Small scale industries are also initiated with other objectives in mind—offering employment to people, using locally available produce. Their long term vision or goals are not clear. There is, therefore, a strong need to promote entrepreneurship backed by sound knowledge. Some of the challenges to be met for large scale manufacture and distribution of traditional foods are:

5.2.1 Indigenous Machinery/Process Development

Indigenous development of machinery is required, and improvements/developments in machinery design and efficiency should be an ongoing activity. This requires complex engineering inputs to be given. The process of manufacturing must also be developed and optimized in order to get optimum product quality as well as desired shelf life.

5.2.2 Availability/Procurement of Good/Consistent Quality Raw Materials

A thorough knowledge of raw materials/ingredients to be used in the manufacturing process is necessary. This demands a thorough investigation of the materials available in the country; their chemical composition and how this would affect product quality.

5.2.3 Standardisation of Product Quality

The product needs to be standardized to avoid batch-to-batch variation in quality. Thus, both raw material specifications and product specifications need to be developed.

5.2.4 Development of Analytical Methods

Appropriate analytical methods have to be developed to ensure compliance with the specifications laid down.

5.2.5 Preservation for Extended Shelf Life

Emerging preservation techniques should aim to interfere with the homeostatic mechanisms that microorganisms have evolved in order to survive environmental adversities, resulting in extension of shelf life of food items.

5.2.6 Packaging and Marketing

Packaging plays an important role in protection of food items against physical, chemical and environmental factors that could possibly contaminate the food if not packaged correctly. Promotional activities (marketing) are required to improve the sales of any product, and so this is true for traditional foods. A greater challenge is to make traditional foods attractive to certain target populations such as children and teenagers.

5.2.7 Process and Product Improvement

Product improvement/development and innovation must be an ongoing endeavor. Application of technologies for improved process economics or improved product quality must be considered. In a society with growing health awareness improvements in nutritional quality of large scale processed foods must also be taken into account. Likewise, issues of safety arise in many instances and must be addressed adequately. All this requires a great deal of scientific knowledge to be made available through research.

6 Nutritive Value of Indian Traditional Foods

Endorsement of traditional foods and effective dissemination of education on nutrition can assist and encourage consumers to select appropriate diet and overcome malnutrition. Under-nutrition, micronutrient deficiencies, and obesity are various forms of malnutrition. Good nutrition is thus not just a result of economic augmentation and social progress but an essential input as well. The challenges of achieving good nutrition, on the other hand, are increasingly obscure. The utilization of processed foods coupled with sedentary lives of an increasingly urban population is amplifying the incidence of life style diseases and metabolic disorders such as obesity, diabetes and cardio- vascular diseases. Other commonly observed disorders are poor sleep quality and diseases associated with gut such as irritable bowel syndrome.

The consumption of hardy millets has declined exponentially since urbanization and green revolution propelled Indians towards a diet centered primiarily around wheat and rice. Millets have been used traditionally for centuries in food preparation. *Sorghum* (*jowar*) and millets [pearl (*bajra*), little (*kutki*), finger (ragi), kodo (*kodon*), foxtail (*kakum*)] are packed with nutrients and being non fastidious crops they can be easily grown, sometimes even in hostile climates of drought. A typical Indian meal or '*thali*' is an amalgamation of all food groups grown locally and processed to cater to all the four basic human tastes viz. salty, sweet, sour and bitter, in a balanced and acceptable manner. In addition, traditional foods such as raw banana have also contributed to a favourable gut microbiome by virtue of the prebiotic inulin present therein (Shalini, Abhinaya, Saranya, & Antony, 2017).

The proximate analysis of the entire range of most Indian foods has been documented recently (Longvah, Ananth, Bhaskarachary, & Venkaiah, 2017) with names given in all vernacular languages. The reader is refereed to this document for detailed composition. Table 6.28 lists the proximate composition and bioactives present in some of the common food ingredients. Table 6.29 lists the proximate composition and bioactives present in some selected Indian traditional food preparations.

7 Overview of Research Work Done So Far in Indian Traditional Foods

The range of research done in traditional Indian foods provides an understanding of the role and optimal levels of traditional nutrients, as well as identifying bioactive components present in foods. Understanding the role of nutrients, science and technology at the molecular level provide specific recommended dietary allowances for different population subgroups. Numerous investigations have been carried out to identify the role of bioactive food components. Prospective and authentic health benefits of bioactive food components signify an analogous area in diet-health research.

Food ingredients in many traditional food preparations perform the function of food additives such as those of acidulants, anti-oxidants, preservatives, stabilizers, colourants and flavorants (Abuajah et al., 2015). Certain food ingredients like *dry mango powder*, tamarind, *kokum*, and *amla* have traditionally been used as acidulatnts in conventional cooking since ages. Researchers have investigated the properties of these acidulatns and their health benfits (Table 6.30). Many of them have been reported to improve the bioavailability of minerals such as iron and copper.

Table 6.28 Proximate composition, bioactive components and health benefits of commonly used food ingredients in Indian traditional foods

Food commodity	Proximate composition, bioactive component and health benefits	Reference
Ragi (*Eleusine coracana*)	Moisture: 10.34 ± 0.17%; Fat: 1.22 ± 0.093%; Protein: 8.17 ± 0.12%; Ash: 2.22 ± 0.13%; Crude fiber: 3.79 ± 0.09%; Carbohydrate: 74.26 ± 0.25% Water soluble fibers like arabinoxylans, β-glucan; Water insoluble fibers like lignin, cellulose, hemi-cellulose; Minerals (calcium, phosphorous and iron); Vitamins (thiamin, pantothanic acid) Reduced LDL and total cholesterol levels; reduced blood sugar and insulin response; increased feeling of fullness; increased growth of good bacteria in the digestive tract	Kudake et al. (2018)
Oat (*Avena sativa* L.)	Moisture: 8.64 ± 0.21%; Fat: 7.75 ± 0.13%; Protein: 13.57 ± 0.18%; Ash: 1.84 ± 0.12%; Crude fiber: 3.46 ± 0. 06%; Carbohydrate: 64.71 ± 0.37% Dietary fiber specifically β-glucan; Polyphenolic compounds like avenanthramides, anthranilic acids, ester coupled alkyl conjugates, ether and ester coupled glycerides Minimize the blood cholesterol level by enhancing the excretion of bile in the body; decreases the risks of colon cancer; antioxidant properties due to presence of several phenolic compounds	Kudake, Pawar, Muley, Parate, and Talib (2017)
Peanuts (*Arachis hypogaea*)	Fat: 49.24%; Protein: 25.80%; Ash: 0.33%; Crude fiber: 8.5%; Carbohydrate: 16.13% Dietary fiber; Vitamins (Folates, niacin, pantothenic acid, pyridoxine, riboflavin, riboflavin, vitamin A, vitamin C, vitamin E); Minerals (calcium, copper, iron, magnesium, manganese, phosphorous, selenium, zinc); Antioxidants (vitamin E in oil or chlorogenic acid, caffeic acid, coumaric acid, ferulic acid, flavonoids and stilbene (resveratrol)) Functioning of the digestive systems, skin, nerves, metabolic reaction function; peptides strengthen the body's immune system, regulates hormones and blood sugar and promotes male fertility; phytosterols help in controlling body's cholesterol level; nutrients can be used to ironically combat malnourished population	Davis and Dean (2016)
Groundnut (*Vigna subterranea* L. Verdc),	Moisture: 5.8%; Fat: 47.0%; Protein: 5.8%; Ash: 3.08%; Crude fiber: 3.70%; Carbohydrate: 4.62% Antioxidants (Vitamin E, tocopherols (α, γ and δ) and tocotrienols (γ and δ)); linoleic acid and α-linolenic acid; good score of essential and non-essential amino acid Positive effect on cardiovascular disease risk factors, such as blood pressure, platelet activity, and inflammation; phytosterols help in controlling body's cholesterol level	Asibuo et al. (2008)

(continued)

Table 6.28 (continued)

Food commodity	Proximate composition, bioactive component and health benefits	Reference
Mung beans (*Vigna radiate*)	Moisture: 9.31%; Fat: 1.02%; Protein: 23.31%; Ash: 2.87%; Crude fiber: 3.64%; Carbohydrate: 59.85% Vitamin C; phenolics; Flavonoids (Vitexin and isovitexin); Antioxidant enzymes (ascorbate oxidase, peroxidase); alpha-amylase inhibitory activity Ability to resist different stress; reduce serum lipids content; maintain proper health of blood vessels; anti diabetic potential	Heller (1927), Liyanage et al. (2018)
Black gram (*Vigna mungo* (L.) Hepper)	Moisture: 10.9%; Fat: 1.4%; Protein: 24.0%; Ash: 3.2%; Crude fiber: 0.9%; Carbohydrate: 59.6% Important source of proteins; minerals (calcium, iron and phosphorus); Vitamins (vitamin B1, vitamin B2, and niacin); Antioxidants (*o*-dihydroxy phenols, flavonols, tannins) Ability to boost energy; protect cardiovascular health; reduce pain and inflammation; improve immunity; aid in maintaining skin health; diabetes management; maintain bone health; strengthen the nervous system; optimize digestion.	Kamboj and Nanda (2017), Suneja, Kaur, Gupta, and Kaur (2011)
Chickpeas (*Cicer arietinum* L.)	Moisture: 8.01 ± 0.38%; Fat: 5.43 ± 0.26%; Protein: 22.83 ± 1.07%; Ash: 3.04 ± 0.0.14%; Crude fiber: 3.50 ± 0.16%; Carbohydrate: 57.19 ± 2.69% Antioxidant peptides (TETWNPNHPEL, FVPH, SAEHGSLH and ALEPDHR); Antioxidant compounds (flavonols, flavone glycosides, oligomeric as well as polymeric proanthocyanidins, cinnamic acid, salicylic acid, hydroxycinnamic acid, *p*-coumaric acid, gallic acid, caffeic acid, vanillic acid, ferulic acid, anise acid, tannic acid, isoferulic acid, piperonyl, and chlorogenic acid) Management of clinical complications associated with diabetes, obesity, cancer, osteoporosis, and cardiovascular diseases; biological roles such as antioxidant, estrogenic, antifungal, and antibacterial activities	Gupta et al. (2017), Rachwa-Rosiak, Nebesny, and Budryn (2015)
Lentil (*Lens culinaris* or *Lens esculenta*)	Moisture: 9.14 ± 0.78%; Fat: 0.81 ± 0.04%; Protein: 31.12 ± 0.68%; Ash: 2.62 ± 0.0.31%; Crude fiber: 15.38 ± 0.22%; Carbohydrate: 52.63 ± 2.22% A good source of proteins; valuable source of total dietary fibers, with insoluble dietary fiber of approximately 93–99.7%; micronutrients (iron, zinc, selenium, copper, manganese, magnesium, molybdenum); vitamins (folate, thiamin, riboflavin, niacin, pantothenic acid and pyridoxine. In addition, vitamin E (α, β, and γ- tocopherols)) Provide essential and non-essential amino acids and carbon skeletons for the metabolic reactions; ability to work as selective promoters for the growth of beneficial gut microbes; muscle growth and development	Faris, Takruri, and Issa (2013)

Tur Pigeon pea (*Cajanus cajan*)	Moisture: 11.20%; Fat:1.7%; Protein: 22.40%; Ash: 3.22%; Crude fiber: 2.5%; Carbohydrate: 58.98% Antioxidants (genistein, genistin, pinostrobin, cajaninstilbene acid, orientin, vitexin); antimicroial agents (cajanuslactone (coumarin), betulinic acid, longistylin A and longistylin C); minerals (calcium, magnessium, zinc, iron, copper) Antibacterial activity; hypocholesterolemic effects; neuroactive properties; antioxidant activities; anticancer activity; hepatoprotective effects; anthelmintic activity	Oke (2014), Pal, Mishra, Sachan, and Ghosh (2011), Saxena, Kumar, and Sultana (2010)
Moth beans (*Vigna aconitifolia*)	Moisture: 9.95%; Fat: 1.48%; Protein: 21.9%; Ash: 3.5%; Crude fiber: 4.5%; Carbohydrate: 58.67% Minerals (calcium, copper, iron, magnesium, manganese, phosphorus, zinc, sodium); Vitamins (thiamine, niacin, pantothenic acid, folate); Potent source of dietary fiber Reduces constipation issues; heart disease management; high blood pressure control; cholesterol balance; osteoporosis control; stress management	Kadam et al. (1985), Kamboj and Nanda (2017)
Rye (*Secale cereale*)	Moisture: 10.95 ± 0.23%; Fat: 1.66 ± 0.20%; Protein: 10.06 ± 0.09%; Ash: 2.01 ± 0.02%; Crude fiber: 15.38 ± 0.22%; Carbohydrate: 59.94 ± 0.47% Minerals (calcium, magnesium, iron,); vitamins (thiamine, riboflavin, niacin, folic acid, pyridoxine); good score of essential amino acids; valuable source of total dietary fibers Prevents gallstones formation; prevent heart diseases; anti-inflammatory activity; anti-cancer activity; helps to improve bowel function	Wrigley and Bushuk (2010)
Fava beans (*Vicia faba*)	Moisture: 12.97 ± 0.71%; Fat: 2.67 ± 0.15%; Protein: 22.61 ± 1.24%; Ash: 2.90 ± 0.0.16%; Crude fiber: 2.46 ± 0.13%; Carbohydrate: 56.39 ± 3.08% Prominent source of proteins; good source of crude fiber and neutral fiber; Vitamins (thiamine, riboflavin, niacin, pyridoxine, pantothanic acid); Minerals (magnesium, copper, iron, calcium, sodium, potassium); Antioxidants (β-carotene) Cardiovascular disease management; obesity control and weight management; boosts up metabolism rate; anti-depression activity due to presence of dopamine; helps in maintaining the balance of fluids	Zee, Boudreau, Bourgeois, and Breton (1988)

(continued)

Table 6.28 (continued)

Food commodity	Proximate composition, bioactive component and health benefits	Reference
Barley (*Hordeum vulgare*)	Moisture: 8.89 ± 0.35%; Fat: 1.78 ± 0.08%; Protein: 14.89 ± 0.70%; Ash: 3.89 ± 0.20%; Crude fiber: 4.89 ± 0.25%; Carbohydrate: 65.66 ± 1.12% Antioxidants (all eight forms of vitamin E and 7-hydroxymatairesinol); Dietary fiber (β-glucan); Vitamins (thiamine, riboflavin, niacin, pyridoxine, pantothanic acid, pyridoxine, folic acid, ascorbic acid, phylloquinone); Minerals (calcium, iron, magnesium, manganese, phosphorus, potassium, sodium, zinc) Cardiovascular disease management; improves regulation of blood sugar; obesity control and weight management; prevents the formation of gall stones; prominent source of antioxidant	Edney (2010), Vilienė, Šašytė, Stupelienė, and Gružauskas (2017)
Pearl Millet (Bajra) (*Pennisetum glaucum*)	Moisture: 10.1±0.23%; Fat: 2.7±0.20%; Protein: 9.1±0.09%; Ash: 1.90±0.0.02%; Crude fiber: 3.8±0.22%; Carbohydrate: 70.00±0.47% Carbohydrates; B complex vitamins (niacin, folic acid); lignins; minerals (calcium, magnesium, phosphorus, iron, potassium, sodium, zinc, manganese, copper and selenium) Good source of starch; diabetes and obesity management; anticancer properties (breast cnacer); reduces constipation issues	Abdalla, El Tinay, Mohamed, and Abdalla (1998)
Wheat (*Triticum aestivum*)	Moisture: 9.72 ± 0.09%; Fat: 1.15 ± 0.10%; Protein: 10.73 ± 0.23%; Ash: 1.52 ± 0.0.11%; Crude fiber: 0.47 ± 0.06%; Carbohydrate: 76.44 ± 0.43 Antioxidants (carotenoids polyphenolics, tocotrienols); dietary fiber (cellulose, pectins, arabinoxylan, glucans, and lignin); Minerals (manganese, phosphorous, magnesium, copper, zinc, Iron); Vitamins (niacin, thiamin and folate) Slow digestibility and suitable for hypoglycemic foods; potential antioxidant properties; PUFA to MUFA ratio >2 shows beneficial effects with respect to low plasma and liver cholesterol levels	Dhanavath and Prasada Rao (2017), Kudake et al. (2017, 2018)
Jowar (*Sorghum bicolor*)	Moisture: 9.40 ± 1.10%; Fat: 3.45 ± 0.36%; Protein: 10.20 ± 0.36%; Ash: 2.43 ± 0.19%; Carbohydrate: 74.49 ± 0.96% Anticancer compound (3-deoxyanthocyanidins); Antioxidants (carotene); Minerals (iron, copper, phosphorous, calcium); Vitamins (pantothenic acid, pyroxidine, thiamin, riboflavin, biotin, folate) Improves digestion and relieves constipation issues; boosts immune system; management of cardiovascular diseases; anti-cancer potential; maintains celiac health	Patil and Arya (2017)

Rice (*Oryza sativa*)	Moisture: 11.54 ± 0.75%; Fat: 2.21 ± 0.22%; Protein: 7.7 ± 0.30%; Ash: 1.20 ± 0.11%; Carbohydrate: 77.33 ± 1.18% Antioxidants as polyphenolics (ferulic acid, p-coumaric acid, isoferulic acid, syringic aicd, vanillic acid, sinapic acid, caffeic acid, (+)-catechin acid, phydroxybenzoic acid, protocatechuic acid); Minerals (magnesium, phosphorus, manganese, selenium, iron, copper); Vitamins (folic acid, thiamine and niacin) Potential antioxidant properties; ability to boost energy	Goufo and Trindade (2014), Shao and Bao (2015)
Kidney beans (*Phaseolus vulgaris*)	Moisture: 9.15 ± 0.24%; Fat: 2.46 ± 0.06%; Protein: 20.09 ± 0.52%; Ash: 3.85 ± 0.0.10%; Crude fiber: 6.78 ± 0.17%; Carbohydrate: 57.67 ± 1.48% Minerals (molybdenum, iron, copper, manganese, phosphorus); Vitamins (Vit. K1, Folate); Antioxidants (isoflavones, anthocyanins) Cardiovascular disease management; diabetes control and management; prevent damage caused by oxidative stress; potential antibacterial activity	Ganesan and Xu (2017), Qayyum, Butt, Anjum, and Nawaz (2012)
Horse gram (*Macrotyloma uniflorum*)	Moisture: 9.5%; Fat: 1.36%; Protein: 23.6%; Ash: 3.3%; Crude fiber: 5.37%; Carbohydrate: 56% Phenolic compounds; highest calcium content among pulses; low in fat and high in carbohydrate content; low in lipid and sodium content Therapeutic agent to treat kidney stones; urinary diseases, piles, common cold, throat infection; astringent and diuretic properties.	Kadam et al. (1985)
Lima bean (*Phaseolus lunatus*)	Moisture: 10.6%; Fat: 0.9%; Protein: 23.0%; Ash: 4.1%; Crude fiber: 5.0%; Carbohydrate: 56.4% Rich source of folate, phosphorus, protein, potassi, vitamin B1, iron, magnesium and vitamin B6. Improve digestion; boost heart health, positively impact energy levels; aid circulation	Moses, Olawuni, and Iwouno (2012)
Common beans (*Phaseolus vulgaris* L.),	Moisture: 12%; Fat: 1.30%; Protein: 22.90%; Ash: 3.20%; Crude fiber: 4.8%; Carbohydrate: 60.6% Excellent source of polyunsaturated fatty acids; main source of vegetable protein; high amounts of isoflavones Diuretic and hypoglycemic action; anticancer	Baptista et al. (2017)

(continued)

Table 6.28 (continued)

Food commodity	Proximate composition, bioactive component and health benefits	Reference
Cowpea (*Vigna unguiculata* L. Walp)	Moisture: 13.4%; Fat: 1.0%; Protein: 24.1%; Ash: 3.2%; Crude fiber: 3.8%; Carbohydrate: 54.5% Rich source of folic acid, an important vitamin that helps prevent neural tube defects in unborn babies Improve digestion; manage diabetes; improves blood circulation	Baptista et al. (2017), Devi, Kushwaha, and Kumar (2015)
Maize (*Zea mays*)	Moisture: 12.8 ± 0.68%; Fat: 3.75 ± 0.17%; Protein: 9.57 ± 0.14%; Ash: 2.08 ± 0.20%; Crude fiber: 7.5 ± 0.60%; Carbohydrate: 64.3 ± 2.69% Anticancer compound (3-deoxyanthocyanidins); Antioxidants (Carotenoids, criptoxanthin, zeaxantin, lutein); Vitamins (thiamin, niacin, pantothanic acid, folate, tocopherols) Anti-cancer potential; antioxidant potential; metabolic reaction management and growth promotion Improves kidney functioning; bone density maintenance; maintenance of heart rate	Bacchetti, Masciangelo, Micheletti, and Ferretti (2013)
Carrot (*Daucus carota*)	Moisture: 86%; Fat: 0.2%; Protein: 0.9%; Ash: 1.1%; Crude fiber: 1.2%; Carbohydrate: 10.6% Minerals (calcium, iron, phosphorus, sodium, potassium, magnesium, copper, zinc); Antioxidants (Carotenes, phenolics—hydroxycinnamic acids and para-hydroxybenzoic acids); Vitamins (Riboflavin, thiamine, ascorbic acid) Maintenance and protection from coronary heart disease and cancer; carotenoids and phenolics may play a significant role in protecting biological systems from the effects of oxidative stress; enhancement of immune system and decreased risk of degenerative diseases such as cancer, cardiovascular disease, age related mascular degeneration and cataract formation	Sharma, Karki, Thakur, and Attri (2012)
Pumpkin (*Telfairia occidentalis*)	Moisture: 4.2 ± 0.4, Fat: 47.0 ± 0.6, Protein: 28.1 ± 0.3, Ash: 4.7 ± 0.1, Crude fiber: 3.1 ± 0.3, Carbohydrate:12.9 ± 05 Minerals (iron, phosphorous)	Giami and Bekebain (1992), Montesano, Rocchetti, Putnik, and Lucini (2018)

Potato (*Solanum tuberosaum*)	Edible portion: 85%; Moisture: 74.70 gm, Protein: 1.60 gm, Fat: 0.10 gm, Minerals: 0.60 gm, Carbohydrates: 22.60 gm Minerals (calcium, iron, phosphorous, potassium); Antioxidant (carotene); Vitamins (thiamine, riboflavin, niacin, vitamin C) The B-complex vitamins are involved in energy metabolism reactions; minerals helps in bone and teeth health; prominent antioxidant properties; constipation relief; blood pressure control management	http://www.potatoes.net.au/nutritional-info/
Jack fruit seed (*Artocarpus heterophyllus Lam)*	Water (g): 51.0–64.5, Protein (g): 6.6–7.04, Fat (g): 0.40–0.43, Fibre (g): 1.0–1.5, Ash (g): 2–2.5, Carbohydrate (g): 25.8–38.4 Minerals (Calcium, magnesium, phosphorus, potassium, sodium, iron); Vitamins (vitamin A, thiamine, riboflavin, vitamin C) ; Phenolics (phenolics acids, phenylpropanoids, flavonoids, lignins, melanins, tannins) Positive effects against lung cancer and prostate cancer control; myocardial infarction and coronary heart disease control and management; hypertension control	Swami, Thakor, Haldankar, and Kalse (2012)

Table 6.29 Proximate components and composition of commonly consumed traditional Indian foods

Food product	Product composition and proximate composition	Reference
Plain wheat noodles	Wheat flour (g): 200; salt (g): 4; vegetable oil (g): 26 Moisture: 6.26 ± 0.25 (%); total fat: 12.76 ± 0.50 (%); protein: 8.83 ± 0.31 (%); crude fiber: 0.41 ± 0.06 (%); ash: 1.31 ± 0.20 (%); Carbohydrate: 70.45 ± 2.85 (%)	Kudake et al. (2017, 2018)
Wheat *chapati*	Wheat flour (g): 100; salt (g): 1.5 Moisture: 30.08 ± 0.08 (%); Total fat: 1.97 ± 0.06 (%); Protein: 11.17 ± 0.01 (%); Ash: 2.66 ± 0.00 (%); Carbohydrate: 84.20 ± 0.06 (%)	Banerji, Anthanarayan, and Lele (2017)
Buffalo milk *gulab jamun*	*Khoa* (treatment wise) 80 parts; refined flour/*Maida* 20 parts; sodium bicarbonate-0.5%; cardamom 1% ; nutmeg 1% Moisture: 34.21 (%); Total fat: 14.98 (%); Protein: 8.49 (%); Ash: 1.10 (%); Carbohydrate: 47.30 (%); TSS: 65.78%	Patil, Padghan, and Patil (2017)
Dairy *rasgulla*	Not specified Moisture: 46.29 ± 0.55 (%); Total fat: 7.86 ± 0.29 (%); Protein: 8.24 ± 0.05 (%); Ash: 1.04 ± 0.03 (%); Carbohydrate: 36.57 ± 0.51 (%); TSS: 53.71 ± 0.60%	Sengupta, Samanta, and Bhowal (2017)
Whole wheat *khakra*	Whole wheat flour (100 g); salt (2% w/w dry flour); water (70 mL/100 g dry weight of flour) Moisture: 5.1 ± 0.0 (%); Total fat: 0.9 ± 0.0 (%); Protein: 10.9 ± 0.2 (%); Ash: 2.3 ± 0.01 (%); Carbohydrate: 80.9 ± 0.0 (%)	Chauhan, Sonawane, and Arya (2017)
Thalipeeth	Not specified Moisture: 39.06 ± 3.04 (%); Total fat: 1.89 ± 0.19 (%); Protein: 21.0 ± 0.87 (%); Crude fiber: 0.64 ± 0.24 (%); Ash: 3.40 ± 0.17 (%); Carbohydrate: 34.01 (%)	Arya and Gaikwad (2017)
Upma	Semolina (95%); black gram dal (5%) Moisture: 40.21 (%); Total fat: 6.30 (%); Protein: 11.82 (%); Ash: 2.05 (%); Carbohydrate– 38.44 (%)	Dhumketi, Singh, and Agrawal (2018)
Dosa	Rice: 85%; black gram: 15% Moisture: 40.61 ± 0.36 (%); Total fat: 15.99 ± 0.15 (%); Protein: 7.76 ± 0.58 (%); Crude fiber: 0.32 ± 0.22 (%); Ash: 2.08 ± 0.41 (%); Carbohydrate: 73.85 ± 1.19 (%)	Beniwal and Jood (2015)

Chikki	Peanuts and jaggery Moisture: 2 ± 1.66 (%); Total fat: 21 ± 0.42 (%); Protein: 12 ± 0.65 (%); Crude fiber: 1.6 ± 0.32 (%); Calcium: 83 ± 3.26 (mg %); Iron: 1.5 ± 1.87 (mg %)	Pallavi, Chetana, and Reddy (2014)
Burfi	Buffalo milk khoa and sugar Moisture: 16.96 ± 0.06 (%); Total fat: 20.41 ± 0.06 (%); Protein: 14.88 ± 0.07 (%); Ash: 2.98 ± 0.01 (%); Total sugars: 44.77 ± 0.06 (%); Total solids: 83.04 ± 0.06 (%)	Navale, Deshmuh, Korake, Narwade, and Mule (2014)
Sooji halva	Wheat semolina (*Bansi* variety) 500 g; cane sugar 100 g; hydrogenated oil 300 g ; *chiranji* 150 g; cardamom whole 7 g Moisture: 29.7 (%); Total fat: 8.5(%); Protein: 2.7 (%); Crude fiber: 0.9 (%); Ash: 0.3 (%); Carbohydrate: 58 (%)	Ghosh, Krishnappa, Srivatsa, Eapen, and Vijayaraghavan (1980), Krishnappa, Srivatsa, Ghosh, Eapen, and Vijayaraghavan (1982)
Aloo chole	Dry bengal gram whole (*Kabuli* variety) 100 g; hydrogenated oil 200 g; potatoes 400 g; garlic whole 10 g; ginger fresh 10 g; tomato (*Madanapalli* variety) 150 g; onion fresh 150 g ; chilli powder 5 g; turmeric powder 5 g; jeera powder 2.5 g; white pepper powder 2.5 g; salt 15 g Moisture: 76.4 (%); Total fat: 5.2 (%); Protein: 3.3 (%); Crude fiber: 1.1 (%); Ash: 1.6 (%); Carbohydrate: 12.4 (%)	Ghosh et al. (1980), Krishnappa et al. (1982)
Potatoes peas curry	Dry peas 100 g; hydrogenated oil 200 g; potatoes 400 g; garlic whole 10 g; ginger fresh 10 g; tomato (*Madanapalli* variety) 150 g; onion fresh 150 g; chilli powder 5 g; turmeric powder 5 g; jeera powder 2.5 g; white pepper powder 2.5 g; salt 15 g Moisture: 72 (%); Total fat: 3.2 (%); Protein: 3.3 (%); Crude fiber: 1.9 (%); Ash: 1.8 (%); Carbohydrate: 16.8 (%)	Ghosh et al. (1980)
Baked beans	Soaked beans (*Rajma*) cooked in tomato puree Moisture: 59.9 (%); Total fat: 0.4 (%); Protein: 9.5 (%); Crude fiber: 1.5 (%); Ash: 2.8 (%); Carbohydrate: 25.9 (%)	Srivatsa et al., 1993

(continued)

Table 6.29 (continued)

Food product	Product composition and proximate composition	Reference
Sorghum laddu	Wheat flour 90 g; germinated sorghum flour 60 g; desi ghee 80 g; powdered sugar 150 g; cardamom 2 g Moisture: 3.82 ± 0.27 (%); Fat: 21.24 ± 0.07 (%); Protein: 13.57 ± 0.02 (%); Crude Fiber: 2.40 ± 0.13 (%); Ash: 2.25 ± 0.13 (%); Carbohydrates: 65.72 ± 0.26 (%)	Singh et al. (2016)
Sorghum chaat	Germinated grain 150 g; onion 50 g; Tomato 80 g; green chilli 10 g; coriander leaves 25 g; salt ½ tsp; refined oil 5 mL; water 150 mL Moisture: 8.51 ± 0.09 (%); Fat: 9.04 ± 0.01 (%); Protein: 14.13 ± 0.04 (%); Crude fiber: 2.07 ± 0.09 (%); Ash: 1.86 ± 0.02 (%); Carbohydrates: 64.39 ± 0.32 (%)	Singh et al. (2016)
Sweet potato pickle	Not specified Organic matter: 236 ± 13.1 (%); Fat:11.2 ± 1.3 (%); Protein: 78.1 ± 4.9 (%); Crude fiber: 3.49 ± 0.10 (%); Ash: 972.5 ± 30.0 (%); Carbohydrates: 21.79 ± 0.29 (%)	Panda, Panda, Sethuraman, and Ray (2009)
Curry leaves chutney powder	Not specified Moisture: 5.0 ± 0.11 (%); Fat: 1402 ± 0.086 (%); Protein: 16.4 ± 0.28 (%); Crude fiber: 7.0 ± 0.14 (%); Ash: 4.0 ± 0.09 (%); Carbohydrates: 44.3 (%)	Balaswamy, Jyothirmayi, and Rao (2004)
Sapota pulp (shrikhand)	Not specified Moisture: 61.96 (%); Fat: 8.50 (%); Protein: 7.60 (%); Ash: 0.50 (%); Carbohydrates: 45.36 (%)	Sharma, Peter, Sharma, Kumar, and Rai (2017)
Kulfi	Not specified Moisture: 57.52 (%); Fat: 10.01 (%); Protein: 3.52 (%); Ash: 0.51 (%)	Siddhu, Broadway, Ali, and Singh (2017)

Table 6.30 Typical ingredients that function as acidulants in traditional Indian foods and their health benefits

Acidulants	Active ingredient, brief research work and health benefits	Reference
Dry mango seed powder/pulp	Active ingredients: Phenolic compounds (mangiferin, isomangiferin, homomangiferin, quercetin, kaempferol, anthocyanins), phenolic acids (gallic, protocatechuic, ferulic, caffeic, coumaric, ellagic, 4-caffeoylquinic acids), and mineral antioxidants (potassium, copper, zinc, manganese, iron, selenium) Food products prepared : Mango *lassi*, Mango yoghurt, *amrakhand*, Beverages *Other food products: Ambavadi, aamras, kairi panha, amba barfi, morabba, pedha, sonpapdi, sheera.* Health benefits: Antioxidant, anti-cancer potential, antimicrobial potential (*Staphylococcus aureus, Escherichia coli, Vibrio vulnificus*)	Torres-León et al. (2016), Ravani and Joshi (2013), Ansari and Kumar (2012)
Amla (*Emblica officinalis*)	Active ingredients: Tannins, mucic acid, amino acids, alkaloids, flavone glycosides, phenolic glycosides, flavonol glycosides, phenolic acids, esquiterpenoids, norsesquiterpenoids and carbohydrates, Vitamin C *Food products: Churna*, juice, pickle, *amla* candy, *morabba.* Health benefits: Antioxidant, anti-diabetic and related metabolic functions, anti-hyperlipidemic and related metabolic syndrome, cardio-protective, anticancer and anti-proliferative, anti-mutagenic, cerebro-protective and anti-brain aging, hepato-protective, analgesic, antipyretic and anti-inflammatory, antimicrobial, GI ailments, antitussive and other respiratory problems, reno-protective, wound healing property	Variya, Bakrania, and Patel (2016)
Tamarind (*Tamarindus indica*)	Active ingredients: Triterpenes, lupanone, lupeol *Food products:* Juice, candy, chutney, curry, sausage for *pani-puri, dhokla, samosa.* Health benefits : Refrigerant, digestive, carminative, laxative, antiscorbutic, febrifuge, ophthalmic, useful in gastropathy, datura poisoning, alcoholic intoxication, scabies, constipation, antimicrobial (*Burkholderia pseudomallei, Klebsiella pneumonia, Salmonella paratyphi, Bacillus subtilis, Salmonella typhi, and Staphylococcus aureus*)	Kurian (2012), Bhadoriya, Ganeshpurkar, Narwaria, Rai, and Jain (2011)
Kokum (*Garcinia indica*)	Active ingredients: Garcicol, xanthochymol, isoxanthochymol,, hydroxycitric acid (hca), garcim-1 *Food products:* Juice, *solkadhi, chutney*, pickle, curry. Health benefits: Cooling, anthelmintic, cardiotonic, astringent, emollient, useful in piles, dysentery, heart complaints, bilious affections	Kurian (2012), Semwal, Semwal, Vermaak, and Viljoen (2015)
Lemon juice (*Citrus limon* L.)	Active ingredient: 5–6% Citric acid *Food products:* Sharbat, syrups, pickle, juice, lemon tea. Health benefits: Flavour enhancer, pH control, microbial control, chelating agent	Yang, Zhang, Bhandari, and Liu (2018)
Vinegars	Active ingredient: 5–10% Acetic acid *Food products:* Pickle, sauces. Health benefits: Flavour enhancer, pHcontrol, microbial control	Sengun and Karapinar (2004)

Spices and spice blends (condiments as well as pastes) are powerful antioxidants and include a wide range of chemically distinct compounds, such as those present in capsicum, garlic (*Allium sativum*), ginger (*Zingiber officinale*), spices like clove, (*Syzygium aromaticum*), coriander (*Coriendrum sativum*), cumin (*Cuminum cyminum*), turmeric (*Curcuma longa*), tamarind (*Tamarindus indica*), saffron (*Crocus sativus*), nutmeg (*Myrestica fragrance*), mustard (*Brassica juncia*), mint (*Mentha* spp.), cardamon (*Electtaria cardamomum*), cassia (*Cinamomum cassia*), cinnamon (*Cinnamomum verum*), curry leaves (*Murraya koeniggini*), fennel (*Foeniculum vulgare*), fenugreek (*Trigonella foenum-graceum*), and *kokum* (*Garcinia indica*). Health benefits of antioxidant food ingredients are listed in Table 6.31. These spice(s) are used either directly or for marination of foods or even through tempering. The latter is almost used for all pulses and many other foods such as *chutneys* and rice preparations. Both whole spices as well as powders (coarse as well as fine, depending on the preparation) are used.

Food manufacturers need to assess shelf life of all products under different storage, transport and distribution conditions. Shelf life studies and design of appropriate packaging are key parameters in facililating successful commercialization of these traditional food items. Researchers have been working in this direction. Preservation for long time storage embodies better commercial scope and viability. Extension of shelf life is being achieved by manipulation in packaging, using additives and/or using their combined roles to reduce the process time. Exploration and assesment of innovative food packaging system is a critical factor for maintaining the organoleptic and gustative characteriscs of food items. Studies related to the evaluation of qualitative degeneration of food (hygienic, microbial, sensorial, nutritional, chemical-physical), texture evaluation, and sensorial characterization are important for food safety during manufacturing and best before consumption dates. Commonly used food ingredients such as oil, salt, sugar, and spices do act as preservatives and extend the shelf life of certain food items. A tentative list of such food ingredients having preservative functions is listed in Table 6.32.

The most commonly used sweetener in India is sucrose that is produced commercially from sugar cane. However, products such as jaggery which is prepared from raw sugarcane juice by incomplete bleaching, is also used extensively in preparation of traditional sweets such as *laddoos*, *barfis*, *syrups*, fruit juices, *kheer*, *chikki*, *puran poli*, *sweet puri*, *shankar pala*, *karanji*, *anarsa*, *bundi*, *halva*, *khobra vadi*, *sheer khurma*, and *sonpapdi*, amongst many others. Jaggery has distinct advantages over sugar in terms of micronuitrients such as iron, calcium, phosphorus and manganese apart from having a typical flavour that is relished by Indians (Singh, Solomon, & Kumar, 2013).

The flavours in Indian traditional foods are a combined attribute of added flavours (spices and herbs) as well those generated during various steps of processing and the nature and proportion of the ingredients used therein. The subject is too complex, and perhaps, not studied as well in depth.

Table 6.31 Typical ingredients that function as antioxidants in traditional Indian foods and their health benefits

Antioxidants	Active ingredient, brief research work and health benefits	Reference
Capsicum (*Capsicum* spp.)	Capsaicin, capsaicinol Digestive, thermogenic, carminative, stimulant, cardiotonic, antipyretic, serdorific, rubefacient and sialagogue	Embuscado (2015), Kurian (2012), Patcharatrakul and Gonlachanvit (2016)
Clove (*Syzygium aromaticum*)	Eugenol, arvacrol, thymol, and cinnamaldehyde Refrigerant, ophthalmic, digestive, carminative, stomachic, stimulant, antispasmodic, expectorant, rubefacient, aphrodisiac, appetizer, emollient, antimicrobial (*Listeria monocytogens, Aeromonashydrophila, Escherichia coli O157: H7, Candida albicans, Staphylococcus aureus*, and other oral pathogens)	Chaieb et al. (2007), D'Souza, Chavannavar, Kanchanashri, and Niveditha (2017), Embuscado (2015), Kurian (2012)
Coriander (*Coriendrum sativum*)	Linalool, g-terpinene, a-pinene, p-cymene, camphor and geranyl acetate Carminative, diuretic, tonic, stimulant, stomachic, refrigerant, aphrodisiac, analgesic, anti-infl amatory Antimicrobial (*Escherichia coli, Salmonella typhi, Pseudomonas aeruginosa, Bacillus cereus, Bacillus subtilis*)	D'Souza et al. (2017), Kurian (2012), Mandal and Mandal (2015)
Cumin (*Cuminum cyminum*)	Cuminaldehyde Digestive, carminative, astringent, anti-inflammatory, constipating, diuretic, revulsive, galactogogue, uterine and nerve stimulant, Antimicrobial (*Bacillus pumilus, Pseudomonas aeruginosa, Staphylococcus aureus*)	D'Souza et al. (2017), Kurian (2012)
Turmeric (*Curcuma longa*)	Curcumin, Demethoxy curcumin, Bisdemethoxycurcumin, Tetrahydrocurcumin, Hexahydrocurcumin, Octahydrocurcumin Thermogenic, emollient, anodyne, anti-inflammatory, vulnerary, depurative, antiseptic, appetizer, carminative, expectorant, stomachic, anthelmintic, stimulant, ophthalmic, tonic, used in skin diseases, dyspepsia, asthma, cough, bronchitis, inflammations, ulcers, worms, skin discolouration, antimicrobial (*Bacillus macerans, Bacillus licheniformis, Escherichia coli, Helicobacter pylori*)	D'Souza et al. (2017), Javeri and Chand (2016), Kurian (2012), Pagano, Romano, Izzo, and Borrelli (2018), Patcharatrakul and Gonlachanvit (2016)
Tamarind (*Tamarindus indica*)	Triterpenes, lupanone, lupeol Refrigerant, digestive, carminative, laxative, antiscorbutic, febrifuge, ophthalmic, useful in gastropathy, datura poisoning, alcoholic intoxication, scabies, constipation, antimicrobial (*Burkholderiapseudomallei, Klebsiella pneumonia, Salmonella paratyphi, Bacillus subtilis, Salmonella typhi, and Staphylococcus aureus*)	Bhadoriya et al. (2011), Kurian (2012)

(continued)

Table 6.31 (continued)

Antioxidants	Active ingredient, brief research work and health benefits	Reference
Saffron (*Crocus sativus*)	Crocin, picrocrocin and safranal Stimulant, tonic, stomachic, aphrodisiac, anodyne, antispasmodic, emmenagogue, diuretic, laxative, used in bronchitis, fever, epilepsy, skin diseases, decolouration of skin	Embuscado (2015), Kurian (2012), Rahmani, Khan, and Aldebasi (2017)
Nutmeg (*Myrestica fragrance*)	Argenteane (lignans), Myristphenone, phenolic volatile oils, phenolic acid (caffeic acid), flavanols (catechin) Astringent, sweet, thermogenic, aphrodisiac, anti-inflammatory, anodyne, deodorant, digestive, expectorant, narcotic, anticonvulsant, antiseptic, constipating, antimicrobial (*Bacillus subtilis, Staphylococcus aureus, Shigelladysentriae*)	D'Souza et al. (2017), Embuscado, 2015, Kurian (2012)
Mustard (*Brassica juncia*)	Tocopherols, phospholipids, sterols, sinalbin, sinigrin Thermogenic, anodyne, anti-inflammatory, carminative, digestive, anthelmintic, sudorphic, tonic, emetic, used in vomiting, abdominal colic, dyspepsia, flatulence, skin diseases	Embuscado (2015), Kurian (2012), Popova and Morra (2014)
Mint (*Mentha spp.*)	Carbone, menthol, Limonene, 1,8-cineol, monoterpenoides, glycosides, spicatosideA, β-D-glucopyranosyloximethylethyl Stimulant, stomachic, carminative, antiseptic, digestive, antispasmodic, contraceptive, used in vomiting, skin diseases, amenorrhoea, dental caries	Kunnumakkara, Chung, Koca, and Dey (2009), Kurian (2012)
Cardamon (*Electtaria cardamomum*)	1,8-cineole and a-terpinyl acetate Stimulant, tonic, diuretic, carminative, digestive, expectorant, cardiotonic and used in several pharmaceutical preparations	Embuscado (2015), Kurian (2012), Sardar, Tarade, & Singhal, 2013
Cassia (*Cinamomum cassia*)	Occidentol-I, Occidentol-II, Vitexin, apigenin, Sennosides Astringent, stimulant, carminative, germicidal, for checking nausea and vomiting	Kurian (2012), Yadav et al. (2010)
Cinnamon (*Cinnamomum verum*)	Cinnamaldehyde, cinnamic acid Astringent, diuretic, carminative, aphrodisiac, deodorant, expectorant, febrifuge, stomachic, antimicrobial (*Staphylococcus aureus, Enterococcus faecalis, Bacillus cereus, Escherichia coli, Klebsiella pneumoniae, Rhizomucorsp*)	D'Souza et al. (2017), Kurian (2012)
Curry leaves (*Murraya koeniggini*)	Linalool, elemol, geranyl acetate, myrcene, allo-ocimene, α-terpinene, and (E)-β-ocimene Astringent, anthelmintic, febrifuge, stomachic, appetizing, carminative, constipating, anti-inflammatory, antiseptic, used in skin diseases, and in diarrhoea, ulcers	Kurian (2012), Rajendran, Pallaiyan, and Selvaraj (2014)

Fennel (*Foeniculum vulgare)*	Gentiobiose, quinic acid, quercitin, kaempferol Stimulant, carminative, stomachic, emmenagogue, refrigerant, cardiac stimulant, antiemetic, aphrodisiac, anthelmintic	Ahmad et al. (2018), Kurian (2012), Pacifico et al. (2018)
Garlic (*Allium sativum)*	Allicin, allyl-alcohols Anticholesterol, antifungal, tonic, rubefacient, stimulant, thermogenic, aphrodisiac, used in cough, asthma, cardiopathy, antimicrobial (*Candida albicans, Escherichia coli, Entamoeba histolytica, Giardia lamblia*, and viruses)	D'Souza et al. (2017), Kurian (2012)
Ginger (*Zingiber officinale)*	Zingiberene, geranial, gingerols Digestive, carminative, emollient, appetizer, stomachic, rubefacient, anodyne, expectorant, anthelmintic, stimulant, antimicrobial (*Bacillus subtilis, Staphylococcus aureus, Listeria monocytogens, Klebsiella pneumonia*)	D'Souza et al. (2017), Kurian (2012)
Fenugreek (*Trigonella foenum-graceum)*	Sapogenins, Coumarin, trimethylamine, neurin, trigonelline, choline, gentianine, carpaine and betain Carminative, tonic, aphrodisiac, emollient, antibacterial, used in vomiting, fever, anorexia, colonitis	Garg (2016), Kurian (2012), Wani and Kumar (2018)
Kokum (*Garcinia indica*)	Garcicol, xanthochymol, isoxanthochymol, hydroxycitric acid (HCA), garcim-1 Cooling, anthelmintic, cardiotonic, astringent, emollient, useful in piles, dysentery, heart complaints, bilious affections	Kurian (2012), Pandey, Kunnumakkara, and Aggarwal (2009)

Table 6.32 Typical ingredients that function as antimicrobial agents in traditional Indian foods and their health benefits

Preservatives	Active ingredient and antimicrobial activity	Reference
Clove oil	Active ingredient: Eugenol, carvacrol, thymol, and cinnamaldehyde Susceptible microorganism: *Bacillus subtilis, Pseudomonas fluorescens, Staphylococcus aureus, Vibrio parahaemolyticus, Aspergillus flavus*	Kadoma, Murakami, Atsumi, Ito, and Fujisawa (2009), Liu et al. (2017)
Thyme oil (*Thymus vulgaris* L.)	Active ingredient: Thymol, *p*-cymene, γ-terpinene, linalool, carvacrol, myrcene, terpinen-4-ol Susceptible microorganism: *Bacillus* spp., *Clostridium*spp., *Bifidobacterium* spp., *Lactobacillus* spp., *Pseudomonas*spp., *Enterococcus* spp., *Lactococcuslactis*, *Streptococcus salivarius*, *Agrobacterium vitis*, *Xanthomonaspruni*, *Erwinia carotovora*	Kalemba, Matla, and Smętek (2012)
Peppermint oil (*Mentapiperita* L.)	Active ingredient: Menthol, menthone, methyl acetate, isomenthone, menthofuran, pulegone, carvone, isopulegol Susceptible microorganism: *S. aureus, C. jejuni, L. monocytogenes, Clostridium spp., Bacillus spp., V. cholera, E. coli, S. aureus, P. aeruginosa, Staphylococcus epidermidis*	Kalemba et al. (2012)
Cornmint oil (*Menthacanadensis* L)	Active ingredient: Menthol, menthone, isomenthone, menthyl acetate, limonene, isopulegol, pulegone, carvone, 1,8-cineole Susceptible microorganism: *S. aureus, B. subtilis, K. pneumoniae, Helicobacter pylori , S. enteritidis, E. coli*	Kalemba et al. (2012)
Cinnamon bark oil (*Cinnamomum zeylanicum*)	Active ingredient: *trans*-Cinnamic aldehyde, eugenol, linalool, β-caryophyllene, 1,8-cineole, benzyl benzoate, *trans*-2-methoxycinnamaldehyde, coumarin, safrole Susceptible microorganism: *B. subtilis, S. typhi, S. pyogene, E. coli, P. aeruginosa, S. aureus, S. pneumoniae, Acinetobacter lwoffii, S. pyogenes , Enterobacter aerogenes , B. cereus, Listeria strains*	Kalemba et al. (2012)
Cassia oil (*Cinnamomum cassia*)	Active ingredient: *Trans*-Cinnamaldehyde, *trans*-2-methoxycinnamaldehyde, cinnamyl acetate, coumarin, eugenol Susceptible microorganism: *E. coli, S. typhimurium , L. monocytogenes, S. aureus, Vibrio parahaemolyticus, E. coli , P. vulgaris, P. aeruginosa, Vibrio cholerae, S. typhymurium, S. aureus E. aerogenes, C. jejuni*	Kalemba et al. (2012)
Berry fruit	Active ingredient: Hydroxycinnamic acids, hydroxybenzoic acid	Seeram & Heber, 2007

Mustards seeds	Active ingredient: Isothiocyanates Susceptible microorganism: *Cytophaga*, *E. coli* O157:H7, *Pseudomonas corrugata*, *Salmonella montevideo*, *S. typhimurium*, and *Serratiagrimesii*), yeast (*Endomyces fibuligera*) as well as fungi (*Aspergillus flavus*, *Aspergillus niger*, *Botrytis cinerea*, *Cladosporium cladosporioides*, *Penicillium citrinum*, *Penicillium commune*, *Penicillium expansum*, and *Penicillium roqueforti*, *L. monocytogenes* and *S. aureus*	Barberis, Quiroga, Barcia, Talia, and Debattista (2018), Davidson, Critzer, and Taylor (2013)
Onion (*Allium cepa*) and garlic (*Allium sativum*)	Active ingredient: Thiosulfinates Susceptible microorganism: *B. cereus*, *Clostridium botulinum* type A, *E. coli Lactobacillus plantarum*, *Salmonella, Shigella* spp., *S. aureus*, *A. flavus*, *C. albicans*, and species of *Cryptococcus*, *Rhodotorula*, *Saccharomyces*, *Torulopsis,* and *Trichosporon*	Davidson et al., 2013, Kyung, 2011

8 Summary and Conclusions

In a country where the taste, texture and flavour changes every five kilometers, it is not surprising that populations are resistant to packaged processed foods, although processed foods from take aways and catering establishments are well accepted. The changing socio-economic conditions of the Indian consumer along with the readiness to try foods from other cultures has resulted in newer flavours, a phenomenon which has been used by multinational companies to full advantage. It is not uncommon to see 'Indianized Chinese' or 'Indianized pizzas with toppings specific to India' being relished by population of all age groups and geographical regions, and infact has become the unifying force of the population. The exposure to other cuisines of the world, new ingredients such as exotic vegetables and spices have garnered attention from the connoisseurs. The average age of an Indian today is a young 29, which makes them open to try foods from all cultures and blend it into their own.

While newer and innovative food processing and preservation techniques are emerging, it is necessary to first use the conventional and readily available techniques for traditional foods. There are lot of lacunae and gaps in the scientific knowledge on traditional foods. The information available is scattered and region-specific.

The willingness of the young Indian to accept processed foods as compared to the resistance of the older generations is opening up opportunities for food processing. The political leadership as well as social conditioning coupled with robust regulations to ensure safe and nutritious foods is creating a path for the processing and marketing within and beyond the boundaries of Indian sub-continent. The role of marketing and positioning of many Indian foods in terms of qualities that are sought by the global community such as 'gluten-free', 'low-fat', 'fibre-rich' or 'low GI' need serious attention and focus from the stake holders. Indian food industry need to take up the challenges of manufacturing traditional foods on a large scale which will require scientific inputs in various domains such as machinery design, process development, raw material selection, end product packaging, and preservation for shelf life extension.

References

Abdalla, A. A., El Tinay, A. H., Mohamed, B. E., & Abdalla, A. H. (1998). Proximate composition, starch, phytate and mineral contents of 10 pearl millet genotypes. *Food Chemistry, 63*(2), 243–246.

Abuajah, C. I., Ogbonna, A. C., & Osuji, C. M. (2015). Functional components and medicinal properties of food: A review. *Journal of Food Science and Technology, 52*(5B2), 2522–2529.

Ahmad, B. S., Talou, T., Saad, Z., Hijazi, A., Cerny, M., Kanaan, H., … Merah, O. (2018). Fennel oil and by-products seed characterization and their potential applications. *Industrial Crops and Products, 111*, 92–98.

Ansari, M. M., & Kumar, D. S. (2012). Fortification of food and beverages with phytonutrients. *Food and Public Health, 2*(6), 241–253.

Arya, S., & Gaikwad, S. (2017). Optimization of ingredients and process formulations on functional, nutritional, sensory and textural properties of *thalipeeth*: Indian multigrain pancake. *Journal of Food Processing and Preservation, 41*, 1–8.

Asibuo, J. Y., Akromah, R., Safo-Kantanka, O., Adu-Dapaah, H. K., Ohemeng-Dapaah, S., & Agyeman, A. (2008). Chemical composition of groundnut, *Arachis hypogaea* (L) landraces. *African Journal of Biotechnology, 7*(13), 2203–2208.

Bacchetti, T., Masciangelo, S., Micheletti, A., & Ferretti, G. (2013). Carotenoids, phenolic compounds and antioxidant capacity of five local Italian corn (*Zea mays* L.) kernels. *Journal of Nutrition & Food Sciences, 3*(6), 1–4.

Balaswamy, K., Jyothirmayi, T., & Rao, D. G. (2004). Studies on preparation of curry leaf (*Murraya koenigii* L.) chutney powder. *Journal of Food Service, 14*(3), 175–187.

Banerji, A., Anthanarayan, L., & Lele, S. (2017). Rheological and nutritional studied of amaranth enriched chapatti (Indian flat bread). *Journal of Food Processing and Preservation, 42*(1), 1–8.

Baptista, A., Pinho, O., Pinto, E., Casal, S., Mota, C., & Ferreira, I. M. (2017). Characterization of protein and fat composition of seeds from common beans (*Phaseolus vulgaris* L.), cowpea (*Vigna unguiculata* L. Walp) and bambara groundnuts (*Vigna subterranea* L. Verdc) from Mozambique. *Journal of Food Measurement and Characterization, 11*(2), 442–450.

Barberis, S., Quiroga, H. G., Barcia, C., Talia, J. M., & Debattista, N. (2018). Natural food preservatives against microorganisms. In *Food Safety and Preservation* (pp. 621–658). Academic Press.

Beniwal, P., & Jood, S. (2015). Development, chemical composition and antioxidant activity of dosa prepared using Bengal gram seed coat. *International Journal of Advanced Nutritional and Health Science, 3*(1), 109–115.

Bhadoriya, S. S., Ganeshpurkar, A., Narwaria, J., Rai, G., & Jain, A. P. (2011). *Tamarindus indica*: Extent of explored potential. *Pharmacognosy Reviews, 5*(9), 73.

Chaieb, K., Hajlaoui, H., Zmantar, T., Kahla-Nakbi, A. B., Rouabhia, M., Mahdouani, K., & Bakhrouf, A. (2007). The chemical composition and biological activity of clove essential oil, Eugenia caryophyllata (*Syzigium aromaticum* L. Myrtaceae): A short review. *Phytotherapy Research, 21*(6), 501–506.

Chauhan, S., Sonawane, S. K., & Arya, S. S. (2017). Nutritional evaluation of multigrain *khakra*. *Food Bioscience, 19*, 80–84.

D'Souza, S. P., Chavannavar, S. V., Kanchanashri, B., & Niveditha, S. B. (2017). Pharmaceutical perspectives of spices and condiments as alternative antimicrobial remedy. *Journal of Evidence-based Complementary & Alternative Medicine, 22*(4), 1002–1010.

Davidson, P. M., Critzer, F. J., & Taylor, T. M. (2013). Naturally occurring antimicrobials for minimally processed foods. *Annual Reviews in Food Science and Technology, 4*, 163–190.

Davis, J. P., & Dean, L. L. (2016). Peanut composition, flavor and nutrition. In *Peanuts* (pp. 289–345). AOCS Press, Academic Publishers.

Devi, C. B., Kushwaha, A., & Kumar, A. (2015). Sprouting characteristics and associated changes in nutritional composition of cowpea (*Vigna unguiculata*). *Journal of Food Science and Technology, 52*(10), 6821–6827.

Dhanavath, S., & Prasada Rao, U. J. S. (2017). Nutritional and nutraceutical properties of *Triticum dicoccum* wheat and its health benefits: An overview. *Journal of Food Science, 82*(10), 2243–2250.

Dhumketi, K., Singh, A., & Agrawal, P. (2018). Formulation and quality evaluation of modified upma mix from Foxtail millet and soy for nutritional security. *International Journal of Current Microbiology and Applied Sciences, 7*(2), 888–896.

Dubey, K. G. (2010). *The Indian Cuisine*. Delhi: PHI Learning Pvt. Ltd.

Edney, M. J. (2010). Barley: Characteristics and quality requirements. In *Cereal Grains* (pp. 141–162). Woodhead Publishing Limited.

Embuscado, M. E. (2015). Herbs and spices as antioxidants for food preservation. In *Handbook of Antioxidants for Food Preservation* (pp. 251–283). Woodhead Publishing Limited.
Faris, M. E. A. I. E., Takruri, H. R., & Issa, A. Y. (2013). Role of lentils (*Lens culinaris* L.) in human health and nutrition: A review. *Mediterranean Journal of Nutrition and Metabolism, 6*(1), 3–16.
Ganesan, K., & Xu, B. (2017). Polyphenol-rich dry common beans (*phaseolus vulgaris* l.) and their health benefits. *International Journal of Molecular Sciences, 18*(11), 2331.
Garg, R. C. (2016). Fenugreek: Multiple health benefits. In *Nutraceuticals* (pp. 599–617). Academic Press.
Ghosh, K. G., Krishnappa, K. G., Srivatsa, A. N., Eapen, K. C., & Vijayaraghavan, P. K. (1980). Pilot plant production of thermostabilized ready-to-eat pouch foods. *Research and Industry, 25*, 140–145.
Giami, S. Y., & Bekebain, D. A. (1992). Proximate composition and functional properties of raw and processed full-fat fluted pumpkin (*Telfairia occidentalis*) seed flour. *Journal of the Science of Food and Agriculture, 59*(3), 321–325.
Goufo, P., & Trindade, H. (2014). Rice antioxidants: Phenolic acids, flavonoids, anthocyanins, proanthocyanidins, tocopherols, tocotrienols, γ-oryzanol, and phytic acid. *Food Science & Nutrition, 2*(2), 75–104.
Gupta, R. K., Gupta, K., Sharma, A., Das, M., Ansari, I. A., & Dwivedi, P. D. (2017). Health risks and benefits of chickpea (*Cicer arietinum*) consumption. *Journal of Agricultural and Food Chemistry, 65*(1), 6–22.
Heller, V. G. (1927). Nutritive properties of the mung bean. *Journal of Biological Chemistry, 75*, 435–442.
Javeri, I., & Chand, N. (2016). Curcumin. In *Nutraceuticals* (pp. 435–445). Academic Press.
Kadam, S. S., Salunkhe, D. K., & Maga, J. A. (1985). Nutritional composition, processing, and utilization of horse gram and moth bean. *Critical Reviews in Food Science and Nutrition, 22*(1), 1–26.
Kadoma, Y., Murakami, Y., Atsumi, T., Ito, S., & Fujisawa, S. (2009). Cloves (Eugenol). In *Molecular Targets and Therapeutic Uses of Spices: Modern Uses for Ancient Medicine* (pp. 117–148). World Scientific.
Kalemba, D., Matla, M., & Smętek, A. (2012). Antimicrobial activities of essential oils. In A. Patra (Ed.), *Dietary Phytochemicals and Microbes*. Dordrecht: Springer.
Kamboj, R., & Nanda, V. (2017). Proximate composition, nutritional profile and health benefits of legumes: A review. *Legume Research, 41*(3), 325–332.
Krishnappa, K. G., Srivatsa, A. N., Ghosh, K. G., Eapen, K. C., & Vijayaraghavan, P. K. (1982). Heat processing of foods in flexible pouches development and future trends in India. *Indian Food Packer, 36*, 11–17.
Kudake, D. C., Bhalerao, P. P., Chaudhari, N. S., Muley, A. B., Talib, M. I., & Parate, V. R. (2018). Fortification of wheat flour with ragi flour: Effect on physical, nutritional, antioxidant and sensory profile of noodles. *Current Research in Nutrition and Food Science, 6*(1), 165–173.
Kudake, D. C., Pawar, A. V., Muley, A. B., Parate, V. R., & Talib, M. I. (2017). Enrichment of wheat flour noodles with oat flour: Effect on physical, nutritional, antioxidant and sensory properties. *International Journal of Current Microbiology and Applied Sciences, 6*(12), 204–213.
Kunnumakkara, A. B., Chung, J. G., Koca, C., & Dey, S. (2009). Mint and its constituents. In *Molecular Targets and Therapeutic Uses of Spices: Modern Uses for Ancient medicine* (pp. 373–401). World Scientific.
Kurian, A. (2012). Health benefits of herbs and spices. In *Handbook of Herbs and Spices, Vol. 2* (2nd ed., pp. 72–88). Woodhead Publishing Limited.
Kyung, K. H. (2011). Antimicrobial activity of volatile sulfur compounds in foods. In M. C. Qian, X. Fan, & K. Mahattanatawee (Eds.), *Volatile Sulfur Compounds in Food* (pp. 323–338). Bangkok: ACS Division of Agricultural and Food Chemistry, Inc..
Liu, Q., Meng, X., Li, Y., Zhao, C. N., Tang, G. Y., & Li, H. B. (2017). Antibacterial and antifungal activities of spices. *International Journal of Molecular Sciences, 18*(6), 1283.

Liyanage, R., Kiramage, C., Visvanathan, R., Jayathilake, C., Weththasinghe, P., Bangamuwage, R., … Vidanarachchi, J. (2018). Hypolipidemic and hypoglycemic potential of raw, boiled, and sprouted mung beans (*Vigna radiata* L. Wilczek) in rats. *Journal of Food Biochemistry, 42*(1), 12457.

Longvah, T., Ananth, R., Bhaskarachary, K., & Venkaiah, K. (2017). *Indian Food Composition Tables*. New Delhi: National Institute of Nutrition, ICMR.

Mandal, S., & Mandal, M. (2015). Coriander (*Coriandrum sativum* L.) essential oil: Chemistry and biological activity. *Asian Pacific Journal of Tropical Biomedicine, 5*(6), 421–428.

Montesano, D., Rocchetti, G., Putnik, P., & Lucini, L. (2018). Bioactive profile of pumpkin: An overview on terpenoids and their health-promoting properties. *Current Opinion in Food Science, 22*, 81–87.

Moses, O., Olawuni, I., & Iwouno, J. O. (2012). The proximate composition and functional properties of full-fat flour, and protein isolate of lima bean (*Phaseolus lunatus*). *Open Access Scientific Reports, 1*(7), 1–5.

Navale, A. S., Deshmuh, B. R., Korake, R. L., Narwade, S. G., & Mule, P. R. (2014). Production profile, proximate composition, sensory evaluation and cost configuration of wood apple burfi. *Animal Science Reporter, 8*, 11–120.

Oke, D. G. (2014). Proximate and phytochemical analysis of *Cajanus cajan* (Pigeon pea) leaves. *Chemical Science Transactions, 3*(3), 1172–1178.

Pacifico, S., Galasso, S., Piccolella, S., Kretschmer, N., Pan, S. P., Nocera, P., … Monaco, P. (2018). Winter wild fennel leaves as a source of anti-inflammatory and antioxidant polyphenols. *Arabian Journal of Chemistry, 11*(4), 513–524.

Pagano, E., Romano, B., Izzo, A. A., & Borrelli, F. (2018). The clinical efficacy of curcumin-containing nutraceuticals: An overview of systematic reviews. *Pharmacological Research, 134*, 79–91.

Pal, D., Mishra, P., Sachan, N., & Ghosh, A. K. (2011). Biological activities and medicinal properties of *Cajanus cajan* (L) Millsp. *Journal of Advanced Pharmaceutical Technology & Research, 2*(4), 207.

Pallavi, B. V., Chetana, & Reddy, S. Y. (2014). Processing, physic-chemical, sensory and nutritional evaluation of protein, mineral and vitamin enriched peanut chikki-an Indian traditional sweet. *Journal of Food Science and Technology, 51*(1), 158–162.

Panda, S. H., Panda, S., Sethuraman, S. P., & Ray, R. C. (2009). Anthocyanin-rich sweet potato lacto-pickle: Production, nutritional and proximate composition. *International Journal of Food Science & Technology, 44*(3), 445–455.

Pandey, M. K., Kunnumakkara, A. B., & Aggarwal, B. B. (2009). Kokum (Garcinol). In *Molecular Targets and Therapeutic Uses of Spices: Modern Uses for Ancient Medicine* (pp. 281–309). World Scientific.

Patcharatrakul, T., & Gonlachanvit, S. (2016). Chili peppers, curcumins, and prebiotics in gastrointestinal health and disease. *Current Gastroenterology Reports, 18*(4), 19.

Patil, S., Padghan, P. V., & Patil, R. A. (2017). Physico-chemical properties of gulabjamun prepared from khoa of buffalo milk blended with sweet corn milk. *Journal of pharmacology and phytochemistry, 6*(6), 69–72.

Patil, S. P., & Arya, S. S. (2017). Nutritional, functional, phytochemical and structural characterization of gluten-free flours. *Journal of Food Measurement and Characterization, 11*(3), 1284–1294.

Popova, I. E., & Morra, M. J. (2014). Simultaneous quantification of sinigrin, sinalbin, and anionic glucosinolate hydrolysis products in *Brassica juncea* and *Sinapis alba* seed extracts using ion chromatography. *Journal of Agricultural and Food Chemistry, 62*(44), 10,687–10,693.

Qayyum, M. M. N., Butt, M. S., Anjum, F. M., & Nawaz, H. (2012). Composition analysis of some selected legumes for protein isolates recovery. *The Journal of Animal and Plant Sciences, 22*(4), 1156–1162.

Rachwa-Rosiak, D., Nebesny, E., & Budryn, G. (2015). Chickpeas—composition, nutritional value, health benefits, application to bread and snacks: A review. *Critical Reviews in Food Science and Nutrition, 55*(8), 1137–1145.

Rahmani, A. H., Khan, A. A., & Aldebasi, Y. H. (2017). Saffron (*Crocus sativus*) and its active ingredients: Role in the prevention and treatment of disease. *Pharmacognosy Journal, 9*(6), 873–879.

Rajendran, M. P., Pallaiyan, B. B., & Selvaraj, N. (2014). Chemical composition, antibacterial and antioxidant profile of essential oil from *Murraya koenigii* (L.) leaves. *Avicenna Journal of Phytomedicine, 4*(3), 200–214.

Ravani, A., & Joshi, D. (2013). Mango and its by product utilization–A review. *Trends in Post Harvest Technology, 1*(1), 55–67.

Sardar, B. R., Tarade, K. M., & Singhal, R. S. (2013). Stability of active components of cardamom oleoresin in co-crystallized sugar cube during storage. *Journal of Food Engineering, 117*(4), 530–537.

Saxena, K. B., Kumar, R. V., & Sultana, R. (2010). Quality nutrition through pigeonpea-a review. *Health, 2*(11), 1335–1344.

Seeram, N. P., & Heber, D. (2007). Impact of berry phytochemicals on human health: Effects beyond antioxidation. *ACS Symposium Series, 956*, 326–336.

Semwal, R. B., Semwal, D. K., Vermaak, I., & Viljoen, A. (2015). A comprehensive scientific overview of *Garcinia cambogia*. *Fitoterapia, 102*, 134–148.

Sengun, I. Y., & Karapinar, M. (2004). Effectiveness of lemon juice, vinegar and their mixture in the elimination of *Salmonella typhimurium* on carrots (*Daucus carota* L.). *International Journal of Food Microbiology, 96*(3), 301–305.

Sengupta, S., Samanta, A., & Bhowal, J. (2017). Effect of different types of coagulating agent on physic-chemical and organoleptic properties of non dairy rasgulla (cheese ball). *Journal of Microbiology, Biotechnology and Food Sciences, 6*(4), 1107–1111.

Shalini, R., Abhinaya, G., Saranya, P., & Antony, U. (2017). Growth of selected probiotic bacterial strains with fructans from Nendran banana and garlic. *LWT- Food Science and Technology, 83*, 68–78.

Shao, Y., & Bao, J. (2015). Polyphenols in whole rice grain: Genetic diversity and health benefits. *Food Chemistry, 180*, 86–97.

Sharma, K. D., Karki, S., Thakur, N. S., & Attri, S. (2012). Chemical composition, functional properties and processing of carrot—A review. *Journal of Food Science and Technology, 49*(1), 22–32.

Sharma, S. K., Peter, S., Sharma, A., Kumar, A., & Rai, V. P. (2017). Effect of incorporation of sapota pulp and cocoa powder on the physico-chemical and microbiological characteristics of shrikhand. *Asian Journal of Dairy and Food Research, 36*(1), 34–36.

Siddhu, D., Broadway, A. A., Ali, M. N., & Singh, A. (2017). Physico-chemical and sensory qualities in *kulfi* prepared by fruit pomace and bura (Khandsari). *The Pharma Innovation Journal, 6*(7), 905–907.

Singh, J., Solomon, S., & Kumar, D. (2013). Manufacturing jaggery, a product of sugarcane, as health food. *Agrotechnology*, (S11), 1–3.

Singh, V., Sati, V., & Agarwal, S. (2016). Potential of germinated sorghum (Sorghum bicolor) for utilization as a health food. *International Journal of Science, Environment and Technology, 5*(6), 4203–4212.

Srivatsa, A. N., Ramakrishna, A., Gopinathan, V. K., Nataraju, S., Leela, R. K., Jayaraman, K. S., & Santhanam, K. (1993). Suitability of indigenously fabricated aluminium cans for canning of Indian foods. *Journal of Food Science and Technology, 30*, 429–434.

Suneja, Y., Kaur, S., Gupta, A. K., & Kaur, N. (2011). Levels of nutritional constituents and antinutritional factors in black gram (*Vigna mungo* L. Hepper). *Food Research International, 44*(2), 621–628.

Swami, S. B., Thakor, N. J., Haldankar, P. M., & Kalse, S. B. (2012). Jackfruit and its many functional components as related to human health: A review. *Comprehensive Reviews in Food Science and Food Safety, 11*(6), 565–576.

Torres-León, C., Rojas, R., Contreras-Esquivel, J. C., Serna-Cock, L., Belmares-Cerda, R. E., & Aguilar, C. N. (2016). Mango seed: Functional and nutritional properties. *Trends in Food Science & Technology, 55*, 109–117.

Variya, B. C., Bakrania, A. K., & Patel, S. S. (2016). Emblica officinalis (Amla): A review for its phytochemistry, ethnomedicinal uses and medicinal potentials with respect to molecular mechanisms. *Pharmacological Research, 111*, 180–200.

Vilienė, V., Šašytė, V., Stupelienė, A. R., & Gružauskas, R. (2017). Nutritional value determination of different varieties of oats and barley using near-infrared spectroscopy method for the horses nutrition. In *World Academy of Science, Engineering and Technology: Conference Proceedings: 19 [International Conference on Animal Nutrition]: Dubai, UAE, January 30–31, 2017 (ICAN 2017)* (pp. 2969–2973). World Academy of Science, Engineering and Technology.

Wani, S. A., & Kumar, P. (2018). Fenugreek: A review on its nutraceutical properties and utilization in various food products. *Journal of the Saudi Society of Agricultural Sciences, 17*, 97–106.

Wrigley, C. W., & Bushuk, W. (2010). Rye and triticale: Characteristics and quality requirements. In *Cereal Grains* (pp. 112–140). Woodhead Publishing Limited.

Yadav, J. P., Arya, V., Yadav, S., Panghal, M., Kumar, S., & Dhankhar, S. (2010). Cassia occidentalis L.: A review on its ethnobotany, phytochemical and pharmacological profile. *Fitoterapia, 81*(4), 223–230.

Yang, F., Zhang, M., Bhandari, B., & Liu, Y. (2018). Investigation on lemon juice gel as food material for 3D printing and optimization of printing parameters. *LWT- Food Science and Technology, 87*, 67–76.

Zee, J. A., Boudreau, A., Bourgeois, M., & Breton, R. (1988). Chemical composition and nutritional quality of faba bean (*Vicia faba* L. Minor) based *tofu*. *Journal of Food Science, 53*(6), 1772–1774.

Chapter 7
Mediterranean Foods and Human Health

Mostafa I. Waly

1 Introduction

Chronic diseases have become the largest contributor to worldwide morbidity and mortality. Healthy diet represents an effective primary intervention measure in combating the onset of chronic diseases, and epidemiological studies had documented that Mediterranean foods are associated with low incidence of chronic diseases. Mediterranean diet is a combination of vegetables, fruits, moderate amounts of fish, white meat and some dairy produce. One key aspect of the Mediterranean diet is the inclusion of olive oil, nuts and fish. This chapter aims to clarify the positive effects of Mediterranean foods, mainly olive oil, polyphenols, nuts and fish in relation to human health and therapeutic properties.

2 Olive Oil

Olives (*Olea europea* L.) is the foremost source of fat in the Mediterranean diet and its composition depends on many factors, including geographical origin, weather and irrigation, ripening and processing after harvesting (Alkhatib, Tsang, & Tuomilehto, 2018; Cai et al., 2018). The main constituents (98%) of olive oil are triglycerides, and the main fatty acids in the triglyceride fraction are monounsaturated fatty acids (oleic acid, 55–83%), a saturated fatty acid (palmitic acid, 7.5–20%) and a polyunsaturated fatty acid (linoleic acid, 2.5–21%) (Cai et al., 2018). The remaining fraction (2%) of olive oil contains: (i) lipophilic phenols (tocopherols),

M. I. Waly (✉)
Department of Food Science and Nutrition, College of Agricultural and Marine Sciences, Sultan Qaboos University, Muscat, Oman

M. Al-Khusaibi et al. (eds.), *Traditional Foods*, Food Engineering Series,
https://doi.org/10.1007/978-3-030-24620-4_7

(ii) sterols, the main sterol being beta-sitosterol, (iii) color pigments, mainly chlorophylls and carotenoids (the most important is beta-carotene), (iv) alcohols, (v) waxes, aldehydes, esters, ketones and (vi) phenolic compounds (hydrophilic phenols) (Alkhatib et al., 2018; Cai et al., 2018). Historically, the beneficial health effects of olive oil intake was first attributed to the high concentration of monounsaturated fatty acids (Crespo, Tomé-Carneiro, Dávalos, & Visioli, 2018). Oleic acid showed its ability to reduce the risk of: cardiovascular diseases, insulin resistance, endothelial dysfunction, inflammation, and proliferation in vascular smooth muscle cells (Visioli et al., 2018a). Several studies demonstrated that phenolic compounds in olive oil have positive effects including anti-oxidant and anti-inflammatory activities (Visioli et al., 2018a). There is an abundant flow of research available in the literature that provides mechanistic insights into the healthful properties of olive oil and its components. Back in 1993, pure hydroxytyrosol was isolated from extra virgin olive oil and in vitro experiments confirmed its antioxidant potential properties against the accumulation of oxidized low-density lipoprotein to atherosclerosis and cardiovascular diseases (Yarla, Polito, & Peluso, 2018). Several human studies indicate that olive oil polyphenols can modulate a variety of surrogate markers of cardiovascular disease (Gouvinhas et al., 2017). From a molecular, mechanistic viewpoint, hydroxytyrosol has anti-inflammatory effect and improved the functionality of high density lipoprotein as supported from human clinical trials and basic science investigations (Chin & Pang, 2017).

2.1 Cardioprotective Effect of Olive Oil

Cardiovascular diseases account for 17.7 million deaths every year and are the leading cause of death worldwide. In the early stages of the development of atherosclerotic plaques, cardiovascular risk factors stimulate vascular endothelial cells, initiating an inflammatory process, fundamental in the pathogenesis of cardiovascular disease. Therefore, primary and secondary prevention strategies aim to reduce traditional risk factors (diabetes mellitus, hypertension, dyslipidemia, obesity) and lifestyle-related disorders, such as inadequate diet, smoking, physical inactivity, and abusive use of alcohol. Cardiovascular risk factors, such as hypertension and dyslipidemia cause endothelial dysfunction, contributing to the onset of the inflammatory process in atherosclerosis (Pedret et al., 2018). Obesity increases the contribution of inflammation to the pathogenesis of cardiovascular diseases (George et al., 2018). Several studies have shown the anti-inflammatory effects of virgin olive oils supplementation at different stages of development of atherosclerosis (George et al., 2018; Nocella et al., 2018). In individuals with endothelial dysfunction, the supplementation of olive oil to the usual diet for four months had a positive effect on the reduction of cell adhesion molecules, platelets, monocytes, and lymphocytes involved in the inflammatory process (Casas et al., 2016). A recent study demonstrated that virgin olive oil supplementation was able to reduce plasma concentrations of inflammatory molecules and atherosclerotic plaque stability in subjects at high cardiovascular risk (Sala-Vila et al., 2016).

In vitro and in vivo studies suggest that olives contain phenolic compounds, such as hydroxytyrosol, tyrosol, and their secoiridoid derivatives, which block the signal and expression of the nuclear factor kappa B, inhibit the action of enzymes responsible to produce eicosanoids, and consequently, decrease circulating levels of inflammatory markers (Sarna et al., 2016). Daily consumption of olive oil seems to modulate cytokines and inflammatory markers related to cardiovascular diseases (Sala-Vila et al., 2016; Sarna et al., 2016).

In the early stages of development of atherosclerotic plaques, risk factors stimulate vascular endothelial cells to express chemotactic and adhesion molecules, initiating the inflammatory process, fundamental in the pathogenesis of cardiovascular diseases (Sarna et al., 2016). In the context of a dietary pattern considered to be healthy, the inclusion of potentially cardioprotective foods, including sources of unsaturated fats, such as olive oil, may modulate the concentration of pro-inflammatory cytokines and markers of inflammation (Sarna et al., 2016), and assist in the control of modifiable risk factors, such as diabetes mellitus, hypertension, dyslipidemia, and overweight (Lee et al., 2015). Moreover, the consumption of 50 mL/day of extra-virgin olive oil may reduce the chance of developing cardiovascular diseases by 37%, and the incidence of major cardiovascular events by 30% (Pedret et al., 2015). The concentration of phenolic compounds in extra-virgin olive oil is influenced, among other factors, by the extraction procedure of the oil, it is obtained by mechanical processes, while refined olive oil is subjected to both physical and chemical procedures (McDonald, O'Kane, McConville, Devine, & McVeigh, 2013; Venø, Nielsen, Lundbye-Christensen, Schmidt, & Handberg, 2013). Although refined olive oil presents a similar composition of fatty acids, due to the low phenolic content, it does not bring the same beneficial effects when compared with extra-virgin olive oil (Nielsen et al., 2012). Phenols such as hydroxytyrosol derivatives (oleuropein complex and tyrosol) are mainly responsible for the beneficial effects of olive oil in the prevention and progression of atherosclerosis, by improving endothelial function, antioxidant effect, and the high-density lipoprotein function, reducing the concentration and the atherogenicity of the low-density lipoprotein and inhibiting platelet aggregation (Manning, Jong, Ryalls, & Sutherland, 2012).

To ensure the cardiovascular benefits of olive oil, the European Food Safety Authority recommends the daily intake of 5 mg of hydroxytyrosol or its derivatives, which can be obtained by the daily consumption of at least 20 g of extra virgin olive oil (Omar, 2010). The potential benefits of olive oil and its phenolic compounds in the reduction of cytokines and inflammatory markers and preventing the mechanisms that may lead to atherosclerosis are well documented. Therefore, the inclusion of olive oil in the daily food consumption might aid in the control of these risk factors, and in the reduction of cardiovascular incidence.

2.2 Anti-Inflammatory Effect of Olive Oil

Evidence has shown that regular consumption of foods rich in phenolic compounds may decrease the risk for the development of chronic diseases, mainly due to their ability to modulate low-grade inflammation (Downie et al., 2018). The mechanisms by which these compounds may exert an anti-inflammatory effect, specifically on cardiovascular diseases, involves: antioxidant activity, blocking the signaling and expression of nuclear factor kappa B, decrease of the adhesion of immune cells to the endothelium, and improvement of endothelial dysfunction (Sabino et al., 2018). In human monocytes, different olive oil polyphenols (Tyrosol and hydroxyl-isocroman derivatives) significantly inhibited the production of superoxide anions (O^{2-}) and inhibited the production of COX-2 mRNA (Serra et al., 2018). Tyrosol and hydroxyl-isocroman compounds also inhibited the release of arachidonic acid and synthesis of metabolites induced by exogenous oxygen-reactive species (Voltes et al., 2018). This action was mediated by the suppression of NF-κB activation, leading to a decrease in COX-2 synthesis (Voltes et al., 2018). In healthy subjects, consumption of a meal (150 g of tomatoes) with extra virgin olive oil (607 mg/kg phenolic content, 300 mg/kg of HT derivatives) reduced the inflammatory markers after 2 and 6 h (Luque-Sierra, Alvarez-Amor, Kleemann, Martín, & Varela, 2018). Oleuropein administration inhibited proliferation of vascular smooth muscle cells in vitro (Lee et al., 2018). Male Sprague Dawly rats that received different concentrations of oleuropein (10, 20, and 30 mg/kg) for 7 days before acute myocardial infarction have lower inflammatory markers, when compared to the group with acute myocardial infarction that received only the vehicle (Rondanelli et al., 2018).

Phenolic compounds in olive oil might decrease hyperhomocysteinemia, which has been associated with a high risk of cardiovascular disease by decreasing vascular endothelial adhesiveness (Pang & Chin, 2018). Recent animal studies have also suggested that protective role of phenolic compounds in olive oil against insulin resistance, non-alcoholic fatty liver disease and atherosclerosis (Currò et al., 2018). The effects of hydroxytyrosol were investigated using a high-fat diet in an animal model using rats as they were divided into three groups: control diet, high fat diet, and high-fat diet + hydroxytyrosol (10 mg/kg/day). After 6 weeks, hydroxytyrosol attenuated, significantly, the metabolic impairment induced by the high-fat diet. It had also reduced hepatic inflammation and nitrosative–oxidative stress through decreased lipid peroxidation, and production of oxygen-reactive species (Currò et al., 2018).

Clinical trials conducted in individuals at risk for the development of cardiovascular diseases show positive effects of daily intake of different amounts of olive oil on inflammatory markers, where the daily supplementation of 50 mL/day of olive oil in cardiac patients decreased proinflammatory cytokines and inflammation markers (Robles-Almazan et al., 2018). Furthermore, dietary supplementation with olive oil is associated with changes in dietary patterns, which may improve the inflammatory profile of cardiac patients.

In vitro and animal model studies have suggested mechanisms of the olive oil action in inflammatory activity at the cardiovascular level, including effects on the arachidonic acid cascade and on signaling pathways and receptors, improvement of vascular function, and reduction of adhesion molecules and chemokines (Lin, Zhong, & Santiago, 2017). The protective effects of olive oil and its phenolic on the inflammatory process needs long-term clinical trials/studies, to evaluate and confirm their beneficial effects in the prevention and treatment of inflammation.

2.3 *Chemopreventive Effect of Olive Oil*

Epidemiological studies show that people of the Mediterranean region have a lower incidence of several cancers when compared to other populations. The consumption of olive oil is an important factor in the Mediterranean diet and is generally believed to be beneficial for health. Olive oil consumption was proven to prevent from colorectal cancer, breast cancer and skin cancer. Carcinogenesis is characterized by a change in the transcriptional activity of many genes, and consequently in the biological function of the proteins that are encoded by those genes. There are many studies pointing out an essential role of polyphenolic compounds as derived from vegetables, fruits, or herbs in the regulation of epigenetic modifications, resulting in the antiproliferative protection. Several epidemiological publications reported an inverse association between adherence to the Mediterranean diet and incidence of different types of cancers (Reboredo-Rodríguez et al., 2018). The anti-cancer effect was attributed to the olive oil content and its phenolic compounds like hydroxytyrosol, tyrosol, and their secoiridoid derivatives (oleuropein, ligstroside aglycon, oleocanthal, and oleuropein) (Visioli et al., 2018b). The anticancer properties of olive oil are correlated to the antioxidant activity of phenolic and polyphenolic compounds, which act as a scavenger for free radicals and reactive oxygen species (Farinetti, Zurlo, Manenti, Coppi, & Mattioli, 2017). Oleuropein, tyrosol, hydroxytyrosol, verboscoside, ligustroide, demethyleuropein displayed antimicrobial and antiviral effects (Griffiths et al., 2016). Antioxidant and anti-atherogenic effects of olive oil polyphenols, like oleuropein and hydroxytyrosol, have been vastly confirmed in the literature (Capurso & Vendemiale, 2017).

The effect of Mediterranean diet on colorectal and prostate cancers have been described in several experimental studies suggest that much of the potential beneficial effects of olive oil can be attributed to its minor components, which are abundant in extra-virgin olive oils (Peluso, Yarla, Ambra, Pastore, & Perry, 2019). The chemo-preventive actions are multiple and include inhibition of cell proliferation and tumor progression as well as increased rates of apoptosis (Shamshoum, Vlavcheski, & Tsiani, 2017). Also, natural plant-derived polyphenols prevent the generation of cancer cells from the initiation of neoplastic transformation due to carcinogenic agents and thus contribute to a lower risk of developing cancer (Ahmad Farooqi et al., 2017). Hydroxytyrosol Treatment of human colon adenocarcinoma cells with olive oil polyphenols significantly inhibited cell proliferation (Fabiani, 2016). Despite the relatively low concentrations of hydroxytyrosol in olive oil, it is in high micromolar concentration in the colon after gastric hydrolysis (Kwan et al., 2017a). In colon cancer cells hydroxytyrosol reduces epidermal growth factor receptor level by promoting its degradation. The epidermal growth factor receptor is one of the key receptors triggering colon carcinogenesis as it regulates the proliferation, apoptosis, angiogenesis, and invasion of cancer cells (Urquiza-Salvat et al., 2018). Hydroxytyrosol was found to be an effective cytotoxic agent in breast cancer cell models. It inhibited cell cycle in the G0/G1 phase by decreasing the level of cyclin D1 in osteosarcoma, neuroblastoma, breast cancer (Corominas-Faja et al., 2018). In addition to that, hydroxytyrosol inhibited lipooxygenase, the enzyme responsible for leukotriene synthesis, and thus modulating the inflammatory response (Corominas-Faja et al., 2018).

The olive leaf extract caused cancer cell death, starting with early apoptosis and completing by the following necrosis in different animal skin cancer models (Garcia-Martinez, Ruiz, Gutierrez-Ibanez, Illescas-Montes, & Melguizo-Rodriguez, 2018). Polyphenols from olive leaf extract showed synergistic effects as combined with standard chemotherapeutic agents (Cusimano et al., 2017). Interestingly, olive oil polyphenols were proven to protect biological membranes against oxidative modification and losing structural integrity (Cusimano et al., 2017).

2.4 Pleiotropic Effect of Olive Oil in Relation to Brain Function

Studies of the Mediterranean diet across the world have observed an association with lower risk of cognitive impairment (Casamenti & Stefani, 2017a). A study in mice who were given extra virgin olive oil and compared with a control diet showed that the mice given olive oil had less Alzheimer's disease neuropathology in their brains (Carito et al., 2016). The effect of the Mediterranean diet and olive oil on brain outcomes does not seem to be specific only to cognition and cerebrovascular disease (Lang, Beglinger, Schweinfurth, Walter, & Borgwardt, 2015). Several studies have shown that higher adherence to a Mediterranean diet, higher intake of olive oil, is associated with lower risk of depression (Yannakoulia, Kontogianni, & Scarmeas, 2015). Observational studies show that high consumption of olive oil, fruit, vegetables, whole grain, fish, and low consumption of animal foods has been associated with a decreased risk of mild cognitive impairment, lower cerebrovascular disease, and lower neurodegeneration (Casamenti & Stefani, 2017b; Lang & Borgwardt, 2013). These associations seem to be mediated by lower cerebrovascular disease and lower neurodegeneration (Casamenti & Stefani, 2017b).

The principle that dietary interventions can prevent the development of cognitive functions of the brain has dominated research on the etiology, prevention, and treatment of Alzheimer's disease (Carito et al., 2016). The individual contribution of olive oil in the treatment of Alzheimer's disease is not clear, but it benefits multiple cerebral disorders and considered as a general protective to the brain (Pase et al., 2015). The inverse relationship between the Mediterranean diet and olive oil and cognitive disorders needs more studies to identify the specific mechanisms and neuropathology that are involved.

2.5 Antidiabetic Effect of Olive Oil

Type 2 diabetes mellitus is increasing all over the world and an effective intervention for the primary prevention is the adoption of Mediterranean diet along with health life modifications. The American Diabetes Association has established a positive association between the Mediterranean diet and a reduced mortality and reduced incidence for type 2 diabetes mellitus (Cao et al., 2018). There is substantial evidence that dietary patterns high in olive oil reduce type 2 diabetes risk, potentially due to its monounsaturated fatty acid content and high amounts of polyphenols (Cao et al., 2018). In a study of US women, total olive oil consumption was inversely associated with type 2 diabetes mellitus risk after adjustment for other dietary and lifestyle factors (Villani, Sultana, Doecke, & Mantzioris, 2018). Weight loss and long-term weight maintenance are important two aspects for the management of type 2 diabetes mellitus and it was reported that low carbohydrate diets and supplemented with olive oil was effective in the prevention of type 2 diabetes mellitus (Galbete et al., 2018). The energy-restricted Mediterranean dietary regimen combats type 2 diabetes mellitus by weight reduction (Galbete et al., 2018; Villani et al., 2018).

Hydroxytyrosol are abundant in olives and acts as a scavenger against free radicals and oxidizing agents (Mandøe et al., 2018). Oxidative stress is involved in the pathogenesis of type 2 diabetes mellitus, and accordingly free radicals and reactive oxygen species increase the risk of type 2 diabetes mellitus (Fujiwara et al., 2017). In experimental animals, hydroxytyrosol acts as antioxidant against alloxan-induced type 2 diabetes mellitus (Torres-Peña et al., 2018). Several studies on different cell lines have shown that hydroxytyrosol increases regulate mitochondrial respiratory chain complexes and this attributed to decreased oxidative-mediated pathogenesis and its associated type 2 diabetes mellitus (Guasch-Ferré, Merino, Sun, Fitó, & Salas-Salvadó, 2017). Hydroxytyrosol modulates activity of the mitochondrial nuclear respiratory factor 1 and transcription factor A, which are associated with oxidative stress (Khan, Iqbal, & Rashid, 2017).

Hydroxytyrosol, the main polyphenol of olive oil, has been shown to improve the lipid profile, glycaemia, and insulin sensitivity, and counteract oxidative and inflammatory processes (Schwingshackl et al., 2017). Therefore, olive oil rich diet has protective benefits in relation to type 2 diabetes mellitus and it is recommended to adopt the Mediterranean diet for the management of type 2 diabetes mellitus.

3 Bioactive Polyphenols

Polyphenols compounds in Mediterranean Foods are a diverse and heterogeneous group of secondary plant metabolites commonly classified as phenolic acids, flavonoids, stilbenes, and lignans. Plant polyphenols have a protective effect against different chronic diseases. Bioactive polyphenols are natural compounds of various chemical structures. Their sources are common in the Mediterranean dietary components and include fruits, vegetables, whole grains, and leaves of different plants. Consumption of polyphenols reduce the risk of the development of human chronic diseases (cardiovascular, type 2 diabetes mellitus, neurodegenerative diseases, and cancer). Biological activity of polyphenols is strongly related to their antioxidant properties as they interact reactive oxygen species and neutralize the active metabolites (Montserrat-de la Paz et al., 2016). Polyphenols possess antioxidant, anti-inflammatory, anti-allergic, anti-atherogenic, and anti-mutagenic effects (Montserrat-de la Paz et al., 2016). In addition, polyphenols improve the immunological defense by increasing the production of white blood cells and cytokines (Santangelo et al., 2016). Polyphenols are classified into lipophilic (α, β, and γ-tocopherols and tocotrienols) and hydrophilic phenolic compounds (InterAct Consortium, 2016). The phenolic compounds are responsible for the aroma and flavor of olive oil (Esposito, Maiorino, Bellastella, Panagiotakos, & Giugliano, 2017). The polyphenols concentration in oils depend on plant variety, maturation conditions, and extraction conditions (Rigacci & Stefani, 2016). In olive oil, the most abundant phenolic compounds are hydroxytyrosol and tyrosol derivatives (secoiridoid, oleacein, oleuropein, and ligstroside (Rigacci & Stefani, 2016). Two flavonoids (luteolin and apigenin) are found in olive oil, but lower than that of other phenolic compounds (Milajerdi, Maghsoudi, & Ghiasvand, 2016).

Epidemiological and experimental studies reported their therapeutic effect in preventing human chronic diseases by suppressing oxidative stress, the upregulation of antioxidant response element gene expression, and activation of the antioxidant enzymes (Kim, Keogh, & Clifton, 2016). Specific phenols have demonstrated ability to inhibit the cortisol producing enzyme in experimental model systems, and influence insulin sensitivity, vascular endothelial function, fat and carbohydrate metabolism, and inflammatory mediators (Guasch-Ferré et al., 2015). These mechanisms have been described in mediating the association between obesity and cardiometabolic risk factors, including hyperinsulinemia and insulin resistance (Guasch-Ferré et al., 2015). The beneficial effect of polyphenols within the Mediterranean foods is promising and is characterized by a broad spectrum of biological activities that play a critical metabolic role in the human metabolism (Guasch-Ferré et al., 2015; Kim et al., 2016). Polyphenols inhibit cancer induction by modulating DNA hypomethylation and activation of the suppressor genes (Nasiri, Fayazi, Jahani, Yazdanpanah, & Haghighizadeh, 2015). Other polyphenols interact with estrogen receptors and influence the regulation of estrogen-dependent gene expression, followed by the inhibition of breast cancer cell growth (Casamenti et al., 2015). In human tumor cell line models, phenols inhibited cell proliferation with no apoptosis induction (Casamenti et al., 2015). Polyphenols show anti-inflammatory properties due to inhibition of cellular nitric oxide production and exhibit in vitro antiproliferative activity against colon cancer cells (Silva et al., 2015). Interestingly, polyphenols decreased metastatic stages of lung and breast cancer, and it is concerned as an efficacious therapeutic agent for cancer treatment by inhibiting DNA damage, lipids peroxidation, and suppresses reactive oxygen species-mediated nuclear factor NF-κB (Storniolo, Roselló-Catafau, Pintó, Mitjavila, & Moreno, 2014). Moreover, polyphenols have been reported to induce epigenetic modifications and control the expression pattern of oncogenic proteins (Storniolo et al., 2014). Recent studies have shown that hydroxytyrosol and oleuropein inhibited angiogenesis (Ceriello et al., 2014).

4 Mediterranean Foods and Human Health

Mediterranean foods include plant-based foods (fruit, vegetables and legumes), moderate intake of fish and dairy products, and low intake of red meat. Epidemiological studies have shown an inverse relationship between Mediterranean foods consumption and incidence of human chronic diseases. Other clinical trials have demonstrated effective management for diabetes, and enhanced metabolic state with high intake of Mediterranean foods and the positive effect was attributed to the high content of polyphenols bioactive compounds as they improve impaired fasting glucose (IFG), impaired glucose tolerance (IGT) and metabolic syndrome (Mattioli et al., 2017). In addition, the glycosylated hemoglobin A1c (HbA1c) levels was reduced in diabetic patients after their reliance on medication and Mediterranean foods rich diet for 5 years as compared with those diabetic patients who had medications only (Capurso & Vendemiale, 2017). Recent meta-analyses studies have reported that adherence to Mediterranean foods (fruit, vegetables, and legumes) reduces incidence of obesity suggesting that Mediterranean foods may attenuate the adverse effects of obesity on the risk of human chronic disease (Pérez-Martínez et al., 2017). Foods present within the Mediterranean diet contains polyphenols, terpenoids, flavonoids, alkaloids, sterols, pigments, and unsaturated fatty acids that all collectively play an important role in maintaining wellness, and contribute to prevent cardiovascular diseases, cancer, depression, diabetes, obesity, and atherosclerosis (Fig. 7.1).

Mediterranean diet is considered one of the healthiest diets for human longevity (Zelber-Sagi, Salomone, & Mlynarsky, 2017). Functional foods prevent or delay onset of chronic diseases by controlling caloric intake and increased physical activity (Kwan et al., 2017b). Nutraceuticals developed form Mediterranean foods are prepared by extracting, purifying, and concentrating ingredients, which are biologically active ingredients and are considered functional because of their association with physiological health benefits related to the prevention of several chronic diseases (Benhammou et al., 2016). Several studies demonstrate that regular intake of Mediterranean foods have impact glycemic control, blood pressure regulation,

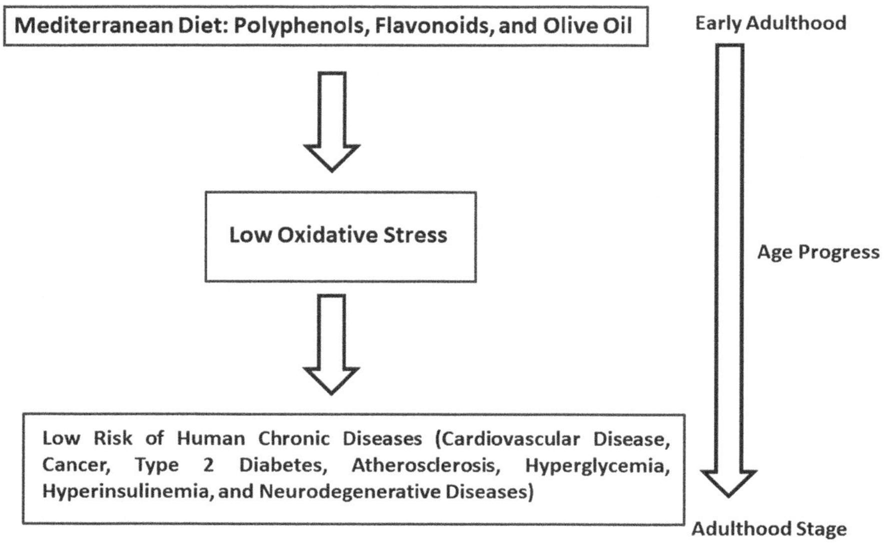

Fig. 7.1 Foods present within the Mediterranean diet contains polyphenols, terpenoids, flavonoids, alkaloids, sterols, pigments, and unsaturated fatty acids that all collectively play an important role in maintaining wellness, and contribute to prevent cardiovascular diseases, cancer, depression, diabetes, obesity, and atherosclerosis

activation of antioxidant enzymes, gut microbiota, and suppress over production of pro-inflammatory cytokines during diabetes (Hadjimbei, Botsaris, Gekas, & Panayiotou, 2016). Additionally, the use of functional foods as a complement therapy for prevention and management of chronic diseases as the reported actions of such foods include an enhanced antioxidant and anti-inflammatory, which considered integral parts of the prevention, management, and treatment of chronic diseases (García-Calzón et al., 2016). Increased consumption of raw vegetables also plays major roles on health, as it has cardioprotective activities, chemopreventive, and neuroprotection (Chatzianagnostou et al., 2015).

5 Nuts and Human Health

Nuts are nutrient-rich, and research on how nuts influence human health has largely focused on the nutritional value of nuts. Although individual nut types differ in nutrient composition, they are all considered to be nutrient-dense. This nutrient profile is likely to contribute to their health properties. Nuts are typically rich sources of cis-unsaturated fatty acids, fibre, and plant protein (Brown et al., 2016). Individual nut types for instance, almonds and hazelnuts are rich sources of vitamin E, peanuts and hazelnuts are high in folate, Brazil nuts are high in magnesium selenium, pine nuts are high in zinc, cashew nuts are rich in iron, and pistachios contain reasonable amounts of vitamin B6. Therefore, the consumption of a variety of nuts can be used to meet the adequate nutrient intake.

The Mediterranean diets are rich in different nuts (pine nuts, almonds, pistachios, walnuts, and cashew nuts); almonds can provide over 100% of the daily value of vitamin E among adults, whereas a serving of cashew nuts can provide nearly three-quarters of the daily value of copper and 25% of the daily value of iron (King, Blumberg, Ingwersen, Jenab, & Tucker, 2008; O'Neil, Keast, Fulgoni, & Nicklas, 2010). A serving of pistachios and pine nuts provide 10-fold the daily value for selenium, a third of the daily value of magnesium and phosphorous, and over 10% of the daily value for potassium and zinc (Jaceldo-Siegl, Joan, Rajaram, & Fraser, 2004).

Several epidemiological studies have shown that nut consumers were less likely to have inadequate intakes of vitamins A, C, E, folate, calcium, iron, magnesium, and zinc compared to non-nut consumers (Tey, Brown, Gray, Chisholm, & Delahunty, 2011). Intervention study carried out in population consuming Mediterranean diets-rich in nuts reported that this population have higher intakes of protein, polyunsaturated fat, and several micronutrients, including magnesium, manganese and copper compared to those who did not consume walnuts (Tey et al., 2011). Collectively, regular consumption of nuts, which are nutrient-dense, is likely to improve diet quality. This will be useful dietary intervention to combat micronutrient insufficiencies among high risk groups.

6 Fish Consumption and Human Health

Fish is recommended as a part of a healthy diet (Mozaffarian, Appel, & Van Horn, 2011) and it is a key component of a cardio-protective diet (Lund, 2013; Mozaffarian et al., 2011). Furthermore, fish is an important source of various nutrients, such as protein, omega-3 fatty acids, vitamin D, iodine, taurine, and selenium, which may contribute to a healthier metabolic profile (Jayedi, Shab-Bidar, Eimeri, & Djafarian, 2018; Raatz, Silverstein, Jahns, & Picklo, 2013).

Meta-analysis showed that there is an inverse association between fish consumption and risk of all-cause and total cardiovascular mortality (Panagiotakos et al., 2007). Mediterranean diet is higher in fish intake, which is a protective measure against higher risk of all-cause and cardiovascular mortality in Western countries, but not in Asians (Strom, Halldorsson, Mortensen, Torp-Pedersen, & Olsen, 2012). These findings may suggest that types of fish/species consumed, and methods of fish preparation, should be considered in addition to other possible confounders, such as regular food items consumed together with the fish in the Mediterranean die (Li et al., 2013). Consumption of diet rich in fatty fish has been suggested to reduce the risk of cerebrovascular disease, coronary calcification, ischemic stroke, and metabolic syndrome (Chowdhury et al., 2012; Heine-Broring et al., 2010; Torris, Molin, & Cvancarova Smastuen, 2014).

7 Conclusion

In conclusion, the most important protection mechanism by which the Mediterranean foods are beneficial for human health is the alleviation of oxidative stress. It is worth mentioning that oleic acid being the major fatty acid present in olive oil is less susceptible to oxidation than other polyunsaturated acids that are in other oils. Polyphenols in olive oil act as a scavenger to oxidant agents, and thus minimizing the risk of formation of free radicals' products and other species of lipid peroxidation. Long term consumption of Mediterranean foods induces apoptosis and arrest of a cellular cycle, leading to a cell proliferation reduction, leading to the reduction of cellular proliferation which prevent malignant cell migration and invasiveness. Mediterranean diet is associated with longevity and wellbeing, and provides health benefits as being a part of balanced diet and improves prognosis of human chronic diseases, Fig. 7.1.

References

Ahmad Farooqi, A., Fayyaz, S., Silva, A. S., Sureda, A., Nabavi, S. F., Mocan, A., ... Bishayee, A. (2017). Oleuropein and cancer chemoprevention: The link is hot. *Molecules, 22*(5), pii: E705. https://doi.org/10.3390/molecules22050705

Alkhatib, A., Tsang, C., & Tuomilehto, J. (2018). Olive oil nutraceuticals in the prevention and management of diabetes: From molecules to lifestyle. *International Journal of Molecular Sciences, 19*(7), pii: E2024. https://doi.org/10.3390/ijms19072024

Benhammou, S., Heras-González, L., Ibáñez-Peinado, D., Barceló, C., Hamdan, M., Rivas, A., ... Monteagudo, C. (2016). Comparison of Mediterranean diet compliance between European and non-European populations in the Mediterranean basin. *Appetite, 1*(107), 521–526. https://doi.org/10.1016/j.appet.2016.08.117. Epub 2016 Sep 3.

Brown, R. C., Tey, S. L., Gray, A. R., Chisholm, A., Smith, C., Fleming, E., & Parnell, W. (2016). Nut consumption is associated with better nutrient intakes: Results from the 2008/09 New Zealand adult nutrition survey. *The British Journal of Nutrition, 115*, 105–112.

Cai, W., Calder, P. C., Cury-Boaventura, M. F., De Waele, E., Jakubowski, J., & Zaloga, G. (2018). Biological and clinical aspects of an olive oil-based lipid emulsion—A review. *Nutrients, 10*(6), pii: E776. https://doi.org/10.3390/nu10060776

Cao, H., Ou, J., Chen, L., Zhang, Y., Szkudelski, T., Delmas, D., ... Xiao, J. (2018). Dietary polyphenols and type 2 diabetes: Human study and clinical trials. *Critical Reviews in Food Science and Nutrition*, 1–19. https://doi.org/10.1080/10408398.2018.1492900. [Epub ahead of print].

Capurso, C., & Vendemiale, G. (2017). The Mediterranean diet reduces the risk and mortality of the prostate cancer: A narrative review. *Frontiers in Nutrition, 24*(4), 38. https://doi.org/10.3389/fnut.2017.00038. eCollection 2017.

Carito, V., Ceccanti, M., Tarani, L., Ferraguti, G., Chaldakov, G. N., & Fiore, M. (2016). Neurotrophins' modulation by olive polyphenols. *Current Medicinal Chemistry, 23*(28), 3189–3197.

Casamenti, F., Grossi, C., Rigacci, S., Pantano, D., Luccarini, I., & Stefani, M. (2015). Oleuropein aglycone: A possible drug against degenerative conditions. in vivo evidence of its effectiveness against Alzheimer's disease. *Journal of Alzheimer's Disease, 45*(3), 679–688. https://doi.org/10.3233/JAD-142850

Casamenti, F., & Stefani, M. (2017a). Olive polyphenols: New promising agents to combat aging-associated neurodegeneration. *Expert Review of Neurotherapeutics, 17*(4), 345–358. https://doi.org/10.1080/14737175.2017.1245617. Epub 2016 Oct 20.

Casamenti, F., & Stefani, M. (2017b). Olive polyphenols: New promising agents to combat aging-associated neurodegeneration. *Expert Review of Neurotherapeutics, 17*(4), 345–358. https://doi.org/10.1080/14737175.2017.1245617. Epub 2016 Oct 20.

Casas, R., Sacanella, E., Urpí-Sardà, M., Corella, D., Castañer, O., Lamuela-Raventos, R. M., ... Estruch, R. (2016). Long-term immunomodulatory effects of a Mediterranean diet in adults at high risk of cardiovascular disease in the PREvención con DIeta MEDiterránea (PREDIMED) randomized controlled trial. *The Journal of Nutrition, 146*(9), 1684–1693. https://doi.org/10.3945/jn.115.229476. Epub 2016 Jul 20.

Ceriello, A., Esposito, K., La Sala, L., Pujadas, G., De Nigris, V., Testa, R., ... Genovese, S. (2014). The protective effect of the Mediterranean diet on endothelial resistance to GLP-1 in type 2 diabetes: A preliminary report. *Cardiovascular Diabetology, 19*(13), 140. https://doi.org/10.1186/s12933-014-0140-9

Chatzianagnostou, K., Del Turco, S., Pingitore, A., Sabatino, L., & Vassalle, C. (2015). The Mediterranean lifestyle as a non-pharmacological and natural antioxidant for healthy aging. *Antioxidants (Basel), 4*(4), 719–736. https://doi.org/10.3390/antiox4040719

Chin, K. Y., & Pang, K. L. (2017). Therapeutic effects of olive and its derivatives on osteoarthritis: From bench to bedside. *Nutrients, 9*(10), pii: E1060. https://doi.org/10.3390/nu9101060

Chowdhury, R., Stevens, S., Gorman, D., Pan, A., Warnakula, S., Chowdhury, S., ... Franco, O. H. (2012). Association between fish consumption, long chain omega 3 fatty acids, and risk of cerebrovascular disease: Systematic review and meta-analysis. *BMJ, 345*, e6698. https://doi.org/10.1136/bmj.e6698

Corominas-Faja, B., Cuyàs, E., Lozano-Sánchez, J., Cufí, S., Verdura, S., Fernández-Arroyo, S., ... Menendez, J. A. (2018). Extra-virgin olive oil contains a metabolo-epigenetic inhibitor of cancer stem cells. *Carcinogenesis, 39*(4), 601–613. https://doi.org/10.1093/carcin/bgy023

Crespo, M. C., Tomé-Carneiro, J., Dávalos, A., & Visioli, F. (2018). Pharma-nutritional properties of olive oil phenols. Transfer of new findings to human nutrition. *Foods, 7*(6), pii: E90. https://doi.org/10.3390/foods7060090

Currò, M., Russo, T., Ferlazzo, N., Caccamo, D., Antonuccio, P., Arena, S., ... Impellizzeri, P. (2018). Anti-inflammatory and tissue regenerative effects of topical treatment with ozonated olive oil/vitamin e acetate in balanitis xerotica obliterans. *Molecules, 23*(3), pii: E645. https://doi.org/10.3390/molecules23030645

Cusimano, A., Balasus, D., Azzolina, A., Augello, G., Emma, M. R., Di Sano, C., ... Cervello, M. (2017). Oleocanthal exerts antitumor effects on human liver and colon cancer cells through ROS generation. *International Journal of Oncology, 51*(2), 533–544. https://doi.org/10.3892/ijo.2017.4049. Epub 2017 Jun 21.

Downie, L. E., Gad, A., Wong, C. Y., Gray, J. H. V., Zeng, W., Jackson, D. C., & Vingrys, A. J. (2018). Modulating contact lens discomfort with anti-inflammatory approaches: A randomized controlled trial. *Investigative Ophthalmology & Visual Science, 59*(8), 3755–3766. https://doi.org/10.1167/iovs.18-24758

Esposito, K., Maiorino, M. I., Bellastella, G., Panagiotakos, D. B., & Giugliano, D. (2017). Mediterranean diet for type 2 diabetes: Cardiometabolic benefits. *Endocrine, 56*(1), 27–32. https://doi.org/10.1007/s12020-016-1018-2. Epub 2016 Jul 9.

Fabiani, R. (2016). Anti-cancer properties of olive oil secoiridoid phenols: A systematic review of in vivo studies. *Food & Function, 7*(10), 4145–4159.

Farinetti, A., Zurlo, V., Manenti, A., Coppi, F., & Mattioli, A. V. (2017). Mediterranean diet and colorectal cancer: A systematic review. *Nutrition, 43*(44), 83–88. https://doi.org/10.1016/j.nut.2017.06.008. Epub 2017 Jul 8.

Fujiwara, Y., Tsukahara, C., Ikeda, N., Sone, Y., Ishikawa, T., Ichi, I., ... Aoki, Y. (2017). Oleuropein improves insulin resistance in skeletal muscle by promoting the translocation of GLUT4. *Journal of Clinical Biochemistry and Nutrition, 61*(3), 196–202. https://doi.org/10.3164/jcbn.16-120. Epub 2017 Oct 3.

Galbete, C., Nicolaou, M., Meeks, K., Klipstein-Grobusch, K., de-Graft Aikins, A., Addo, J., ... Danquah, I. (2018). Dietary patterns and type 2 diabetes among Ghanaian migrants in Europe and their compatriots in Ghana: The RODAM study. *Nutrition & Diabetes, 8*(1), 25. https://doi.org/10.1038/s41387-018-0029-x

García-Calzón, S., Martínez-González, M. A., Razquin, C., Arós, F., Lapetra, J., Martínez, J. A., ... Marti, A. (2016). Mediterranean diet and telomere length in high cardiovascular risk subjects from the PREDIMED-NAVARRA study. *Clinical Nutrition, 35*(6), 1399–1405. https://doi.org/10.1016/j.clnu.2016.03.013. Epub 2016 Apr 1.

Garcia-Martinez, O., Ruiz, C., Gutierrez-Ibanez, A., Illescas-Montes, R., & Melguizo-Rodriguez, L. (2018). Benefits of olive oil phenolic compounds in disease prevention. *Endocrine, Metabolic & Immune Disorders Drug Targets, 18*(4), 333–340. https://doi.org/10.2174/1871530318666180213113211

George, E. S., Marshall, S., Mayr, H. L., Trakman, G. L., Tatucu-Babet, O. A., Lassemillante, A. M., ... Marx, W. (2018). The effect of high-polyphenol extra virgin olive oil on cardiovascular risk factors: A systematic review and meta-analysis. *Critical Reviews in Food Science and Nutrition, 30*, 1–138. https://doi.org/10.1080/10408398.2018.1470491. [Epub ahead of print].

Gouvinhas, I., Machado, N., Sobreira, C., Domínguez-Perles, R., Gomes, S., ... Barros, A. I. R. N. A. (2017). Critical review on the significance of olive phytochemicals in plant physiology and human health. *Molecules, 22*(11), pii: E1986. https://doi.org/10.3390/molecules22111986

Griffiths, K., Aggarwal, B. B., Singh, R. B., Buttar, H. S., Wilson, D., & De Meester, F. (2016). Food antioxidants and their anti-inflammatory properties: A potential role in cardiovascular diseases and cancer prevention. *Diseases., 4*(3), pii: E28. https://doi.org/10.3390/diseases4030028

Guasch-Ferré, M., Hruby, A., Salas-Salvadó, J., Martínez-González, M. A., Sun, Q., ... Hu, F. B. (2015). Olive oil consumption and risk of type 2 diabetes in US women. *The American Journal of Clinical Nutrition, 102*(2), 479–486. https://doi.org/10.3945/ajcn.115.112029. Epub 2015 Jul 8.

Guasch-Ferré, M., Merino, J., Sun, Q., Fitó, M., & Salas-Salvadó, J. (2017). Dietary polyphenols, mediterranean diet, prediabetes, and type 2 diabetes: A narrative review of the evidence. *Oxidative Medicine and Cellular Longevity, 2017*, 6723931. https://doi.org/10.1155/2017/6723931. Epub 2017 Aug 13.

Hadjimbei, E., Botsaris, G., Gekas, V., & Panayiotou, A. G. (2016). Adherence to the Mediterranean diet and lifestyle characteristics of university students in Cyprus: A cross-sectional survey. *Journal of Nutrition and Metabolism, 2016*, 2742841. https://doi.org/10.1155/2016/2742841. Epub 2016 May 16.

Heine-Broring, R. C., Brouwer, I. A., Proenca, R. V., van Rooij, F. J., Hofman, A., Oudkerk, M., ... Geleijnse, J. M. (2010). Intake of fish and marine *n*-3 fatty acids in relation to coronary calcification: The Rotterdam Study. *The American Journal of Clinical Nutrition, 91*, 1317–1323. https://doi.org/10.3945/ajcn.2009.28416

InterAct Consortium. (2016). Investigation of gene-diet interactions in the incretin system and risk of type 2 diabetes: The EPIC-InterAct study. *Diabetologia, 59*(12), 2613–2621. Epub 2016 Sep 13.

Jaceldo-Siegl, K., Joan, S., Rajaram, S., & Fraser, G. E. (2004). Long-term almond supplementation without advice on food replacement induces favourable nutrient modifications to the habitual diets of free-living individuals. *The British Journal of Nutrition, 92*, 533–540.

Jayedi, A., Shab-Bidar, S., Eimeri, S., & Djafarian, K. (2018). Fish consumption and risk of all-cause and cardiovascular mortality: A dose-response meta-analysis of prospective observational studies. *Public Health Nutrition, 21*, 1297–1306. https://doi.org/10.1017/S1368980017003834

Khan, T. M., Iqbal, S., & Rashid, M. A. (2017). Comparison of lipid lowering effect of extra virgin olive oil and atorvastatin in dyslipidaemia in type 2 diabetes mellitus. *Journal of Ayub Medical College, Abbottabad, 29*(1), 83–86.

Kim, Y., Keogh, J. B., & Clifton, P. M. (2016). Polyphenols and glycemic control. *Nutrients, 8*(1), pii: E17. https://doi.org/10.3390/nu8010017

King, J. C., Blumberg, J., Ingwersen, L., Jenab, M., & Tucker, K. L. (2008). Tree nuts and peanuts as components of a healthy diet. *The Journal of Nutrition, 138*, S1736–S1740.

Kwan, H. Y., Chao, X., Su, T., Fu, X., Tse, A. K., Fong, W. F., & Yu, Z. L. (2017a). The anticancer and antiobesity effects of Mediterranean diet. *Critical Reviews in Food Science and Nutrition, 57*(1), 82–94.

Kwan, H. Y., Chao, X., Su, T., Fu, X., Tse, A. K., Fong, W. F., & Yu, Z. L. (2017b). The anticancer and antiobesity effects of Mediterranean diet. *Critical Reviews in Food Science and Nutrition, 57*(1), 82–94.
Lang, U. E., Beglinger, C., Schweinfurth, N., Walter, M., & Borgwardt, S. (2015). Nutritional aspects of depression. *Cellular Physiology and Biochemistry, 37*(3), 1029–1043. https://doi.org/10.1159/000430229. Epub 2015 Sep 25.
Lang, U. E., & Borgwardt, S. (2013). Molecular mechanisms of depression: Perspectives on new treatment strategies. *Cellular Physiology and Biochemistry, 31*(6), 761–777. https://doi.org/10.1159/000350094. Epub 2013 May 31.
Lee, S. M., Chung, S. H., Park, Y., Park, M. K., Son, Y. K., Kim, S. E., & An, W. S. (2015). Effect of omega-3 fatty acid on the fatty acid content of the erythrocyte membrane and proteinuria in patients with diabetic nephropathy. *International Journal of Endocrinology, 2015*, 208121. https://doi.org/10.1155/2015/208121. Epub 2015 May 18.
Lee, Y., Kim, S., Yang, B., Lim, C., Kim, J. H., Kim, H., & Cho, S. (2018). Anti-inflammatory effects of Brassica oleracea Var. capitata L. (cabbage) methanol extract in mice with contact dermatitis. *Pharmacognosy Magazine, 14*(54), 174–179. https://doi.org/10.4103/pm.pm_152_17. Epub 2018 Apr 10.
Li, Y. H., Zhou, C. H., Pei, H. J., Zhou, X. L., Li, L. H., Wu, Y. J., & Hui, R. T. (2013). Fish consumption and incidence of heart failure: A meta-analysis of prospective cohort studies. *Chinese Medical Journal, 126*, 942–948.
Lin, T. K., Zhong, L., & Santiago, J. L. (2017). Anti-inflammatory and skin barrier repair effects of topical application of some plant oils. *International Journal of Molecular Sciences, 19*(1), pii: E70. https://doi.org/10.3390/ijms19010070
Lund, E. K. (2013). Health benefits of seafood; is it just the fatty acids? *Food Chemistry, 140*, 413–420. https://doi.org/10.1016/j.foodchem.2013.01.034
Luque-Sierra, A., Alvarez-Amor, L., Kleemann, R., Martín, F., & Varela, L. M. (2018). Extra-virgin olive oil with natural phenolic content exerts an anti-inflammatory effect in adipose tissue and attenuates the severity of atherosclerotic lesions in Ldlr-/-.leiden mice. *Molecular Nutrition & Food Research, 62*(13), e1800295. https://doi.org/10.1002/mnfr.201800295. Epub 2018 Jun 12.
Mandøe, M. J., Hansen, K. B., Windeløv, J. A., Knop, F. K., Rehfeld, J. F., Rosenkilde, M. M., … Hansen, H. S. (2018). Comparing olive oil and C4-dietary oil, a prodrug for the GPR119 agonist, 2-oleoyl glycerol, less energy intake of the latter is needed to stimulate incretin hormone secretion in overweight subjects with type 2 diabetes. *Nutrition & Diabetes, 8*(1), 2. https://doi.org/10.1038/s41387-017-0011-z
Manning, P. J., Jong, S. A., Ryalls, A. R., & Sutherland, W. H. (2012). Paraoxonase 1 activity in chylomicrons and VLDL: The effect of type 2 diabetes and meals rich in saturated fat and oleic acid. *Lipids, 47*(3), 259–267. https://doi.org/10.1007/s11745-011-3640-3. Epub 2011 Dec 10.
Mattioli, A. V., Palmiero, P., Manfrini, O., Puddu, P. E., Nodari, S., Dei Cas, A., … Ciccone, M. M. (2017). Mediterranean diet impact on cardiovascular diseases: A narrative review. *Journal of Cardiovascular Medicine (Hagerstown, Md.), 18*(12), 925–935. https://doi.org/10.2459/JCM.0000000000000573
McDonald, D. M., O'Kane, F., McConville, M., Devine, A. B., & McVeigh, G. E. (2013). Platelet redox balance in diabetic patients with hypertension improved by n-3 fatty acids. *Diabetes Care, 36*(4), 998–1005. https://doi.org/10.2337/dc12-0304. Epub 2012 Dec 13.
Milajerdi, A., Maghsoudi, Z., & Ghiasvand, R. (2016). Different consumed oils and metabolic parameters in type 2 diabetes patients in diabetes society of Natanz. *Diabetes and Metabolic Syndrome: Clinical Research and Reviews, 10*(2 Suppl 1), S11–S15. https://doi.org/10.1016/j.dsx.2016.01.010. Epub 2016 Jan 14.
Montserrat-de la Paz, S., Bermudez, B., Cardelo, M. P., Lopez, S., Abia, R., & Muriana, F. J. (2016). Olive oil and postprandial hyperlipidemia: Implications for atherosclerosis and metabolic syndrome. *Food & Function, 7*(12), 4734–4744.

Mozaffarian, D., Appel, L. J., & Van Horn, L. (2011). Components of a cardioprotective diet: New insights. *Circulation, 123*, 2870–2891. https://doi.org/10.1161/CIRCULATIONAHA.110.968735

Nasiri, M., Fayazi, S., Jahani, S., Yazdanpanah, L., & Haghighizadeh, M. H. (2015). The effect of topical olive oil on the healing of foot ulcer in patients with type 2 diabetes: A double-blind randomized clinical trial study in Iran. *Journal of Diabetes and Metabolic Disorders, 14*, 38. https://doi.org/10.1186/s40200-015-0167-9. eCollection 2015.

Nielsen, M. S., Gammelmark, A., Madsen, T., Obel, T., Aardestrup, I., & Schmidt, E. B. (2012). The effect of low-dose marine n-3 fatty acids on the biosynthesis of pro-inflammatory 5-lipoxygenase pathway metabolites in overweight subjects: A randomized controlled trial. *Prostaglandins, Leukotrienes, and Essential Fatty Acids, 87*(1), 43–48. https://doi.org/10.1016/j.plefa.2012.05.009. Epub 2012 Jun 27.

Nocella, C., Cammisotto, V., Fianchini, L., D'Amico, A., Novo, M., Castellani, V., … Carnevale, R. (2018). Extra virgin olive oil and cardiovascular diseases: Benefits for human health. *Endocrine, Metabolic & Immune Disorders Drug Targets, 18*(1), 4–13. https://doi.org/10.2174/1871530317666171114121533

O'Neil, C. E., Keast, D. R., Fulgoni, V. L., & Nicklas, T. A. (2010). Tree nut consumption improves nutrient intake and diet quality in us adults: An analysis of national health and nutrition examination survey (nhanes) 1999–2004. *Asia Pacific Journal of Clinical Nutrition, 19*, 142–150.

Omar, S. H. (2010). Cardioprotective and neuroprotective roles of oleuropein in olive. *Saudi Pharmaceutical Journal, 18*(3), 111–121. https://doi.org/10.1016/j.jsps.2010.05.005. Epub 2010 May 31.

Panagiotakos, D. B., Zeimbekis, A., Boutziouka, V., Economou, M., Kourlaba, G., Toutouzas, P., & Polychronopoulos, E. (2007). Long-term fish intake is associated with better lipid profile, arterial blood pressure, and blood glucose levels in elderly people from Mediterranean islands (MEDIS epidemiological study). *Medical Science Monitor, 13*, CR307–CR312.

Pang, K. L., & Chin, K. Y. (2018). The biological activities of oleocanthal from a molecular perspective. *Nutrients, 10*(5), pii: E570. https://doi.org/10.3390/nu10050570

Pase, C. S., Teixeira, A. M., Roversi, K., Dias, V. T., Calabrese, F., Molteni, R., … Burger, M. E. (2015). Olive oil-enriched diet reduces brain oxidative damages and ameliorates neurotrophic factor gene expression in different life stages of rats. *The Journal of Nutritional Biochemistry, 26*(11), 1200–1207. https://doi.org/10.1016/j.jnutbio.2015.05.013. Epub 2015 Jun 20.

Pedret, A., Catalán, Ú., Fernández-Castillejo, S., Farràs, M., Valls, R. M., Rubió, L., … Solà, R. (2015). Impact of virgin olive oil and phenol-enriched virgin olive oils on the HDL proteome in hypercholesterolemic subjects: A double blind, randomized, controlled, cross-over clinical trial (VOHF study). *PLoS One, 10*(6), e0129160. https://doi.org/10.1371/journal.pone.0129160. eCollection 2015.

Pedret, A., Fernández-Castillejo, S., Valls, R. M., Catalán, Ú., Rubió, L., Romeu, M., … Solà, R. (2018). Cardiovascular benefits of phenol-enriched virgin olive oils: New insights from the virgin olive oil and HDL functionality (VOHF) study. *Molecular Nutrition & Food Research, 62*(16), e1800456. https://doi.org/10.1002/mnfr.201800456. Epub 2018 Jul 30.

Peluso, I., Yarla, N. S., Ambra, R., Pastore, G., & Perry, G. (2019). MAPK signalling pathway in cancers: Olive products as cancer preventive and therapeutic agents. *Seminars in Cancer Biology, 56*, 185–195. https://doi.org/10.1016/j.semcancer.2017.09.002. [Epub ahead of print].

Pérez-Martínez, P., Mikhailidis, D. P., Athyros, V. G., Bullo, M., Couture, P., Covas, M. I., … López-Miranda, J. (2017). Lifestyle recommendations for the prevention and management of metabolic syndrome: An international panel recommendation. *Nutrition Reviews, 75*(5), 307–326. https://doi.org/10.1093/nutrit/nux014

Raatz, S. K., Silverstein, J. T., Jahns, L., & Picklo, M. J. (2013). Issues of fish consumption for cardiovascular disease risk reduction. *Nutrients, 5*, 1081–1097. https://doi.org/10.3390/nu5041081

Reboredo-Rodríguez, P., Varela-López, A., Forbes-Hernández, T. Y., Gasparrini, M., Afrin, S., Cianciosi, D., … Giampieri, F. (2018). Phenolic compounds isolated from olive oil as nutraceutical tools for the prevention and management of cancer and cardiovascular diseases.

International Journal of Molecular Sciences, 19(8), pii: E2305. https://doi.org/10.3390/ijms19082305

Rigacci, S., & Stefani, M. (2016). Nutraceutical properties of olive oil polyphenols. an itinerary from cultured cells through animal models to humans. *International Journal of Molecular Sciences, 17*(6), pii: E843. https://doi.org/10.3390/ijms17060843

Robles-Almazan, M., Pulido-Moran, M., Moreno-Fernandez, J., Ramirez-Tortosa, C., Rodriguez-Garcia, C., Quiles, J. L., & Ramirez-Tortosa, M. (2018). Hydroxytyrosol. Bioavailability, toxicity, and clinical applications. *Food Research International, 105*, 654–667. https://doi.org/10.1016/j.foodres.2017.11.053. Epub 2017 Nov 21.

Rondanelli, M., Faliva, M. A., Miccono, A., Naso, M., Nichetti, M., Riva, A., ... Perna, S. (2018). Food pyramid for subjects with chronic pain: Foods and dietary constituents as anti-inflammatory and antioxidant agents. *Nutrition Research Reviews, 31*(1), 131–151. https://doi.org/10.1017/S0954422417000270. Epub 2018 Apr 22.

Sabino, M., Cappelli, K., Capomaccio, S., Pascucci, L., Biasato, I., Verini-Supplizi, A., ... Trabalza-Marinucci, M. (2018). Dietary supplementation with olive mill wastewaters induces m0odifications on chicken jejunum epithelial cell transcriptome and modulates jejunum morphology. *BMC Genomics, 19*(1), 576. https://doi.org/10.1186/s12864-018-4962-9

Sala-Vila, A., Guasch-Ferré, M., Hu, F. B., Sánchez-Tainta, A., Bulló, M., ... PREDIMED Investigators. (2016). Dietary α-linolenic acid, marine ω-3 fatty acids, and mortality in a population with high fish consumption: Findings from the PREvención con DIeta MEDiterránea (PREDIMED) study. *Journal of the American Heart Association, 5*(1), pii: e002543. https://doi.org/10.1161/JAHA.115.002543. Erratum in: J Am Heart Assoc. 2016 Feb;5(2). pii: e002077. doi: 10.1161/JAHA.116.002077.

Santangelo, C., Filesi, C., Varì, R., Scazzocchio, B., Filardi, T., Fogliano, V., ... Masella, R. (2016). Consumption of extra-virgin olive oil rich in phenolic compounds improves metabolic control in patients with type 2 diabetes mellitus: A possible involvement of reduced levels of circulating visfatin. *Journal of Endocrinological Investigation, 39*(11), 1295–1301. Epub 2016 Jun 25.

Sarna, L. K., Sid, V., Wang, P., Siow, Y. L., House, J. D., & O K. (2016). Tyrosol attenuates high fat diet-induced hepatic oxidative stress: Potential involvement of cystathionine β-synthase and cystathionine γ-lyase. *Lipids, 51*(5), 583–590. https://doi.org/10.1007/s11745-015-4084-y. Epub 2015 Oct 30.

Schwingshackl, L., Lampousi, A. M., Portillo, M. P., Romaguera, D., Hoffmann, G., & Boeing, H. (2017). Olive oil in the prevention and management of type 2 diabetes mellitus: A systematic review and meta-analysis of cohort studies and intervention trials. *Nutrition & Diabetes, 7*(4), e262. https://doi.org/10.1038/nutd.2017.12

Serra, G., Incani, A., Serreli, G., Porru, L., Melis, M. P., Tuberoso, C. I. G., ... Deiana, M. (2018). Olive oil polyphenols reduce oxysterols -induced redox imbalance and pro-inflammatory response in intestinal cells. *Redox Biology, 17*, 348–354. https://doi.org/10.1016/j.redox.2018.05.006. Epub 2018 May 16.

Shamshoum, H., Vlavcheski, F., & Tsiani, E. (2017). Anticancer effects of oleuropein. *BioFactors, 43*(4), 517–528. https://doi.org/10.1002/biof.1366. Epub 2017 Jun 14.

Silva, S., Bronze, M. R., Figueira, M. E., Siwy, J., Mischak, H., Combet, E., & Mullen, W. (2015). Impact of a 6-wk olive oil supplementation in healthy adults on urinary proteomic biomarkers of coronary artery disease, chronic kidney disease, and diabetes (types 1 and 2): A randomized, parallel, controlled, double-blind study. *The American Journal of Clinical Nutrition, 101*(1), 44–54. https://doi.org/10.3945/ajcn.114.094219

Storniolo, C. E., Roselló-Catafau, J., Pintó, X., Mitjavila, M. T., & Moreno, J. J. (2014). Polyphenol fraction of extra virgin olive oil protects against endothelial dysfunction induced by high glucose and free fatty acids through modulation of nitric oxide and endothelin-1. *Redox Biology, 2*, 971–977. https://doi.org/10.1016/j.redox.2014.07.001. Epub 2014 Jul 21.

Strom, M., Halldorsson, T. I., Mortensen, E. L., Torp-Pedersen, C., & Olsen, S. F. (2012). Fish, *n*-3 fatty acids, and cardiovascular diseases in women of reproductive age: A prospective study in a large national cohort. *Hypertension, 59*, 36–43. https://doi.org/10.1161/HYPERTENSIONAHA.111.179382

Tey, S. L., Brown, R., Gray, A., Chisholm, A., & Delahunty, C. (2011). Nuts improve diet quality compared to other energy-dense snacks while maintaining body weight. *Journal of Nutrition and Metabolism, 2011.*

Torres-Peña, J. D., Garcia-Rios, A., Delgado-Casado, N., Gomez-Luna, P., Alcala-Diaz, J. F., Yubero-Serrano, E. M., … Lopez-Miranda, J. (2018). Mediterranean diet improves endothelial function in patients with diabetes and prediabetes: A report from the CORDIOPREV study. *Atherosclerosis, 269*, 50–56. https://doi.org/10.1016/j.atherosclerosis.2017.12.012. Epub 2017 Dec 8.

Torris, C., Molin, M., & Cvancarova Smastuen, M. (2014). Fish consumption and its possible preventive role on the development and prevalence of metabolic syndrome—A systematic review. *Diabetology and Metabolic Syndrome, 6*, 112. https://doi.org/10.1186/1758-5996-6-112

Urquiza-Salvat, N., Pascual-Geler, M., Lopez-Guarnido, O., Rodrigo, L., Martinez-Burgos, A., Cozar, J. M., … Rivas, A. (2018). Adherence to Mediterranean diet and risk of prostate cancer. *The Aging Male, 15*, 1–7. https://doi.org/10.1080/13685538.2018.1450854. [Epub ahead of print].

Venø, S. K., Nielsen, M. R., Lundbye-Christensen, S., Schmidt, E. B., & Handberg, A. (2013). The effect of low-dose marine n-3 fatty acids on plasma levels of sCD36 in overweight subjects: A randomized, double-blind, placebo-controlled trial. *Marine Drugs, 11*(9), 3324–3334. https://doi.org/10.3390/md11093324

Villani, A., Sultana, J., Doecke, J., & Mantzioris, E. (2018). Differences in the interpretation of a modernized Mediterranean diet prescribed in intervention studies for the management of type 2 diabetes: How closely does this align with a traditional Mediterranean diet? *European Journal of Nutrition*. https://doi.org/10.1007/s00394-018-1757-3. [Epub ahead of print].

Visioli, F., Franco, M., Toledo, E., Luchsinger, J., Willett, W. C., Hu, F. B., & Martinez-Gonzalez, M. A. (2018a). Olive oil and prevention of chronic diseases: Summary of an International conference. *Nutrition, Metabolism, and Cardiovascular Diseases, 28*(7), 649–656. https://doi.org/10.1016/j.numecd.2018.04.004. Epub 2018 Apr 26.

Visioli, F., Franco, M., Toledo, E., Luchsinger, J., Willett, W. C., Hu, F. B., & Martinez-Gonzalez, M. A. (2018b). Olive oil and prevention of chronic diseases: Summary of an international conference. *Nutrition, Metabolism, and Cardiovascular Diseases, 28*(7), 649–656. https://doi.org/10.1016/j.numecd.2018.04.004. Epub 2018 Apr 26.

Voltes, A., Bermúdez, A., Rodríguez-Gutiérrez, G., Reyes, M. L., Olano, C., Fernández-Bolaños, J., & Portilla, F. (2018). Anti-inflammatory local effect of hydroxytyrosol combined with pectin-alginate and olive oil on trinitrobenzene sulfonic acid-induced colitis in wistar rats. *Journal of Investigative Surgery, 15*, 1–7. https://doi.org/10.1080/08941939.2018.1469697. [Epub ahead of print].

Yannakoulia, M., Kontogianni, M., & Scarmeas, N. (2015). Cognitive health and Mediterranean diet: Just diet or lifestyle pattern? *Ageing Research Reviews, 20*, 74–78. https://doi.org/10.1016/j.arr.2014.10.003. Epub 2014 Oct 18.

Yarla, N. S., Polito, A., & Peluso, I. (2018). Effects of olive oil on TNF-α and IL-6 in humans: Implication in obesity and frailty. *Endocrine, Metabolic & Immune Disorders Drug Targets, 18*(1), 63–74. https://doi.org/10.2174/1871530317666171120150329

Zelber-Sagi, S., Salomone, F., & Mlynarsky, L. (2017). The Mediterranean dietary pattern as the diet of choice for non-alcoholic fatty liver disease: Evidence and plausible mechanisms. *Liver International, 37*(7), 936–949. https://doi.org/10.1111/liv.13435. Epub 2017 Jun 1.

Chapter 8
Safety, Nutrition and Functionality of the Traditional Foods

Abul Hossain and Md. Jiaur Rahman

1 Introduction

Traditional foods are made from traditional raw materials using traditional processing techniques and have a long history of supporting good health. Traditional foods are simply prepared from fruits, vegetables, meat, fish, milk, eggs, nuts, legumes, and seeds. Moreover, traditional foods are indigenous in nature, unique from other similar foods from the same groups, and not usually processed or packaged as modern products. According to the European Commission (EU, 2006), "*traditional*" *related to foods: "Traditional means proven usage in the community market for a time period showing transmission between generations; this time period should be the one generally ascribed as one human generation, at least 25 years*". Furthermore, Italian Ministry of Agriculture defines the traditional foods as "*Agrifood products whose methods of processing, storage and ripening are consolidated with time according to uniform and constant local use*" (Ministero Agricoltura, 1999). Therefore, traditional foods are the part of a set of traditions, which are generally free from chemicals, additives, and artificial preservatives compared to modern foods.

Safety of food refers to the safety of raw materials, processing technique, transportation, distribution and handling, and storage condition. According to the World Health Organization (WHO), five key principles of food safety are stopping contaminating food with pathogens (keep clean) and cross-contamination by parting raw and cooked foods, using the proper length of time and appropriate temperature for cooking food and storing them at safe temperature, and using safe water and

A. Hossain (✉)
Department of Biochemistry, Memorial University of Newfoundland, St. John's, NL, Canada

M. J. Rahman (✉)
Department of Food Processing and Preservation, Hajee Mohammad Danesh Science and Technology University, Dinajpur, Bangladesh

M. Al-Khusaibi et al. (eds.), *Traditional Foods*, Food Engineering Series,
https://doi.org/10.1007/978-3-030-24620-4_8

cooked materials. In particular, International Organization for Standardization (ISO) 22000, Hazard Analysis and Critical Control Points (HACCP) principles, and Good Manufacturing Practice (GMP) enhancing and securing food safety. Moreover, in 2003, International Food and Agriculture Organization (FAO) under the United Nations and WHO published the Codex Alimentarius which acts as a general guideline to food safety mainly food hygiene through Codex standards. Generally, traditional foods considered to be completely safe under all circumstances. However, traditional foods may contain natural toxicants, allergens, or anti-nutrients that can produce toxicity under specific circumstances. The safety of traditional foods is an ultimate challenge, mainly potential hazards (physical, chemical, and microbial) arising from the processing, handling, and storage sectors. Therefore, including new technologies such as food irradiation can be a good way to eliminate the biological hazard from the food production, processing, storage, and distribution chain.

Traditional foods have a long history of supporting food security (i.e. availability) with health and wellness due to the presence of valuable nutrients. Due to the presence of functional ingredients such as antioxidants, body-healing chemicals, dietary fibers, and probiotics; traditional foods always considered as functional foods. For example, dairy products (mainly fermented milk products) and fish are the rich sources of probiotics, long chain omega-3, 6, and 9 polyunsaturated fatty acids, conjugated linolenic acid, bioactive peptides, and vitamins as well as minerals. These functional components are responsible for improving the health, medicinal, and nutraceutical values of traditional foods. Traditional foods have attracted much attention because of their high antioxidative, anti-inflammatory, anti-carcinogenic, anti-mutagenic, and immune modulatory properties in cells and demonstrate positive effects against gastrointestinal, obesity, diabetes, blood cholesterol, oxidative stress, cardiovascular, and neurodegenerative diseases (Srinivasan, 2010). However, the functional activities of foods are boosted by different processing methods including malting, sprouting, and fermentation (Hotz & Gibson, 2007). Additionally, incorporation of additional nutraceuticals to the traditional food products boosts the functional properties of foods in terms of health benefits. Hence, there is a realistic possibility to improve the functionality of traditional foods, though the safety is still a big challenge. Therefore, this chapter aims to investigate the safety and functionality of traditional foods. An overview of the beneficial effects of traditional foods on human health promotion is also provided.

2 Categories of Traditional Foods

Traditional foods are the food products that people have eaten for ages and become an important part of traditions and cultures. Traditional foods can be categorized as follows: (1) grain based traditional foods, (2) pulses and legumes based traditional foods, (3) fruits and vegetable-based traditional foods, (4) milk based traditional foods, (5) meat and fish based traditional foods, and (6) spice based traditional foods. The common traditional preservation methods are fermentation, curing/brining (salting), syruping (sugaring), drying, and smoking. The main objective of all the techniques is to control the water activity for inhibiting the microbial growth, providing antibacterial activity, and reducing biochemical as well as chemical reactions.

Fermentation is one of the oldest and most common food preservation techniques. Fermentation is responsible for improving flavor, digestibility, and nutritional as well as pharmacological values. Lactic acid bacteria (LAB) plays an important role during fermentation. Fermented food can be classified into different groups, mainly cereal based, pulse/legume based, cereal/pulse and milk based, milk based, fruit and vegetable based, and fish and meat based as well as alcohol based fermented foods. Fermentation helps to enhance color, flavor, aroma, texture, essential amino acid (mainly methionine, lysine, and tryptophan), essential fatty acid, vitamin, and detoxification of foodstuffs. However, the use of salt (sodium chloride/sodium nitrate) for preserving foods has a long history. Salting is applied to meat (mainly pork, beef, mutton, lamb, and llama) and fish since ancient time. Interestingly, salted hilsha (*Tenualosa ilisha*) fish is still a very popular dish in Bangladesh, whereas salted cod is very common in Portugal, Italy, Nigeria, and Jamaica. On the other hand, addition of sugar helps to lower the water activity, which ultimately controls the microbial growth and inhibits the enzymatic activity. Fruits and vegetables are the major traditional foods, where sugar used as a preservative. Besides, people used different techniques to reduce the moisture content of foods such as sun drying, wind, and a smoky fire since ancient time. Sun drying is the most common method of preserving foods, mainly fruit, vegetable, meat, fish, grain, and herb. For example, codfish has been dried on racks with sun drying in Northern European countries, whereas legumes have been sun-dried during Greek as well as Roman times (Flint-Hamilton, 1999). In addition, smoking is another oldest method for preserving foodstuffs, mainly fish and meat. The reason for smoking is to dry the surface of the foodstuffs for lowering the microbial growth and lipid oxidation caused by phenolic antioxidants. Additionally, it improves the sensory acceptability of foods including color, texture, flavor, and aroma. The most common traditional smoked foods are fish (mainly salmon, cod, mackerel, and trout), meats (mostly dry-cured ham, speck, bacon, German salami, and sausages), cheeses (mainly ricotta and Italian caciocavallo), cereals (mostly wheat and barley), vegetables (mostly potato and leeks), and beverages (mainly smoked beer and lapsang souchong tea) (Prakash et al., 2015).

3 Traditional Foods and Their Functionality

Traditional foods are rich in bioactive compounds due to most of the foodstuffs are made by the addition of spices and herbs. Traditional foods contain a wide range of phenolics (phenolic acids, flavonoids, tannins, and quinones), bioactive peptides, protein, vitamins, minerals, organic acids, and bacteriocins. Fermented traditional foods mainly soy products show a strong antioxidant activity, which inhibits lipid peroxidation and chelates metal ion. For example, fermented soy foods (tofuyo and sufu) show high antioxidant property (Wang, Saito, Tatsumi, & Li, 2003). This is because fermentation improves the total antioxidant activity, radical scavenging activity, and reducing power. Moreover, fermentation helps to produce the low-molecular-weight short chain components, which is responsible for the production of bioactive peptides (Xiao et al., 2015). This is because the antioxidant property is correlated with the type of amino acids and molecular weight distribution in the chain. However, the nutritional value and functionality of traditional foods are highly variable due to their geographical origin, raw material, processing method, and storage condition. Therefore, this chapter intends to summarize the most common traditional foods found in Asian, European, American, and African countries. A summary of nutritional value and potential health benefits of common traditional foods is given in Table 8.1.

3.1 Cereal-Based Traditional Foods

Cereal-based traditional foods are generally made by rice (mainly parboiled rice, flaked rice, hand-pounded rice, puffed rice, and rice wine), whole wheat flour (mostly chapati, puri, bread, naan, biscuit, idli, dosa, noodles, vermicelli, and sweet food), maize (mainly ogi, kenkey, and mawe), finger millet, and sorghum. Cereals are the major source of calorific value, which comes from mainly starch and protein. For example, idli is a fermented product made by rice and black gram, which has a broad range of nutritional value because of fermentation. The major microorganisms involved in the leavening process of idli are *Streptococcus faecalis*, *Lactobacillus mesenteroides*, *Lactobacillus plantarum* and *Lactobacillus lactis* (Mukherjee, Albury, Pederson, Vanveen, & Steinkraus, 1965). These microorganisms are responsible for producing beta-galactosidase enzyme and vitamin B_{12}, which improve the probiotic property (Iyer, Singhal, & Ananthanarayan, 2013). Besides, it has been reported that two Japanese traditional rice (vinegars and komesu) have a potential effect to suppress lipid peroxidation, prevents hypertension, lowers blood pressure, and reduces tumor growth (Murooka & Yamshita, 2008). Moreover, ambali is another fermented millet-based traditional food found in India, which is rich in thiamin, tryptophan, and riboflavin, as well as minerals. The major compounds that are formed during the cereal fermentation are organic acids (mainly butyric, heptanoic, formic, propionic, lactic, caprylic, palmitic, and myristic), alcohols (mostly ethanol, amyl alcohol, and isoamyl alcohol), aldehydes and ketones (mainly formaldehyde and acetaldehyde), and carbonyl compounds (mostly furfural, methional, and glyoxal) (Ray & Montet, 2014). Furthermore, there are few volatile compounds produced during cereal fermentations, which are responsible for providing flavor in fermented products. Due to the presence of acetic acid, butyric acid, and diacetyl, fermented cereal-based products provide aroma

Table 8.1 Functional ingredients of common traditional foods, their sources, and potential benefits

Food category	Traditional foods (ingredients)	Country/origin	Functional components	Potential benefits	References
Fermented drink and beverages	Rice wine and palm wine (coconut and talipot palm), as well as shalgam (fermented carrot drink)	India, China, Korea, Japan, Thailand, Sri Lanka, Nigeria, and Turkey	Amino acids, oligosaccharides, phenolics, and mineral elements	Anti-cancer and anti-bacterial activity, and also reduce the coronary heart disease	He et al. (2013), Kristbergsson and Ötles (2016), Law, Bakar, Hashim, and Hamid (2011)
Fermented cereal foods	Bread, idli, ambali, dhokla, koozhu, nan, dosa, bulgur, etc.	West Africa (mainly Nigeria, Ghana, Benin, and Guinea), India, Bangladesh, Pakistan, Indonesia, and Turkey	Phenolic compounds (catechins, rutin, daidzin, etc.), crude fiber, vitamins, and minerals	Lower the risk of heart disease, type 2 diabetes, obesity, and intestinal cancer, as well as and prevent degenerative diseases	Kristbergsson and Ötles (2016); Oguntoyinbo (2014); Sarkar, Lohith, Dhumal, Panigrahi, and Choudhary (2015); Satish Kumar et al. (2013); Srinivasan (2010)
Fermented fruits and vegetables foods	Chutney and pickles, olives, cucumbers, gundruk, sinki, and sauerkrauts (cabbage), etc.	Greece, Germany Austria, and South Asian countries	Phenolic compounds, sulphur compounds, dietary fibre, beta-carotene, lycopene, vitamins, and minerals	Anticancer, antidiabetic, and cardioprotective properties	Lücke and Zangerl (2014); Panagou et al. (2013); Satish Kumar et al. (2013)
Fermented root foods	Cassava roots (sweet and sour snack (tapai/tape/tempeh)	Malaysia, Indonesia, and African countries	Phenolics and starch	Anti-cancer, arthritis, and improve immune system	Law et al. (2011); Yi et al. (2010)
Fermented spicy foods	Kimchi (Chinese cabbage or radish with spice) and gochujang (spicy red pepper paste)	Korea	Allyl compounds, capsaicin, gingerol, isothiocyanate, chlorophyll, vitamins (mainly B_{12}), and minerals	Anti-cancer, anti-obesity, and anti-diabetic effects	Kwon, Jang, Yang, and Chung (2014); Park et al. (2014)

(continued)

Table 8.1 (continued)

Food category	Traditional foods (ingredients)	Country/origin	Functional components	Potential benefits	References
	Tsukemono (cabbage leaf and salt)	Japan	Organic acids, bacteriocins, bioactive peptides, vitamins, and minerals	Digestive stimulant and reduce chronic diseases	Murooka and Yamshita (2008)
Fermented tea	Persimmon leaf tea, alfalfa leaf, and Aster root	Korea, Japan, and China	Flavonoids, tannins, bioactive peptides, vitamin C, and minerals (Mg, Ca, K, and Mn)	Antimicrobial, anticancer, anti-allergic, and anti-inflammatory activity	Hossain et al. (2017), (2018a), (2018b); Yu et al. (2015)
Fermented milk-based foods	Cheese, butter (ghee), yoghurt (dahi), ayran/lassi (yogurt-based beverage), rabdi (milk with pearl millet), kefir (mildly alcoholic milk with kefir grain), etc.	South East Asia, Middle East, Europe (mainly Turkey, Germany, and Austria), North America, and Africa	Bacteriocins, bioactive peptides, vitamins (riboflavin, B_{12}, A, D, and K), and minerals (Ca, K, P, and Mg)	Improve digestibility, enhance micronutrient bioavailability, and eliminate foodborne pathogens	Kristbergsson and Ötles (2016); Lücke and Zangerl (2014); Ray and Montet (2014); Sarkar et al. (2015); Satish Kumar et al. (2013); Srinivasan (2010)
Fermented meat-based foods	Raw ham (bacon), sausages (lean meat and fat with salt), nham (round red pork with spice), sucuk (fermented dry meat), and kargyong (sausage)	Europe (mainly Greece and Turkey), Middle East, North America, Thailand, and Indi	Protein, fat, vitamins, and minerals	Maintain lean muscle mass and hormone balance, improve immune response modulation, and support red blood cell function	Kristbergsson and Ötles (2016); Panagou et al. (2013); Satish Kumar et al. (2013)
Fermented fish-based Foods	Som-fug (minced fish cake) and nam-pla (fish and salt)	Thailand	Omega-3 fatty acids, protein, vitamins, and minerals	Show positive effects against hypertension	Kristbergsson and Ötles (2016)

Fermented legume-based foods	Soy sauce, doenjang, miso, natto, douche, sufu, tempeh, dawadawa, tarhana, kinema etc.)	Japan, Korea, Thailand, Indonesia, India, and Turkey	Flavonoids, quercetin, protein, vitamin (B and K2), and minerals (Ca, Mg, Mn, Fe, K, Zn, Cu, and Na)	Prevent brain hemorrhages and osteoporosis, and also reduce cholesterol levels	Kristbergsson and Ötles (2016); Li, Yin, and Saito (2004); Liu, Han, and Zhou (2011); Murooka and Yamshita (2008); Satish Kumar et al. (2013)
Meat and meat products	Raw meat ball (red meat, bulgur, and spices), kokorec (lamb Intestines with spices), doner kebab (red meat, poultry, yogurt, and spices), pastirma (dry-cured meat), brühwurst and kochwurst (cooked meat), khouzi (lamb stuffed with chicken, eggs, and rice), kibbeh (lamb with cracked wheat), Kochpökelwaren (cured meat), and roast duck	Qatar, Saudi Arabia, United Arab Emirates, Syria, Turkey, Germany, Austria, and China	Protein, vitamin B, vitamin E, and minerals	Boost immunity, promote muscle growth, and eliminate skin diseases	Cagri-Mehmetoglu (2018); Kristbergsson and Ötles (2016); Lücke and Zangerl (2014); Tajkarimi, Ibrahim, and Fraser (2013); Zhou et al. (2016)
Cereals and grains	Parboiled rice, flaked rice, puffed rice, chapati, tortilla, puri, noodles, vermicelli, Jilebi, laddu, chikki, mansaf, etc.	India, Pakistan, Bangladesh, Sri Lanka, Jordan, Mexico, and Guatemala	Fiber, beta-glucan, vitamin B (thiamin, niacin riboflavin, and folate), and minerals (Fe, Ca, Mg, Na, and Zn).	Reduce the risk of some chronic diseases and minimize the blood cholesterol levels	Kristbergsson and Ötles (2016); Srinivasan (2010); Tajkarimi et al. (2013)

(continued)

Table 8.1 (continued)

Food category	Traditional foods (ingredients)	Country/origin	Functional components	Potential benefits	References
Pulses and legumes	Fried, roasted, or boiled red lentil (masoor), red gram (pigeon pea), chickpea (Bengal gram), black gram (urad), and green gram (mung) as well as papads (pulse flour), wadis (black gram/Bengal gram), and hummus (chickpeas with lemon juice)	India, Pakistan, Bangladesh, Sri Lanka, and Middle East	Protein, fiber, vitamins, and minerals (Fe, Zn, and Mg) as well as isoflavones, saponins, and tannins	Improves serum lipid profile, control diabetes, and reduce the risk of cardiovascular disease	Mudryj, Yu, and Aukema (2014); Tajkarimi et al. (2013); Srinivasan (2010)
Milk and milk products	Sweet products (peda, rasgolla, burf, sandesh, kheer, halwa khoa, rabri, chhana, paneer, lokum, etc.)	India, Pakistan, Bangladesh, Sri Lanka, and Turkey	Bioactive peptides, vitamins (riboflavin, niacin, vitamin D, B, and A), and minerals (Ca and K)	Reduce the risk of osteoporosis, maintain bone formation, control blood pressure and diabetes	Kristbergsson and Ötles (2016); Srinivasan (2010)
Spices and condiments	Spice soup (rasam) and spice powders (turmeric, black pepper, ginger, garlic, fenugreek, cumin, coriander, cardamom, nutmeg, etc.)	India, Pakistan, Bangladesh, Sri Lanka, China, and Mediterranean countries	Alkaloids, glycosides, saponins, phenolic compounds, and organic acids	Antimicrobial, anti-cancer, cardioprotective, anti-mutagenic, and anti-inflammatory properties	Devarajan and Mohanmarugaraja (2017)
Fruits and vegetables	Bale dandu palya (stem of banana plants and spice), banana stem juice, vazhai poo poriyal (flower of banana plant and seasoning), Jackfruit seed curry, mango peel chutney, papaya salad, bitter gourd, and sweet gourd curry, etc.	India and Bangladesh	Phenolics, flavonoids, carotenoids, bioactive peptides, vitamins, and minerals	Improve skin health, intestinal microbial balance, digestibility, chronic diarrhea, and lower the risk of cancer and diabetes	Jamuna, Mahadevamma, Vishwanatha, and Paramahans (2010); Sarkar et al. (2015)

(Kohajdová, 2010). However, some cereals-based foods may contain trypsin inhibitors, but traditional processing methods mainly soaking, cooking, and fermentation help to remove them (Egounlety & Aworh, 2003).

3.2 Pulse and Legume Based Traditional Foods

The most common pulses in South Asia are masoor (red lentil), Bengal gram (chickpea), urad (black gram), pigeon pea (red gram), and mung (green gram). The world's biggest producers of pulses are India, Bangladesh, China, Russia, Canada, USA, Brazil, and Australia. Pulses are a good source of polyphenols, protein, vitamins, and minerals. It consumed as roasted, fried, or boiled with spices, and are used for preparing bread, snacks, purees, soups, chutneys, and sweets. Traditionally, germination and malting are the very common pulse processing methods, which are responsible for increasing of iron absorption. This is because it helps to improve the content of vitamin C or decrease the level of tannin and phytic acid (Tontisirin, Mantel, & Battacharjee, 2002). Moreover, sprouting of legumes enhances the bioaccessibility of iron due to a reduction in tannin content (Hemalatha, Platel, & Srinivasan, 2007). Furthermore, fermentation of legumes helps to reduce the oligosaccharide fractions (mainly stachyose, raffinose, and verbascose), which improves diarrhea, gas, and digestive problems. During fermentation, microorganisms degrade these oligosaccharide fractions to usable mono- and disaccharides (Leroy & De Vuyst, 2004). For example, the fractions of several oligosaccharides are reduced, and the trypsin inhibitor is inactivated during tempe (beans) fermentation (Hesseltine, 1983). It has been reported that the Japanese traditional miso (soybean-barley paste) has the potential to reduce the risk of gastric cancer, who eat the miso soup every day (Hirayama, 1982). Similarly, natto (fermented soybeans) is a common traditional soy food in Japan, which contains saponin, isoflavone, vitamin K_2, and dipicolinic acid. Natto has a potential effect against stomach ulcers, osteoporosis, and heart disease (Murooka & Yamshita, 2008).

3.3 Fruit and Vegetable Based Traditional Foods

Fruit and vegetables are a rich source of flavonoids, carotenoids, vitamins, minerals, and other phytochemicals, which have a positive impact on human health. For example, mulberry fruit juice contains tannins, anthocyanin, organic acids (malic acid, citric acid, oxalic acid, tartaric acid, and fumaric acid), geraniol, eucalyptol, essential oils, vitamins, and minerals (K, Cu, Zn, and Mg), which have the potential effect against coronary diseases (Kristbergsson & Ötles, 2016). Moreover, persimmon and its leaf used as a traditional fruit and tea, respectively in Korea, Japan, and China. This is because persimmon and its leaf contain flavonoids, tannins, vitamins, and minerals, and are used as a traditional medicine (Hossain, Moon, & Kim, 2017,

2018b). Furthermore, jackfruit is an ancient fruit commonly consumed in Asian countries, which contains antioxidants, phenolic compounds, carotenoids, vitamins, and minerals. On the other hand, cucumber is one of the most common vegetables found in all around the world, which contains cucurbitacins, lignans, glucosides, and apigenin (Mukherjee, Nemaa, Maitya, & Sarkar, 2013). Fermentation is one of the most common methods that has been used in cucumber for preserving for a long time. Similarly, kimchi (fermented cabbage) is another fermented vegetable, which is very common in Korea due to its anticancer, anti-constipation, anti-obesity, anti-oxidative, antiaging, and probiotic properties (Park, Jeong, Lee, & Daily, 2014). Kimchi contains vitamins (ascorbic acid, B-complex, and carotene), minerals (Ca, Fe, and K), dietary fiber, and amino acids (lysine, arginine, glutamic acid, aspartic acid, and alanine).

3.4 Milk Based Traditional Foods

Milk-based traditional foods are mainly fermented milk such as curds (yogurt/dahi), cheese, chhena (curds or cheese curds), buttermilk, ghee (clarified butter), and sweet foods, which exhibit tremendous nutritional and therapeutic values. The species and strain of lactic acid bacteria (LAB) play an important role for the flavor and texture of fermented milk. Fermented milk products are probiotics and have a positive effect against gastroenteritis, cholesterol and triacylglycerols, colon cancer, and pathogens. For example, yogurt improves lactose intolerance, enhances appetite and vitality, reduces cholesterol, and stimulates the immune system (Srinivasan, 2010). Numerous volatile compounds have been identified in yogurt, namely carbonyl compounds, alcohols, hydrocarbons, acids, esters, sulfur-containing compounds, aromatic compounds, and heterocyclic compounds, among others (Cheng, 2010). These compounds mainly acetaldehyde, diacetyl and acetoin, and lactic acid are responsible for the final aroma of yogurt. Moreover, butter contains conjugated linoleic acid, butyric acid, vitamins (mainly A, D, E, and K), and minerals. Traditionally butter (ghee) used for the treatment of many diseases in South Asia including antiaging, asthma, cough, digestive problems, dermatitis, hysteria, leprosy, and piles (Sharma, Zhang, & Dwivedi, 2010).

3.5 Meat and Fish Based Traditional Foods

Sausages are the most common traditional meat products in the world. For example, Kochwurst is a type of traditional sausages found in Germany which is made from pre-cooked meat (Lücke & Zangerl, 2014). Sausages are prepared from meat with the addition of salt, nitrites, phosphates, monosodium glutamate (tasting salt), and various spices. Moreover, salted/cured meats are also very common traditional meat products in European countries. However, drying, salting, fermentation, and smoking are the very common methods that have been used to preserve traditional fish products. For example, Shidal, lonailish, and shutki are the common fermented fish products in Northeast India, which are rich in nutritional value. Dried fish contains high protein with low calories and reduces the risk of heart attacks and strokes due to the presence of omega 3 fatty acid. Furthermore, traditional fish and chips is a hot dish of English origin, which is also very popular in Canada, particularly in Newfoundland.

3.6 Spice Based Traditional Foods

Spices have widely been used as flavor enhancers, colorants, preservatives, and potential therapeutic agents since ancient times. Spices contain a wide range of phytochemicals, mainly curcumin, piperine, capsaicin, cinnamaldehyde, thymol, estragole, anethole, 1,8-cineole, carvacrol, eugenol, linalool, limonene, sabinene, myrcene, allicin, gingerol, and safranal (Fig. 8.1) (Devarajan & Mohanmarugaraja, 2017). Most of the traditional foods are treated with spices, particularly South Asian meat and fish curries are marinated with various spices, namely turmeric, black pepper, chili pepper, garlic, ginger, coriander seeds, cumin seeds, fenugreek seeds, nutmeg, and cardamom. Traditionally, spices are used for curing cough, cold, intermittent fever, diseases of the throat, dyspnea, wounds, colic, nosebleed, vomiting, indigestion, allergies, digestive problems, bronchitis, dysentery, piles, tuberculosis, and worms. Active principles of spices are exhibited to confer a broad range of therapeutic effects, such as antioxidant, anti-microbial, anti-cancer, anti-diabetic, anti-obesity, anti-inflammatory, and cardioprotective properties. For example, curcuminoids and essential oils of turmeric demonstrated antimicrobial property against *Bacillus licheniformis, B. macerans, B. subtilis, Candida albicans, C. tropicalis, Trichophyton rubrum*, and *T. mentagrophytes* (Naz et al., 2010). Moreover, curcumin of turmeric has the potential capacity to inhibit different types of cancer, mainly oral epithelial, cervical, gastric, hepatic, prostate, leukemia, ovarian, pancreatic, and colorectal cancer (Villegas, Sánchez-Fidalgo, & Alarcón de la Lastra, 2008). Additionally, cumin has the potential to reduce the high blood sugar, oxidative stress, and also the formation of advanced glycated end products (Jagtap & Patil, 2010).

4 Safety of Traditional Foods

Food is considered as safe when it is properly handled in every step of production, processing, and distribution from retailers to consumers. During the evolution, human has developed various food processing techniques in terms of quality and safety; and how these can be kept for a long time. Fermentation, sun drying, smoking, brining, and syruping are the main processing methods of traditional foods.

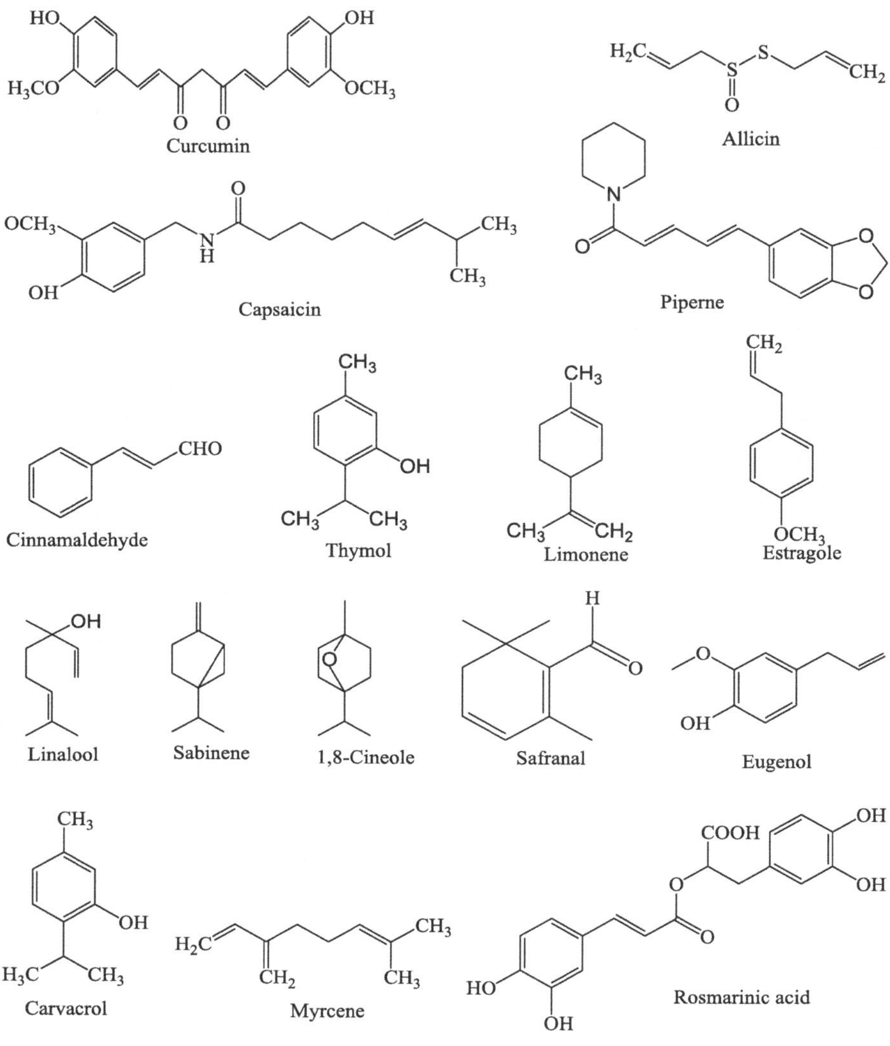

Fig. 8.1 Major phytochemicals of spices

However, physical, chemical, and microbial contaminants are the main challenges of traditional foods. For example, mycotoxins can be produced by molds in crops during storage, diethyl carbamate can be formed during fermentation, acrylamide can be formed during heat treatment, heterocyclic aromatic amines can be formed during cooking, and polycyclic aromatic hydrocarbons can be formed during smoking or cooking (Nedović, Raspor, Lević, Tumbas Šaponjac, & Barbosa-Cánovas, 2016).

Every country has their own food safety regulations, particularly for traditional foods. For example, in the USA, US Department of Agriculture (USDA) regulates the safety of meat and poultry products, whereas US Food and Drug Administration (FDA) deals with other foods. Moreover, Canadian Food Inspection Agency maintains the Canadian food standards system. Furthermore, Australia and New Zealand follow Australia New Zealand Food Standards Code. In China, China Food and Drug Administration, Entry-Exit Inspection and Quarantine Bureaus, Ministry of Agriculture, and National Health and Family Planning identify risk assessment and food safety. Additionally, Ministry of Health & Family Welfare has been established in India for food safety and standards.

4.1 Safety of Cereal and Legume Based Traditional Foods

Generally, cereal-based foods are considered to be safe due to they have comparatively low moisture content and long shelf life. However, molds can grow in cereals when stored under unfavorable weather conditions, particularly humid conditions and mycotoxins can be formed. Microbial counts of cereal grains should be in the range 10^2–10^6 g^{-1} for aerobic bacterial plate counts, 10^2–10^4 g^{-1} for molds and yeasts, 10^2–10^4 g^{-1} for coliform groups, 10^3–10^6 g^{-1} for actinomycetes, and 10–10^3 g^{-1} for *E. coli* (Prakash et al., 2015). Cereals or legume-based products can be contaminated under high-moisture environment, thus aflatoxin can be generated by *Aspergillus flavus*, which is carcinogenic. For example, soybean pastes can be a source of aflatoxin, but the content of aflatoxin can be reduced by fermentation. Moreover, increasing the concentrations of CO_2 or N_2 can reduce the growth and formation of aflatoxin by minimizing the concentration of O_2, which is required for the growth of aerobic organisms (Park, 1984). During storage, ochratoxin A can be produced in all types of cereals under a moist environment, whereas Fusarium toxins can be produced during maturing of cereals in the field. For example, maize and maize-based products can be contaminated during growth, storage, and processing by *Fusarium verticillioides,* which produces fumonisins (mycotoxins) (Nedović et al., 2016). Moreover, acrylamide can be formed in traditional dry baked products, which is a produced from L-asparagine and reducing sugars by Maillard-type reactions. Additionally, several large-scale outbreaks of *Salmonella* also identified in cereals; these outbreaks come from rice straw, which is contaminated by rat feces (Ray & Montet, 2014). Proper handling, processing, storage, and packaging (vacuum, modified atmosphere, and active) can prevent toxin formation, particularly fermentation has the ability to inhibit aflatoxin producing organisms.

4.2 Safety of Fruit- and Vegetable-Based Traditional Foods

Generally, traditional fruit and vegetable are considered as safe, but it may contain small quantities of food toxins, particularly when it consumed as a fresh. Proper handling, processing, packing, and storage can be minimized the common hazards in fruit and vegetables, mainly physical, chemical, and biological hazard. For example, in the early 1960s, poisonous coumarin was isolated from the peel of the mamey fruit (Finnegan, Morris, & Djerassi, 1961). Similarly, Hassal, Reyle, and Feng (1954) isolated two toxic compounds (hypoglycin A and hypoglycin B) from ackee fruit. Selection of cultivar, harvesting time, fruit size, and weather condition are the major factors in terms of toxicity. For example, bitter cassava has a higher amount of cyanogenic glucosides (methyl-linamarin) compared to sweet cassava, particularly when grown in drought environments (Nzwalo & Cliff, 2011). Therefore, it has been recommended that cassava should be carefully processed before consuming. Moreover, mycotoxigenic molds can grow in cassava during improper drying and fermentation, as well as storage in a humid condition (Prakash et al., 2015). Generally, pickles are prepared from various fruit and vegetables with the help of lactic acid fermentation. Physical hazards such as glass, metal, or plastic objects can make pickles unsafe, particularly stones can come with raw materials or brine solutions. Moreover, kimchi can be contaminated during fermentation. Appropriate washing before the fermentation of kimchi could prevent contamination from physical hazards, fermentation of kimchi could kill parasites, while washing cabbage before and after salting could eliminate parasite eggs (Choi, Youn, Yu, Choi, & Lee, 2007). It has been reported that the fermentation of kimchi reduces the level of nitrite, nitrate, biogenic amines, and secondary amines (Lee, Lee, & Lee, 2012). In kimchi, lactic acid bacteria grow first and that is why the possibility of growing pathogenic organisms is very low. This is because organic acids (acetic acid and lactic acid) are formed during fermentation, which are act as an antimicrobial agent due to the lower pH. Similarly, mushrooms can be contaminated by toxic metals, radioactive isotopes, pesticides, process contaminants, microbial contamination, and allergens.

4.3 Safety of Milk-Based Traditional Foods

Pathogenic and enterotoxin-producing bacteria such as *Staphylococcus aureus, Salmonella* spp., *Listeria monocytogenes, Staphylococcus aureus, Listeria monocytogenes, Salmonella* spp., *Escherichia coli*, and *Campylobacter* spp. are the most significance to dairy products. For example, most camel milk in Saudi Arabia is consumed as raw, which might cause foodborne diseases. El-Ziney and Al-Turki (2007) examined the raw camel milk in Qassim region (middle Saudi Arabia); 70% of samples were contaminated with *Staphylococcus aureus*, whereas this percentage was 24 with *Salmonella*. Pasteurization, sterilization, and ultra-high temperature processing can reduce around 99.99% of the microbial population in milk. However, cheese can be contaminated from various sources such as raw milk, floor, brine, packaging material, cheesecloth, cheese vat, curd cutting utensil, and storage room (Temelli, Anar, Sen, & Akyuva, 2006). The major foodborne pathogens in cheese are *Listeria monocytogenes, Salmonella* spp., *Staphylococcus aureus,*

Aeromonas hydrophila, E. coli, Lactococcus spp., *Enterococcus* spp., *lactic acid bacteria, Pseudomonas* spp., *Enterobacteriaceae, Brochothrix thermosphacta*, and yeasts or molds. Additionally, Al-Haddad (2003) reported that the *Salmonella infantis* was able to survive up to 10 days during storage at 4 °C in traditional yogurt as well as bio-yogurt when the yogurt was at pH values of 4.5–4.6. Therefore, there is a need to prevent contamination of fermented milk during processing with pH values.

4.4 Safety of Meat and Fish Based Traditional Foods

Meat and fish are highly perishable due to their high-water content and availability of nutrients for the growth and development of microorganisms. The safety of meat- and fish-based foods is determined by the microbial load in the products and spices. Fermentation is the very common methods and has been used in meat- and fish-based traditional foods since ancient time. Fermentation has the ability to prevent these products from microbial attack due to fermentation produces lactic acid and low pH. For example, Greek traditional sausages have a lower pH value and biogenic amines than the similar products (Panagou, Nychas, & Sofos, 2013). The major foodborne pathogens in fermented sausages are *Salmonella, Staphylococcus aureus*, enterohemorrhagic *E. coli* (EHEC), and *Listeria monocytogenes* (Skandamis & Nychas, 2007). Moreover, Turkish raw meatball (cig kofe) contains various pathogens such as *Salmonella* spp., *Staphylococcus aureus, Listeria monocytogenes, Clostridium perfringens, Escherichia coli* O157:H7, *Yersinia enterocolitica*, and *Bacillus cereus* (Cagri-Mehmetoglu, 2018). It could be due to no cooking, less amount of spices in fresh meat, and low pH. A number of microorganisms have been identified in Turkish doner Kebab including *Staphylococcus aureus, Clostridium perfringens, Bacillus cereus, E. coli* O157:H7, *Listeria monocytogenes*, and *Salmonella* spp. (Elmali, Ulukanli, Tuzcu, Yaman, & Cavli, 2005). Additionally, chemical hazards may generate in cooked meat products. For example, Bordeleau et al. (2016) reported that the traditional snowshoe hare meat is not safe when consumed frequently. Results showed that the concentration of As, Co, Cr, Ni, and Pb was below the detection limit, whereas Cd and Hg detected in few meat samples. However, traditional fish-based foods are associated with the new foodborne diseases that challenge the safety of fish and fish products. It has become a major challenge to deal with the food poisoning microorganisms in fermented, dried, and salted fish. The major foodborne pathogens in fish and fish products are *Salmonella, Vibrio Parahaemolyticus* O3:K6, and *Listeria* momocytogenes (Ray & Montet, 2014). Besides, shrimp and prawn can contain a number of pathogens including *E. coli* and *Salmonella* spp. For example, EU banned Bangladeshi shrimp import due to the presence of pathogenic microorganisms, poor sanitary quality, and foreign matters (European Economic Commission (EEC), 1997). Smoking, salting, drying, cooking, canning, curing, and fermentation can be used to control the microbial load. However, salted fish can be the cause of cancer, particularly nasopharyngeal cancer which is associated to the upper part of the throat behind the nose (Yu, Ho, Lai, & Henderson, 1986). This is because N-nitroso components can be produced during the salting of fish which is highly carcinogenic. Yu, Ho, Ross, and Henderson (1981) reported that nasopharyngeal carcinoma (NPC) are the most consistent with the consumption of Cantonese salted fish.

4.5 Safety of Spice Based Traditional Foods

Traditionally, meat and fish are marinated in cinnamon, cardamom, garlic, onion, and turmeric. Spices-based traditional foods usually considered as safe, but sometimes they can be carcinogenic under specific circumstances. For example, consuming cinnamon oil may cause vomiting, lightheadedness, and central nervous system depression (Barceloux, 2009). Moreover, allicin, a sulfur-containing compound found in garlic can interact with –SH groups of enzymes, which can produce toxic action. For example, dermatitis and occupational allergy can be occurred due to the presence of di (prop-2-enyl) disulfide in garlic. Similarly, bay leaf may cause allergic contact dermatitis due to the presence of one or more sesquiterpene lactone (Peter, 2001). Additionally, life-threatening anaphylaxis may occur due to the consumption of saffron (Wüthrich, Schmid-Grendelmeyer, & Lundberg, 1997). Therefore, it has been recommended that the using of spices in the foods should be in a certain limit, because it may cause of toxicity and allergy under specific conditions. Moreover, spices can be contaminated easily by *salmonella* and *E. coli*, since most of the spices are dried naturally (sun drying). Moreover, sometimes dried spices contained "filth" mainly rodent hairs and insect fragments.

5 Conclusion

Traditional foods with bioactive compounds provide the potential health benefits, particularly they reduce the risk of some chronic diseases mainly cancers and cardiovascular disease. However, the safety of traditional foods is a major concern, predominantly physical, chemical, and biological hazards are the major causes of toxicity. The main hazards for traditional foods are coming from processing technological steps or from the outside of the processing area. During processing, hazards can come in different ways such as inappropriate technological parameters (temperature, time, and pressure) and contamination with lubricants, machine part, cleaning and sanitizing agents. Moreover, hazards also come through raw materials and ingredients, food contact materials, and water. Therefore, GMP (good manufacturing practice) implementation in traditional food sectors could prevent the traditional foods contamination. Besides, using combined preservation methods can be a good way to reduce the microbial risk.

References

Al-Haddad, H. S. K. (2003). Survival of salmonellae in bio-yoghurt. *International Journal of Dairy Technology, 56*, 199–202.
Barceloux, D. G. (2009). Cinnamon (*Cinnamomum* species). *Disease-a-Month, 55*, 327–335.
Bordeleau, S., Asselin, H., Mazerolle, J. M., Imbeau, L. (2016). "Is it still safe to eat traditional food?" Addressing traditional food safety concerns in aboriginal communities. *Science of the Total Environment, 565*, 529–538.
Cagri-Mehmetoglu, A. (2018). Food safety challenges associated with traditional foods of Turkey. *Food Science and Technology, 38*, 1–12.
Cheng, H. (2010). Volatile flavor compounds in yogurt: A review. *Critical Reviews in Food Science and Nutrition, 50*, 938–950.
Choi, I. U., Youn, Y. N., Yu, Y. M., Choi, M. H., & Lee, Y. H. (2007). Comparative evaluation of washing methods of Chinese cabbages for eliminating the parasite eggs in the preparing Kimchi. *Journal of Food Hygiene and Safety, 22*, 192–198.
Devarajan, A., & Mohanmarugaraja, M. K. (2017). A comprehensive review on rasam: a south Indian traditional functional food. *Pharmacognosy Reviews, 11*, 73–82.
Egounlety, M., & Aworh, O. C. (2003). Effect of soaking, dehulling, cooking and fermentation with *Rhizopus oligosporus* on the oligosaccharides, trypsin inhibitor, phytic acid and tannins of soybean (*Glycine max* Merr.), cowpea (*Vigna unguiculata* L. Walp) and groundbean (*Macrotyloma geocarpa* Harms). *Journal of Food Engineering, 56*, 249–254.
Elmali, M., Ulukanli, Z., Tuzcu, M., Yaman, H., & Cavli, P. (2005). Microbiological quality of beef doner kebabs in Turkey. *Archiv fur Lebensmittelhygiene, 56*, 32–34.
El-Ziney, M. G., & Al-Turki, A. I. (2007). Microbiological quality and safety assessment of camel milk (*Camelus dromedaries*) in Saudi Arabia (Qassim region). *Applied Ecology and Environmental Research, 5*, 115–122.
EU. (2006). Council Regulation (EC) No 509/2006 of 20 March 2006 on agricultural products and foodstuffs as traditional specialities guaranteed. *Official Journal of the European Union L, 93*, 1.
European Economic Commission (EEC). (1997). Commission Decision 97/513/EC of 30 July 1997 concerning certain protective measures with regard to certain fishery products originating in Bangladesh. *Official Journal of the European Communities, 214*, 6–8.
Finnegan, R. A., Morris, M. P., & Djerassi, C. (1961). Naturally occurring oxygen heterocyclics X. *Journal of Organic Chemistry, 26*, 1180–1184.
Flint-Hamilton, K. B. (1999). Legumes in ancient Greece and Rome: Food, medicine or poison? *Hesperia, 68*, 371–385.
Hassal, C. H., Reyle, K., & Feng, P. (1954). Hypoglycin A, B: Biologically active polypeptides from Blighia sapida. *Nature, 173*, 356–357.
He, S., Mao, X., Liu, P., Lin, H., Du, Z., Lv, N., … Qiu, C. (2013). Research into the functional components and antioxidant activities of North China rice wine (Ji Mo Lao Jiu). *Food Science & Nutrition, 1*, 307–314.
Hemalatha, S., Platel, K., & Srinivasan, K. (2007). Influence of germination and fermentation on bioaccessibility of zinc and iron from cereals and pulses. *European Journal of Clinical Nutrition, 61*, 342–348.
Hesseltine, C. W. (1983). The future of fermented foods. *Nutrition Reviews, 41*, 293–301.
Hirayama, T. (1982). Relationship of soybean paste soup intake to gastric cancer risk. *Nutrition and Cancer, 3*, 223–233.
Hossain, A., Moon, H. K., & Kim, J. K. (2017). Effect of pre-treatment and extraction conditions on the antioxidant properties of persimmon (*Diospyros kaki*) leaves. *Bioscience, Biotechnology and Biochemistry, 81*, 2079–2085.
Hossain, A., Moon, H. K., & Kim, J.-K. (2018a). Antioxidant properties of Korean major persimmon (*Diospyros kaki*) leaves. *Food Science and Biotechnology, 27*, 177–184.

Hossain, A., Moon, H. K., & Kim, J.-K. (2018b). Effect of drying and harvest time on the physicochemical properties of the most common Korean persimmon leaves. *Korean Journal of Food Preservation, 25*, 428–435.

Hotz, C., & Gibson, R. S. (2007). Traditional food-processing and preparation practices to enhance the bioavailability of micronutrients in plant-based diets. *The Journal of Nutrition, 137*, 1097–1100.

Iyer, B. K., Singhal, R. S., & Ananthanarayan, L. (2013). Characterization and in vitro probiotic evaluation of lactic acid bacteria isolated from idli batter. *Food Science and Technology, 50*, 1114–1121.

Jagtap, A. G., & Patil, P. B. (2010). Antihyperglycemic activity and inhibition of advanced glycation end product formation by *Cuminum cyminum* in streptozotocin induced diabetic rats. *Food and Chemical Toxicology, 48*, 2030–2036.

Jamuna, J. B., Mahadevamma, S., Vishwanatha, S., & Paramahans, V. S. (2010). Effect of banana (Musa Sp. Cultivar Elakki Bale) flower and stem on enzyme activities of intestinal and renal disaccharidases in streptozotocin-induced diabetic rats. *Journal of Food Biochemistry, 34*, 564–580.

Kohajdová, Z. (2010). Fermented cereal products. In A. Pandey, C. R. Soccol, E. Gnansounou, C. Larroche, P. Singh Nigam, & C. G. Dussap (Eds.), *Comprehensive Food Fermentation Biotechnology* (pp. 57–82). New Delhi: Asiatech.

Kristbergsson, K., & Ötles, S. (2016). *Functional properties of traditional foods* (Integrating Food Science and Engineering Knowledge into the Food Chain, ISEKI-Food Series). New York: Springer.

Kwon, Y. D., Jang, D.-J., Yang, H. J., & Chung, K. R. (2014). History of Korean *gochu, gochujang*, and kimchi. *Journal of Ethnic Foods, 1*, 3–7.

Law, S. V., Bakar, A. F., Hashim, D. M., & Hamid, A. (2011). Popular fermented foods and beverages in Southeast Asia. *International Food Research Journal, 18*, 475–484.

Lee, G.-I., Lee, H.-M., & Lee, C.-H. (2012). Food safety issues in industrialization of traditional Korean foods. *Food Control, 24*, 1–5.

Leroy, F. D. R., & De Vuyst, L. (2004). Lactic acid bacteria as functional starter culture for the food fermentation industry. *Trends in Food Science and Technology, 15*, 67–78.

Li, L. T., Yin, L. J., & Saito, M. (2004). Review: Function of traditional foods and food culture in China. *Japan Agricultural Research Quarterly, 38*, 213–220.

Liu, S., Han, Y., & Zhou, Z.-J. (2011). Lactic acid bacteria in traditional fermented Chinese foods. *Food Research International, 44*, 643–651.

Lücke, F. K., & Zangerl, P. (2014). Food safety challenges associated with traditional foods in German-speaking regions. *Food Control, 43*, 217–230.

Ministero Agricoltura (1999). Decreto Legislativo 30 Aprile 1998 n. 173. Decreto Ministero Agricoltura 8 settembre 1999 n. 350.

Mudryj, A. N., Yu, N., & Aukema, H. M. (2014). Nutritional and health benefits of pulses. *Applied Physiology, Nutrition, and Metabolism, 39*, 1197–1204.

Mukherjee, K. P., Nemaa, K. N., Maitya, N., & Sarkar, K. B. (2013). Phytochemical and therapeutic potential of cucumber. *Fitoterapia, 84*, 227–236.

Mukherjee, S. K., Albury, M. N., Pederson, C. S., Vanveen, A. G., & Steinkraus, K. H. (1965). Role of *Leuconostoc mesenteroides* in leavening the batter of idli, a fermented food of India. *Journal of Applied Microbiology, 13*, 227–231.

Murooka, Y., & Yamshita, M. (2008). Traditional healthful fermented products of Japan. *Journal of Industrial Microbiology & Biotechnology, 35*, 791–798.

Naz, S., Jabeen, S., Ilyas, S., Manzoor, F., Aslam, F., & Ali, A. (2010). Antibacterial activity of *Curcuma longa* varieties against different strains of bacteria. *Pakistan Journal of Botany, 42*, 455–462.

Nedović, V., Raspor, P., Lević, J., Tumbas Šaponjac, V., & Barbosa-Cánovas, G. V. (Eds.). (2016). *Emerging and traditional technologies for safe, healthy and quality food* (Food Engineering Series). Cham, Heidelberg, New York, Dordrecht, London: Springer.

Nzwalo, N., & Cliff, J. (2011). Poverty, cassava, and cyanogen intake to toxico-nutritional neurological diseases. *PLoS Neglected Tropical Diseases, 5*, e1051.
Oguntoyinbo, F. A. (2014). Safety challenges associated with traditional foods of West Africa. *Food Reviews International, 30*, 338–358.
Panagou, E. Z., Nychas, G. J. E., & Sofos, J. N. (2013). Types of traditional Greek foods and their safety. *Food Control, 29*, 32–41.
Park, K.-Y. (1984). Aflatoxin: factors affecting aflatoxin production. *Journal of the Korean Society of Food and Nutrition, 13*, 117–126.
Park, K.-Y., Jeong, J.-K., Lee, Y.-E., & Daily, J. W. (2014). Health benefits of kimchi (Korean fermented vegetables) as a probiotic food. *Journal of Medicinal Food, 17*, 6–20.
Peter, K. V. (2001). *Handbook of herbs and spices* (Vol. 1, pp. 573–562). Cambridge: Woodhead. ISBN, 85.
Prakash, V., Martin-Belloso, O., Keener, L., Astley, S., Braun, S., McMahon, H., & Lelieveld, H. (2015). *Regulating safety of traditional and ethnic foods* (1st ed.). Waltham, MA: Academic Press.
Ray, R. C., & Montet, D. (2014). *Microorganisms and fermentation of traditional foods* (pp. 341–369). Boca Raton, London, New York: CRC Press.
Sarkar, P., Lohith, K. D. H., Dhumal, C., Panigrahi, S. S., & Choudhary, R. (2015). Traditional and ayurvedic foods of Indian origin. *Journal of Ethnic Foods, 2*, 97–109.
Satish Kumar, R., Kanmani, P., Yuvaraj, N., Paari, K. A., Pattukumar, V., & Arul, V. (2013). Traditional Indian fermented foods: A rich source of lactic acid bacteria. *International Journal of Food Sciences and Nutrition, 64*, 415–428.
Sharma, H., Zhang, X., & Dwivedi, C. (2010). The effect of ghee (clarified butter) on serum lipid levels and microsomal lipid peroxidation. *Ayu, 31*, 134–140.
Skandamis, P. N., & Nychas, G. J. E. (2007). Pathogens: risks and control. In F. Ed Toldra (Ed.), *Handbook of fermented meat and poultry* (pp. 427–454). Oxford: Blackwell.
Srinivasan, K. (2010). Traditional Indian functional foods. In J. Shi, C. T. Ho, & F. Shahidi (Eds.), *Functional foods of the East* (Nutraceutical Science and Technology. Series 10) (pp. 51–76). Florida: CRC Press.
Tajkarimi, M., Ibrahim, S. A., & Fraser, A. M. (2013). Food safety challenges associated with traditional foods in Arabic speaking countries of the Middle East. *Trends in Food Science and Technology, 29*, 116–123.
Temelli, S., Anar, S., Sen, C., & Akyuva, P. (2006). Determination of microbiological contamination sources during Turkish white cheese production. *Food Control, 17*, 856–861.
Tontisirin, K., Mantel, G., & Battacharjee, L. (2002). Food based strategies to meet the challenges of micronutrient malnutrition in the developing world. *Proceedings of the Nutrition Society, 61*, 243–250.
Villegas, I., Sánchez-Fidalgo, S., & Alarcón de la Lastra, C. (2008). New mechanisms and therapeutic potential of curcumin for colorectal cancer. *Molecular Nutrition & Food Research, 52*, 1040–1061.
Wang, L., Saito, M., Tatsumi, E., & Li, L. (2003). Antioxidative and angiotensin I-converting enzyme inhibitory activities of sufu (fermented tofu) extracts. *Japan Agricultural Research Quarterly, 37*, 129–132.
Wüthrich, B., Schmid-Grendelmeyer, P., & Lundberg, M. (1997). Anaphylaxis to saffron. *Allergy (EAACI), 52*, 476–477.
Xiao, Y., Xing, G., Rui, X., Li, W., Chen, X., Jiang, M., & Dong, M. (2015). Effect of solid-state fermentation with *Cordyceps militaris* SN-18 on physicochemical and functional properties of chickpea (*Cicer arietinum* L.) flour. *LWT-Food Science and Technology, 63*, 1317–1324.
Yi, B., Hu, L., Mei, W., Zhou, K., Wang, H., Luo, Y., ... Dai, H. (2010). Antioxidant phenolic compounds of cassava (*Manihot esculenta*) from Hainan. *Molecules, 16*, 10,157–10,167.
Yu, C. M., Ho, H. C. J., Lai, S.-H., & Henderson, E. B. (1986). Cantonese-style salted fish as a cause of nasopharyngeal carcinoma: report of a case-control study in Hong Kong. *Cancer Research, 46*, 958–961.

Yu, C. M., Ho, H. C. J., Ross, K. R., & Henderson, E. B. (1981). Nasopharyngeal carcinoma in Chinese-Salted fish or inhaled smoke? *Preventive Medicine, 10*, 15–24.
Yu, P., Cheng, S., Xiang, J., Yu, B., Zhang, M., Zhang, C., & Xu, X. (2015). Expectorant, anti-tussive, anti-inflammatory activities and compositional analysis of *Aster tataricus*. *Journal of Ethnopharmacology, 164*, 328–333.
Zhou, Y., Xie, F., Zhou, X., Wang, Y., Tang, W., & Xiao, Y. (2016). Effects of Maillard reaction on flavor and safety of Chinese traditional food: Roast duck. *Journal of the Science of Food and Agriculture, 96*, 1915–1922.

Chapter 9
Laws and Regulations of Traditional Foods: Past, Present and Future

Jamie Gallagher and Aideen McKevitt

1 Introduction

As a symbol of identity, traditional food can constitute a significant element of the culture and heritage of a country regardless of geographical, social, or political differences separating the population (DeSoucey, 2010). This 'food identity' can contribute to international identification and reputation (Takaki, 2012). As well as having a role in generating a sense of identity, the unique characteristics of certain food products can also be leveraged in order to influence consumer behavior and encourage the transfer of cultural heritage across generations and territories (Albayrak & Gunes, 2010).

While traditional food remains subject to the same rules and regulatory requirements designed to ensure safety, geographical Indications are the main tool that the public sector has developed as a means of safeguarding the integrity of products that have strong and long-standing links to a particular territory. However, this legal construct is also a way to add economic value to these traditional products (Sgarbi & Menasche, 2015). In the area of multilateral trade regulation, geographical indications issues are regulated mainly under the Trade Related Aspects of Intellectual Property Rights Agreement (TRIPS). However, TRIPS only establishes a common framework, which is then specified, and in some cases elaborated on, in national regulations, bilateral treaties, and regional pacts.

The purpose of this chapter is to trace the emergence of traditional food as an object of regulatory attention, to explore how concepts informing policies designed

J. Gallagher
School of Agriculture and Food Science, University College Dublin, Dublin 4, Ireland

School of Biomedical Science, Ulster University, Coleraine, Northern Ireland

A. McKevitt (✉)
School of Agriculture and Food Science, University College Dublin, Dublin 4, Ireland
e-mail: aideen.mckevitt@ucd.ie

M. Al-Khusaibi et al. (eds.), *Traditional Foods*, Food Engineering Series,
https://doi.org/10.1007/978-3-030-24620-4_9

to regulate traditional foods have evolved, how these concepts are currently expressed in regulatory policies in various jursidictions around the world, and how these regulatory policies may evolve into the future.

2 What Constitutes Traditional Food?

> There is a growing belief that rural populations should no longer underrate the features which make them unique. Rather, they should at least consider local specialties and resources as potential sources of pride... Changes in consumer tastes and preferences have increased demand for goods... which are in some way special and out of the ordinary... The question is how to best exploit these resources for successful rural development.
>
> *Niche Markets as a Rural Development Strategy*, Paris, OECD (1995):13

When establishing a working definition of what is meant by the term "traditional food", it is important to recognise that food culture is extremely heterogeneous, with noticeable differences existing at global, national, regional and local points of view in food preferences, habits, food-related behavior, and attitudes (Askegaard & Madsen, 1998). Added to this, a variety of terms—local, original, artisanal, farmhouse, typical, specialty, etc.—are used to convey several concepts associated with traditional foods (Verbeke, Guerrero, Almli, Vanhonacker, & Hersleth, 2016). The range of foods that can be considered as traditional foods also continually evolves over time. What is more, traditional foods are continually being re-invented and re-adapted according to new needs and requirements as time goes by. What is perceived as traditional can change over time. The term traditional food is therefore made up of a complex set of meanings associated to a concept that is inevitably relative, and shaped according to a variety of different interests. However, the term 'traditional' can arguably be said to convey a sense of continuity and an impression that a food is made in a time-honoured way or to a time-honoured recipe. Using the Oxford English Dictionary as a starting point, it has been noted that traditional foods can be taken as food products that are created according to a "long established custom that has been passed from generation to generation" (Galli, 2018).

As a means of protecting the intellectual property and goodwill vested in traditional foods, the EU (EC, 2012) has defined traditional foods as food products with "proven usage on the domestic market for a period that allows transmission between generations; this period is to be at least 30 years." Adopting these general statements, one can begin to work towards a concept of traditional food that focuses on historical and geographical features.

One can also look to a body of academic literature in relation to traditional foods that has been growing steadily as the issues and challenges surrounding traditional foods continue to change and evolve (Belletti, Marescotti, & Touzard, 2017; Bramley & Kirsten, 2007; Herrmann & Marauhn, 2009; Quinones-Ruiz, Penker, Belletti, Marescotti, & Scaramuzzi, 2016; Rangnekar, 2010). One of the key groups who have become the focus of academic attention in relation to defining and characterising traditional foods are consumers (Almli, Verbeke, Vanhonacker, Næs, & Hersleth, 2011; Balogh, Békési, Gorton, Popp, & Lengyel,

2016; Guerrero et al., 2009; Pieniak, Verbeke, Vanhonacker, Guerrero, & Hersleth, 2009; Verbeke et al., 2016).

By focussing on the meaning that the word "traditional" has in consumers' minds, and what values they relate to this concept using a free word association method, Guerrero et al. (2009) defined traditional food (from the European consumer's perspective) as:

> a product frequently consumed or associated with specific celebrations and/or seasons, normally transmitted from one generation to another, made accurately in a specific way according to the gastronomic heritage, with little or no processing/manipulation, distinguished and known because of its sensory properties and associated with a certain local area, region or country.

It has also been observed that traditional foods can evoke strong childhood memories (Cerjak, Haas, Brunner, & Tomić, 2014; Rudawska, 2014), possess distinctive positive sensory attributes (Molnár, Gellynck, Vanhonacker, Gagalyuk, & Verbeke, 2011) and are assessed by consumers in relation to their authenticity (Tregear, Kuznesof, & Moxey, 1998). Traditional food can also be perceived culturally, as an example of the cooperation that exists amongst members of communities with links to a particular territory (Bertozzi, 1998).

Because a vast array of foods can arguably possess some if not all of the features described above, one further point to note is that what does and does not constitute a traditional food deserving of legal protection, to the detriment of its imitators, continues to be a source of debate, especially given the price premiums that such food products can command and the potential for consumers placed at the end of ever-lengthening global supply chains to be misled as to the provenance and authenticity of the food they buy.

3 The Development of Food Regulations and Food Law

Laws and rules providing for how foods are to be produced and marketed have existed since ancient times, with laws governing the manufacture of foods and beverages dating back thousands of years. Regulations governing the manufacture of beer can be traced back to the Code of Hammurabi, written in 1700 BC during the Babylonian era. In ancient Egypt, there were laws regulating how wine was labelled and consumer protection laws designed to prevent adulteration of foods were in force in ancient Rome. Echols (2001) noted that the ancient Assyrians established a formal system of weights and measures for grains as early as 200 BC, and the adulteration of grains and oils was punishable by law in India. Meanwhile in ancient China, officials were trying to prevent consumer fraud (Crawford, 1954) and in ancient Athens purity standards were in place for beer and wine.

In Europe, one of the earliest pieces of legislation dealing with food was the English *Assize of Bread and Ale* 51 Hen 3 Stat 1, which was introduced during the reign of Henry III in 1266 and regulated the quality, weight and price of bread and beer. This was followed by other pieces of legislation such as *the Adulteration of*

Tea and Coffee Act, 1724, *the Adulteration of Tea Acts*, 1730 and 1776, *the Bread Acts* of 1822 and 1836, and *the Corn, Peas, Beans or Parsnips and Cocoa Act*, 1822.

Meanwhile, from the fifteenth century onwards the *Reinheitsgebot* provided for a series of regulations limiting the ingredients in beer in Germany and the states of the former Holy Roman Empire. Countries across Europe also regulated the safety and quality of foods such as eggs, cheese, beer, wine, bread, and sausages through networks of self-regulating guilds. One example of this, was the guild of spice traders—the Pepperers—who obtained a royal warrant during the middle ages in England to create the Grocer's Company, which established a code to protect the quality and integrity of the foods under their authority and a body of inspectors to enforce it (Echols, 2001).

Although this chapter focuses on secular legislation and regulatory policies that apply to traditional foods, it is important to recognise the religious origin of rules governing the preparation and consumption of various foods that are now recognised as traditional foods.

In Islam, food rules are contained within the concepts of *Halal* and *Haram*. Halal means lawful, permitted and recommended by Sharia. *Haram* is the opposite, meaning unlawful or prohibited. Between these two categories lies the concept of *mushbooh*, which means doubtful or suspect. Several *hadith* (collections of the Prophet's tradition) prohibit the consumption of the flesh of "animals with fangs" (carnivorous animals), "animals who live on carrion" (insects) and "creeping animals without ears" (reptiles). Edible animals in verses of the Old Testament are those that have cloven hooves and chew their cud. As in Judaism, blood is often alluded to in Islam as signifying the presence of life. It must therefore be discarded before consuming any type of flesh and rules for *Halal* slaughter are designed to effectively drain the animal's blood without inflicting unnecessary suffering.

The Bible also includes various rules identifying foods that should not be eaten such as camels, hares, pigs, shellfish and the books of Leviticus and Deuteronomy also give rise to many of the religious dietary laws making up the Jewish *kashrut*. It should be noted however, that these dietary laws are characterised more by the foods that adherents are not permitted to eat, as opposed to identifying specific foods and recipes that are traditional to the Jewish faith, i.e. any food can be Kosher if prepared correctly, and traditional Ashkenazic Jewish foods like knishes, bagels, blintzes, and matzah ball soup can all be non-kosher if not prepared in accordance with the laws of Kashrut. Viewed in this way, the practice of keeping Kosher can be seen more as a traditional way of eating rather than an example of regulation of a traditional food in the strict sense.

For Hindus, the cow is sacred and the eating of beef is forbidden. Many Hindus go further and are vegetarians for religious reasons. In China, the influence of Taoism has led to the deeply-rooted belief in food as a way to long life. Food is thus deemed to be medicinal, and foods are classified according to their medicinal properties. In Japan, through the influence of the Buddhist proscription on killing, meat was not commonly eaten until the latter part of the nineteenth century.

The above examples serve to illustrate the emergence of rules regulating the production and consumption of foods, some of which may be regarded today as traditional foods with unique characteristics and connection to a particular region.

However, as we begin to transition from the past to the present, it is necessary to reflect on the emergence of the concept of 'traditional food' as a relatively recent phenomenon that has come about as a result of the marketing of food products as packages of values (and associations with quality) that travel through increasingly globalized food systems. Put another way, foods traditionally eaten in any given region have always needed to be safe and free from adulterants. However, the idea of 'traditional food' as a term whose use requires regulation has only come about relatively recently, with the increased trade in food as a commodity with a built-in set of values and associations that can attract a price premium. This protection does not relate to the safety of a food product, but its provenance and perceived quality. A consumer can buy a perfectly safe piece of hard, dry, cheese, made from cow's milk in any number of places and according to whatever process or recipe, however that cheese cannot and should not be marketed as *Parmigiano-Reggiano* unless manufactured in a specific way in the Italian provinces of Parma, Reggio Emilia, Modena, Bologna and Mantua, between the rivers Po and Reno. This is because of the host of characteristics that consumers associate with that name, and the effect that those associations, present in the minds of consumers, have on the quality they perceive and the price they are willing to pay for this 'special' cheese.

4 From Past to Present: A Global Perspective

> A powerful force drives the world toward a converging commonality ... The result is a new commercial reality – the emergence of global markets for standardized consumer products on a previously unimagined scale of magnitude... Gone are accustomed differences in national or regional preference... The world's needs and desires have been irrevocably homogenized.
>
> T. Levitt, 'The globalization of markets'. *Harvard Business Review* 61 (1983): 92–93

During the second half of the twentieth century, agriculture in developed market economies underwent a profound process of industrialization (Bowler, 1992; Goodman & Watts, 1997). On an individual farm level this resulted in intensification, specialization of production and increased reliance on external inputs (Morgan & Murdoch, 2000). In nearly all cases this was accompanied by growth in output and farm size and a decline in labour requirements (Allanson, 1992; Lowe, Cox, MacEwan, O'Riordan, & Winter, 1986). These transformations were accompanied by unparalleled growth in food processing and distribution industries, which have now come to assume a pivotal role in mediating between the realms of (primary) production and consumption (Atkins & Bowler, 2000). Taken together these processes have reshaped both the rural environment and the way in which food finds its way from 'farm to fork' (Fine & Leopold, 1994). In the wake of increased liberalization in international trade it was anticipated that these trends would intensify and this has proven to be the case (Marsden et al., 1993).

Meanwhile, the expansion of national and international markets during the nineteenth century had led to growing geographical separation between producers and consumers, although Parrot, Wilson, and Murdoch (2002) noted that the extent to which this pattern has been played out is far from uniform. Taking food in Europe as an example, southern European countries have shown significant cultural and political resistance to these processes and have sought to defend their long-established agricultural and culinary traditions against the incursions of an increasingly mechanized and standardized food production system (Miele, 2001; Mulholland, 2000). On the other hand, in northern European countries, productivist agriculture and centralized food supply chains are firmly established. Globalization and its counter-movement have thus given rise to what Whatmore and Thorne (1997) described as an 'alternative geography of food' where shifting production and consumption relations have given rise to new regional and local food 'complexes' (Murdoch, Marsden, & Banks, 2000). On the one hand, there are 'hotspots' of globalized food production and consumption (Whatmore, 1994) while, on the other hand, 'quality' production areas have emerged (Ilbery & Kneafsey, 1998). The concept of traditional food can therefore be said to have emerged as a result of a globalising process of standardisation and homogenisation of food production and consumption, and the perceived quality and added-value that consumers attach to traditional foods can be viewed as a type of reaction to the homogenising effect that globalisation has had on the activity of eating and drinking, an activity that has been historically rich in identity and meaning at a more local level.

Outside of the realm of traditional food, the growth of long distance trade in other food and non-food goods can also be seen to exacerbate the problem of quality uncertainty, with producers selling to a set of predominantly anonymous buyers in a global market. In this relatively new globalized market for increasingly homogenised and standardised consumer products, the values of history and heritage have therefore taken on a greater perceived importance, due to their ability to imbue products of all sorts with a set of unique characteristics in the eyes of consumers. This growing interest in heritage towards the end of the twentieth century has been termed a 'nostalgia boom' (Brown, Kozinets, & Sherry, 2003), during which time a new group of consumers emerged whose attitudes, aspirations and purchasing patterns were unlike any before them—because their basic needs were quickly and easily satisfied, these new consumers tended to reject mass-produced and mass-marketed commodities in favour of products and services that can claim to be in some way authentic (Lewis & Bridger, 2001). Herein lies the value and significance of traditional foods, and consequently, the manner in which they are regulated.

A specific example of this can be found again in the European experience. Prompted by a series of food-safety scares in the 1990s culminating in the BSE crisis and a furore over the introduction of GM foods, European consumers began to become wary of the increasingly industrialized agri-food industry and those who regulated it (Shaw, 1999). In consequence, increasing numbers of consumers began to seek out alternatives in the form of organic, environmentally or animal-welfare friendly and/or regionally and locally produced foods (Murdoch & Miele, 1999; Nygard & Storstad, 1998). The way that European consumers assessed the quality of the foods they bought was changing.

Despite the importance of the issue of food quality for food consumers in the wake of various food scares in the 1990s, no clear theorization of food quality was forthcoming (Dixon, 1999; Murdoch et al., 2000). However, the advent of more complex and fragmented 'worlds of production' in the food sector (Murdoch & Miele, 1999) highlighted the fact that a broad range of issues required incorporation into any accurate analysis of what food 'quality' means (Ilbery & Kneafsey, 1998, 1999a, 1999b; Morris & Young, 2000). In the wake of the 'cultural turn' (or 'turn to quality'), consumption and the substance of consumer behaviour therefore began to be taken more seriously in academic circles (Miller, 1998; Miller et al., 1998). It was in this environment that scholars stimulated by a desire to strengthen local production systems in the face of threats from the global food sector became interested in the continued resilience of local production systems in the food sector (Sylvander, 1995, 1998; Thevenot, 1998). It was in this context that regulatory policies designed to protect the integrity of traditional food products and processes began to increase in significance, with the link with territory becoming one of the primary determinative characteristics of traditional food. Indeed, Boza and Munoz (2016) have noted how the interest of consumers for traditional and local food has increased in recent years in many countries, especially in Europe (Albayrak & Gunes, 2010; Balogh et al., 2016; Kühne, Vanhonacker, Gellynk, & Verbeke, 2010; Pieniak et al., 2009). Moreover, the positive attitude of consumers for traditional and local food appear to increase when products have some kind of indication and/or are sold under collective trademarks (Guerrero et al., 2009; Pieniak et al., 2009; Verbeke & Roosen, 2009). Champredonde and González (2016) have attributed this increased perceived value that consumer attribute to traditional foods to an *integral valorization*, based on technical, economic, social, patrimonial, cultural, and environmental characteristics.

Having emerged as a manifestation of the forces of globalisation, it may come as little surprise that legal frameworks providing for the regulation of traditional foods can also best be explained at the international level, before addressing regional and national differences. In recognition of the fact that traditional food is in no way exempt from rules providing for food safety, the next section provides a brief summary of the international agreements that relate to food safety, before describing further agreements that apply to traditional foods as a form of intellectual property. Some examples of the development of regulatory regimes in various countries and regions will then be offered as a means of providing an introduction to the global regulatory landscape that traditional foods inhabit. This cursory survey is in no way intended as an exhaustive global survey, or even a detailed analysis in respect of the regions and countries considered. Instead, the snapshots on food safety provided in these sections are included simply as a means of introducing some important issues in food regulation from around the world. This is done with a view to stimulating further discussion as to how the regulation of traditional foods can form part of debate and discussion regarding wider issues such as food safety, patterns of food consumption and production, development, international cooperation and coordination, intellectual property, and international trade.

5 Current Frameworks and Issues

5.1 Food Safety and Traditional Foods

Although the intellectual property vested in traditional food is perhaps its unique identifying feature from a regulatory perspective, the safety of traditional food products is also a major issue for food producers across the globe. Consumer protection from consumption of unsafe food is a major food policy issue for countries and consumer confidence is essential for the development of the traditional foods sector. However, the globalisation of food has increased the potential for food to become contaminated, and has exposed gaps in national and international capacity to assure adequate levels of food safety and quality for all foods including traditional foods. Food safety concerns are frequently cited as justification for legislation and regulatory frameworks as well as barriers to free trade and other measures curtailing international trade. Therefore, improvements in the safety of traditional foods can serve to increase their presence across global markets. Because the economic value of traditional foods often increase the further they travel from their place of origin—exotic ingredients such as certified Japanese *Wagyu* beef have become even more valuable by the time they arrives at the door of a kitchen in London—the integrity and safety of long, sophisticated supply chains that require high levels of coordination and cooperation at the international level become even more important. Authenticity may imply quality, but quality always presupposes cleanliness and above everything, safety.

The risk of foodborne diseases is most severe in low and middle-income countries, linked to preparing food with unsafe water; poor hygiene and inadequate conditions in food production and storage; lower levels of literacy and education; and insufficient food safety legislation or implementation of such legislation (WHO, 2015). The WHO's "Estimates of the global burden of foodborne diseases" shows that as many as 600 million people, almost 1 in 10 people, fall ill every year from eating contaminated food and 420,000 die as a result. Children under 5 years of age are at particularly high risk, with 125,000 children dying from foodborne diseases every year. WHO African and South-East Asia Regions have the highest burden of foodborne diseases. Diarrhoeal diseases are responsible for more than half of the global burden of foodborne diseases, Diarrhoea is often caused by eating raw or undercooked meat, eggs, fresh produce and dairy products contaminated by norovirus, *Campylobacter*, non-typhoidal *Salmonella* and pathogenic *E. coli*.

Food standards must therefore be developed in a way that avoids barriers to trade, whilst simultaneously affording adequate protection for consumers of the trading countries. The baseline levels of food safety required to conduct international trade are formulated by an international inter-governmental organisation, the Codex Alimentarius Commission (CAC).

5.2 International Standards: Codex Alimentarius Commission

In the 1950s food regulators, traders, consumers, and experts were looking increasingly to the FAO and WHO for leadership about the plethora of food regulations that were impeding trade and that for the most part were not providing adequate protection for consumers.

In 1961, the 11th FAO Conference, on foot of a resolution of the Council of Codex Alimentarius Europaeus proposing that its work on food standards be taken over by FAO and WHO, established the Codex Alimentarius Commission (CAC) and requested endorsement by the WHO of a joint FAO/WHO Food Standards Programme. In 1962, the joint FAO/WHO Food Standards Conference requested the CAC to implement a joint FAO/WHO food standards programme and create the Codex Alimentarius. The inaugural meeting of the CAC took place in Rome in 1963, where the Sixteenth World Health Assembly (the governing body of the WHO, composed of representatives from each member state) approved the establishment of the Joint FAO/WHO Food Standards Programme and adopted the statutes of the Codex Alimentarius Commission (CAC).

The CAC is the pre-eminent global food standards organization, whose aim is to protect consumers' health; ensure fair practices in the food trade through the development of science-based food quality and safety standards, guidelines, and recommendations; and to promote coordination of all food standards work undertaken by governmental and international organizations. These aims are pursued on the basis that harmonization of food standards facilitates trade between countries, with said international trade benefitting from a form of guarantee that food that is traded will be as safe as a similar product made elsewhere. Membership of the CAC is open to all member nations and associate members of the FAO and/or WHO. The CAC currently has 188 Member Countries and one Member Organization (the European Union). In recent years there has been a significant increase in the membership. Developing countries now constitute a significant proportion of total membership. However, many countries are still faced with resource constraints to effective participation in CAC activities. FAO and WHO technical assistance programmes support the efforts of developing countries to strengthen their national food safety systems to protect local consumers and to take advantage of international food trade opportunities. In addition, the FAO/WHO Codex Trust Fund supports the participation of countries in Codex technical committee meetings, and countries have been funded to attend sessions of the CAC.

CAC meetings are held yearly and alternately at the FAO headquarters in Rome and the WHO headquarters in Geneva. At these meetings draft and final standards, guidelines, and codes of practice are adopted. Each member of the Commission has one vote. Decisions of the Commission are taken by a majority of votes cast. Representation is on a country basis. National delegations are led by senior officials appointed by their governments. Delegations may include representatives of industry, consumers' organizations, and academic institutions. Countries not members of the Commission sometimes attend in an observer capacity. A number of international governmental organizations and international NGOs also attend in an observer capacity. These organizations may put forward their points of view at every stage except in the final decision, which is taken by member governments. The Commission and member governments have established country Codex Contact Points and many member countries have National Codex Committees to coordinate activities nationally.

The main standard setting work of the CAC is carried out in more than 20 Codex Committees and Task Forces. These include committees dealing with "vertical" and "horizontal" standards; task forces dedicated to a particular task of limited duration; and regional coordinating committees. This work is also carried out in combination with, and the support of the work of a number of joint FAO/WHO expert scientific groups:

- The Joint FAO/WHO Expert Committee on Food Additives (JECFA) was established in 1956, whose remit now covers the evaluation of contaminants, naturally occurring toxicants and residues of veterinary drugs in food.
- The Joint FAO/WHO Meeting on Pesticide Residues (JMPR) has met regularly since 1963 and was set up to provide independent scientific advice to the FAO and WHO with recommendations from panels of independent experts on the use of pesticides in agriculture and safe levels of residues in foods.
- The Joint FAO/WHO Meeting on Microbiological Risk Assessment (JEMRA) began in 2000. The aim of JEMRA is to optimize the use of microbiological risk assessment as the scientific basis for risk management decisions that address microbiological hazards in foods.

At first, the CAC concentrated on commodity standards called "vertical standards," standards that apply to specific categories of foods such as cereals; fats and oils; fish and fish products; fresh fruits and vegetables; processed and quick frozen fruits and vegetables; fruit juices; meat and meat products; milk and milk products; sugars, cocoa products, and chocolate. However, by the 1980s it was generally agreed that the diversification of food products was accelerating so rapidly that the setting of these types of specific standards was in fact hindering trade. Thus, a move toward "horizontal" standards began. "Horizontal standards" are general standards and principles that have application across a wide range of foods, for example, rules relating to food additives; contaminants; food labeling; food hygiene, methods of analysis and sampling; pesticide residues, residues of veterinary drugs in foods; food import and export inspection and certification systems; nutrition and foods for special dietary uses, etc. These standards are then published in one of the Codex's 13 volumes. Codex standards pass through various stages of ratification by members—the eight-step process—the final one being that of acceptance. When members accept a Codex standard they are committed to allowing products conforming to that standard on to their market. This is an important issue for the global traditional food sector.

5.3 *WTO: Sanitary and Phytosanitary Measures and Technical Barriers to Trade*

The World Trade Organization (WTO) was established in 1995 on the conclusion of the Uruguay Round of multilateral trade negotiations, conducted within the framework of the General Agreement on Tariffs and Trade (GATT). The WTO Agreements relevant to international food safety regulation (and thus the safety of traditional foods) are the Agreement on the Application of Sanitary and Phytosanitary Measures (SPS) and the Agreement on Technical Barriers to Trade (TBT).

The SPS Agreement allows governments to take scientifically justified sanitary and phytosanitary measures to protect human health. The agreement commits members to base these measures on internationally established guidelines and risk assessment procedures. The SPS Agreement has chosen the standards guidelines and recommendations developed by the CAC for food additives, veterinary drug and pesticide residues, contaminants, methods of analysis and sampling, and codes and guidelines of hygienic practice. A national standard that provides a greater level of protection than Codex is considered to be a trade barrier unless the WTO decides that the stricter national standard is based on a risk assessment that demonstrates that the Codex standard, guideline, or recommendation provides insufficient protection or that the country maintaining the stricter standard has other scientific justification.

The TBT agreement seeks to ensure that technical regulations and product standards including packaging, marking and labelling requirements, and analytical procedures for assessing conformity with technical regulations and standards do not create unnecessary obstacles to trade. The importance of Codex standards is also stated in the Technical Regulations and Standards provisions contained in Article 2 of the TBT Agreement. So, although CAC standards are not enshrined in international law, WTO endorsement of these standards through the SPS and TBT agreements has effectively made them mandatory, and Codex standards are the benchmarks standards against which national measures and regulations are evaluated.

Adherence to Codex standards has also become critical as they are used as guidelines for the resolution of disputes under the enhanced WTO dispute settlement procedure. A dispute is triggered when a WTO member complains that another member(s) has failed to live up to the obligations of the GATT or WTO agreements, i.e., a benefit guaranteed under one or other of these agreements has been "nullified or impaired" by another member(s).

5.4 *European Union (EU): The Path to the General Food Law*

The EU can be considered as an example of the evolution toward a modern system of traditional food regulation, providing for both safety and intellectual property protections. In terms of traditional food it worth noting that the EU, when taken as a geographic whole, is made up of a highly diverse array of traditional food products and practices. The European project, viewed through the lens of traditional food regulation, can therefore offer valuable insights for large countries tasked with regulating a diverse internal traditional food landscape (e.g. China, India, or Brazil), or other blocs made up of distinct but related food traditions that would benefit from recognition, definition, and protection (e.g. MERCOSUR, ASEAN, The African Union, etc.).

The EU is currently an association of 28 Member States that have agreed to integrate and coordinate much of their economic policy and some other policy areas. The original European Economic Community (EEC) was formed following the signing of the Treaty of Rome in 1957 and consisted of six Member States, increasing over time to 9, 12, 15, 25, 27 and then 28 Member States in 2013 following the accession of Croatia. At first, Member States concentrated on the free movement of foodstuffs through the common market.

Three institutions, the European Commission, the Council of the European Union, and the European Parliament—the inter-institutional triangle—take decisions in the legislative field. The prompt for political action in any given policy area can come from particular Member States, the Council of Ministers, the European Parliament, lobbying by trade associations, research on risks and hazards, and/or technical developments, etc.

During the first 40 years of what came to become the EU, European food regulation consequently developed in an uncoordinated fashion. This resulted in a fragmented framework of rules and regulations with different national policies relating to food overlapping across a variety of related but separate policy areas such as trade and agriculture to which food production, trade and consumption was linked.

In 1974, a Scientific Committee for Food (SCF) was established by the European Commission "to advise the Commission on any problem relating to the protection of the health and safety of persons arising or likely to arise from the consumption of food, in particular on nutritional, hygienic and toxicological issues." However, various scientific committees established by the European Commission (including the SCF) were continually criticized on a number of grounds by the European Parliament and by industry and consumer NGOs. Following the bovine spongiform encephalopathy (BSE) crisis in the UK, there was a further decrease in confidence in these scientific committees and, with the new powers in public health granted by the Maastricht Treaty, the European Parliament forced the Commission to totally revise their structure. A major reorganization of the Commission's services ensued.

The responsibility for monitoring the implementation of food safety legislation and for providing scientific advice, hitherto jointly share by the Commissioners for Agriculture and Industry, was transferred to the Commissioner for Consumer Affairs. The rationale at the time was that it was necessary to separate monitoring, compliance with and enforcement of the law, from the law-making function itself. This latter function remained for a time with the Agriculture and Industry Commissioners, however the legislation function on food safety was later transferred to the Health and Consumer Protection Commissioner. The Commission also announced that the way in which scientific advice on food safety was provided at European level would be reorganized and strengthened. A Scientific Steering Committee to oversee the work of the eight regrouped scientific committees was created. In 1997, The Green Paper on the General Principles of Food Law was published in order to launch a debate on the future development of EU food law. Its aim was to provide the Commission with a solid background for a major programme of new or amending legislation that it would later propose in the 2000 White Paper on Food Safety.

In January 2002, Regulation (EC) No. 178/2002, otherwise known as the General Food Law, was adopted. The Regulation set out the general principles of EU food law and established a European Food Safety Authority. It also established a rapid alert system (RASFF) for the notification of direct or indirect risks to human health deriving from feed or food, and set down clear procedures for the handling by the Commission and the Member States of food safety emergencies and crises.

This Regulation, for the first time, set out the following main principles of EU food law, which were intended to cover all food and feed at all stages of production, processing, and distribution:

- A requirement that food law must be based on a system of risk analysis founded on risk assessment, risk management, and risk communication.
- A requirement that the precautionary principle will be applied in the case of a potential risk to human health where there is scientific uncertainty as to what measures to take.
- A requirement that public authorities at all levels will apply the principle of transparency in consulting with and informing the public on actual or potential risks and the actions that are taken or proposed to deal with them.

The Regulation also provided for a system to ensure 'farm to fork' traceability of all food and feed at all stages of the food and feed chain. The primary responsibility of the European Food Safety Authority (EFSA) is to provide independent scientific advice on all matters with a direct or indirect impact on food safety. The Authority has been given a wide brief, so that it can cover all stages of food production and supply, from primary production to the safety of animal feed, right through to the supply of food to consumers. It gathers information from all parts of the globe, keeping an eye on new developments in science. Although the Authority's main "customer" is the Commission, it is open to respond to scientific questions from the European Parliament and the Member States and it can also initiate risk assessments on its own behalf. The Authority carries out assessments of risks to the food chain and indeed can carry out scientific assessment on any matter that may have a direct or indirect effect on the safety of the food supply, including matters relating to animal health, animal welfare, and plant health. The Authority also gives scientific advice on non-food and feed, genetically modified organisms (GMOs), and on nutrition in relation to Community legislation.

6 Traditional Food as Intellectual Property: Geographical Indications (GIs)

Until the late twentieth century, traditional knowledge was developed through trial and error and passed from generation to generation without anyone legally owning any of this knowledge. In an era of multilateral trade however, a key feature of the establishment of the WTO was an agreement for the enforcement of 'trade-related intellectual property rights' which allow for an individual or organization to own ideas including trademarks comprising sufficiently distinctive names, words, logos, shapes (of the food product itself or its packaging), slogans, specific colours and sounds.

GIs are used as a generic term for names or signs used on products that correspond to a specific geographical location and were developed to protect product names linked to a particular territory or to a production method. They are addressed in the Annex 1C, Part II, Section 3 of the Agreement on Trade-Related Aspects of Intellectual Property Rights (TRIPs). This sets the basic international regulatory framework for GIs, not only providing a definition but also aligning the standards of protection and providing for access to international dispute settlement mechanisms (ITC, 2009). Various authors view TRIPS as the first multilateral agreement that explains the definition of a GI and related aspects (EFOW, 2016; Gervais, 2010; O'Connor, 2004; Rai, 2009; Zografos, 2010). This is particularly important given the broad membership of TRIPS, thus constituting the basis of many countries' domestic regulations (Errázuriz, 2010).

The text of TRIPs addressed to GIs is composed of three articles: (1) article 22, which describes the basic definition and general standards of protection for GIs, (2) article 23, about additional protection for geographical indications for wines and spirits, and (3) article 24, which establishes some important exceptions and details for international negotiations.

Before TRIPS, there were some approximations to GIs in other international agreements, but they did not have the same scope, either because of their generality (Paris Convention) or the reduced number of signatory parties (Madrid Agreement and Lisbon Agreement) (Zografos, 2010). However, the concept of GIs proposed in those agreements did begin to emerge via terms such as "indication of origin" and "appellation of origin".

Inspired by previous international agreements, but with some significant differences, Article 22 of TRIPs provides the agreed WTO definition for GI. It establishes that GIs are:

> indications which identify a good as originating in the territory of a (WTO) Member, or a region or locality in that territory, where a given quality, reputation or other characteristic of the good is essentially attributable to its geographical origin.

It is important to observe that, in its definition for geographical indications, the WTO refers to "goods" and not to "products". Products are understood as goods and services; therefore, services are excluded (Rangnekar, 2003).

In Article 22, TRIPs also specifies that members will decide on legal forms to prevent the misuse of their protected GI, as well as to avoid "any use which constitutes an act of unfair competition within the meaning of Article 10bis of the Paris Convention (1967)". Additionally, a WTO member, if its legislation allows it or an interested party requires it, can refuse or invalidate the registration of a trademark when it has the same denomination of a GI, when the good that has such a mark is not originally from the same place of the GI, and this may induce consumers to confusion. This protection is also applicable in case of deceptive GI, namely: "geographical indication which, although literally true as to the territory, region or locality in which the goods originate, falsely represents to the public that the goods originate in another territory". Therefore, TRIPs allows members to choose the most appropriate legal strategy to protect their GI, such as laws on business practices, trademark law, and owner systems specifically designed to protect GI, i.e. *sui generis* systems (WTO, 2016).

Article 23 refers to the specific protection for geographical indications in wines and spirits. It establishes a considerably higher level of protection regarding the rest of the goods contemplated in Article 22. In fact, countries can protect GI for wines and spirits even if there is no risk of misleading consumers or unfair competition. This differential treatment is not based on economics or any other justified reason but, rather, is the result of negotiations and specific circumstances that were particular to the wine sector. The Agreement assures entire protection of members' GI on wines and spirit from producers who use them, either in translation or accompanied by the expressions "kind", "type", "style", "imitation", or similar by the legal means established by each country. In this sense, if a new trademark for wines or spirits contains or consists of an existing GI that does not correspond to its origin, it shall be refused or invalidated if the legislation of the Member allows it (WTO, 2016).

To grant this extra protection in the case of a wine or spirit, the Agreement establishes a multilateral system of notification and registration of related GIs. Some developed and developing countries have proposed to extend that system to a higher number of products, such as crafts, agricultural products, and other drinks. That proposal has been largely discussed, and it was included in the working program for the Doha Round. The debate continues as to how many products should be included as part of this registration system, an issue which is especially relevant for those Members who link the GI with the access to new market segments, thereby increasing differentiation and enhancing competitiveness. In contrast, other groups of WTO members, e.g. Chile, Argentina, New Zealand, and the United States, disagree with the extension of the GI registration system for diverse reasons; the most important being a dearth of evidence to support the claim that the existing protection for geographical indications under Article 22 of the TRIPS is in some way insufficient, and the conviction that further protection would constitute a barrier that disrupts current, legitimate market practices (WTO, 2005, 2016).

Finally, Article 24 specifies some relevant exceptions to previous articles and further details for international negotiations. It establishes that, in some cases, GI does not need protection, or it can be limited—for instance, when a denomination has become a common or generic term (e.g., cheddar now is referring to a type of cheese, beyond which is made in Cheddar, UK), when the geographical indications have been used similarly and continuously for many years, or when a trademark has been obtained before the registration of the GI, with the exception that such trademark has not been used or had been registered in bad faith.

GIs became the subject of the WTO dispute settlement procedure when an EU system for the registration of protected geographical indications and protected designations of origin in 1992 became the subject of complaints brought under Articles

22 and 24 of TRIPs by the USA and Australia, the *Geographical Indications Dispute*. The Dispute Settlement Body Panel found that although the EU's system was designed to fulfill a legitimate objective, it had failed to provide for national treatment in that it had provided a privileged status for applicants within its own member states, owing to the problems faced by applicants from third countries gaining recognition under the EU system. The EU subsequently changed their system by adopting two new regulations in 2006 which recognised the significance of TRIPs for the content and application of EU food laws.

TRIPS can be said to exist as the global keystone agreement in terms of GI's and by extension regulation of traditional foods on the global level. As has been noted above however, members are permitted to choose the most appropriate legal strategy to protect their GI. This has given rise to a patchwork of regulatory regimes and policies that have TRIPS as their common denominator. What follows is a sample of some approaches that have been developed and which are in use today. It is also important to note that the various means by which members go about protecting their GIs can often give rise to disputes, and GIs are becoming an ever more important issue in the context of both multilateral and bilateral trade agreements.

6.1 Geographical Indications in the EU: A Case Study

In relation to traditional foods in particular, the EU also has its own system of registration for GIs. In post-World War 2 Europe, the decline of the agricultural sector compared to manufacturing and services had created several problems in rural areas such as depopulation and increasing unemployment and poverty. In recognition of the fact that traditional price-support measures were of little benefit, the OECD began to emphasise the need for structural adjustment within the agricultural sector and the need for better targeting of support. In particular, local initiatives such as the formation of 'product specialisation areas' and 'local productive systems' were identified as beneficial. Meanwhile, the European Common Agricultural Policy (CAP) was established in 1962 with the aim of increasing productivity, stabilising markets and prices, and securing food supplies. CAP policies were also designed with the goal of securing a fair standard of living for agricultural communities across Europe, taking into account the socio-economic disparities present across the various European agricultural regions. However, the pricing and subsidy schemes which comprised the main part of CAP resulted in substantial overproduction, artificially high food prices in Europe, and soaring agricultural subsidies in Europe in comparison to other developed countries. An adjustment between supply and demand was required. An EU GI registration system was part of the EU response, a further factor being the belief that the promotion of products possessing geographically defined characteristics would improve the income of farmers.

Although rural development and the wellbeing of European farmers and agricultural communities provided some impetus for a European GI scheme, free movement of goods and the interests of consumers were also key factors. MacMaolain

(2015) has noted that when the European Court of Justice developed the principle of mutual recognition in 1979, declaring that a product 'lawfully produced and marketed in one of the Member States' should be marketable in all of the Member States, it essentially created an inter-State market without restrictions as to compositional requirements or methods of manufacture. One consequence of this was that a producer in one Member State could make a food product, call it whatever she liked (and include a place name in the name with which his product had no connection) and sell it in another Member State. The emergence of such an unrestricted market therefore led to an increased risk that consumers could be misled in relation to the provenance (and associated quality) of the food that they were purchasing. The introduction of a register of geographical food names also sought to remedy this risk to consumers as well as the related problem faced by regional producers who sought to protect the reputation of products built up over generations and which attached a price premium in many cases.

As alluded to above, in 2005 the WTO found that elements of the EU system for registering and protecting geographical food names were contrary to TRIPs in that the equivalence and reciprocity conditions for access to the system established by Regulation 2081/1992 violated the national treatment obligations set out in TRIPs and the General Agreement on Tariffs and Trade (GATT). A 2006 Regulation was therefore introduced. The law as it relates to geographical food names remains largely unchanged in the 2012 Regulation which is the current legislation in force. The preamble to Regulation (EU) No. 1151/2012 (Official Journal L343 of 2012) states that 'citizens and consumers in the Union increasingly demand quality as well as traditional products'. This then 'generates a demand for agricultural products for food-stuffs with identifiable specific characteristics, in particular those linked to their geographical origin'. Traditional foods in the EU may or may not possess some form of origin label, although they are often sold under different collective trademarks, i.e., quality labels. Registration affords producers in the designated area protection against imitation and use of the name outside the region.

The 2012 Regulation sets out how applications for entry on the register must show how the characteristics of the region—topographic, savoir faire, reputation, natural resources—have an impact on the characteristics of the particular applicant product. An applicant product need not be unique, but the characteristics of the region—topographic, savoir-faire, reputation, and natural resources—must have an impact on the characteristics of the product. It is also important to note that application for GI registration in the EU is more often than not a collective effort on the part of trade associations and farmer groups. GI producers must maintain the quality of the product, comply with scheme specifications and meet control requirements and producers within the defined geographic area who meets the specifications set out by the Commission can avail of the GI, even if not a member of the applicant group. There are three types of Geographic Indication:

- **PDO (Protected Designation of Origin)** for products with a strong link to the defined geographical area where they are produced.
- **PGI (Protected Geographical Indication)** for agricultural products and foods linked to a geographical area where at least one production step has taken place.

Famous PGIs across Europe include Champagne, Camembert, Feta cheese and Parma ham.
- **TSG (Traditional Speciality Guaranteed)** for traditional products (25 years/ handed down through generations) or established by custom.

As of the end of 2016, there are almost 1400 PDO, PGI and TSG registrations, which include some of Europe's most famous food products. Examples of PDOs are French cheeses from Cantal, and Munster; dry-cured hams from Teruel, Parma and San Daniele; as well as Cornish clotted cream, Jersey Royal potatoes, and Orkney beef and lamb. Typical PGI products include German beers originating from Dortmund, Cologne and Munich. Indeed, cheese and beer have been identified as the two most important products covered by the EU scheme, accounting for 33% and 20% of the total value of cheese and beer production respectively, and taken together, 51% of total EU GI sales based on 2012 figures (European Commission, 2012).

Since its inception, the effectiveness of the EU scheme has been studied in detail. Although PDO/PGI certification is a useful means of reaffirming the link between locality and produce which affords protection against the global supply of generic produce, the benefits of PDO/PGI status are ambiguous and unclear owing primarily to the ambiguous link between quality and origin, and the related failure of the EU to successfully market products with PDO/PGI status. Within the EU the challenge of devising a response to the shift in consumer preference towards products for which cultural heritage, artisanal production, and provenance were emphasised and guaranteed has at all times been exacerbated by both the diversity of foodstuffs and the number of associated brands, logos, consortia labels, and other product information that consumers are exposed to. Moreover, between 1996 and 2016 over 60% of total registrations were attributable to four countries: France, Italy, Spain and Portugal. Between 1995 and 2005, agriculture as a proportion of national income in these four countries comfortably exceeded the average for EU-25 countries, and was particularly pronounced in relation to Northern European countries such as Belgium and Britain, a trend which continues. Higgins (2018) noted that even allowing for the fact that agriculture as a proportion of national income has declined in all EU Member States, a fair generalisation is that those in which agriculture remains relatively important have engaged most strongly with the EU scheme. Viewed in this way, producers in these Member States can be viewed as the 'winners' from the EU scheme. Meanwhile, doubt remains as to whether this particular registration system operates to protect the viability of EU farmers and the EU agricultural sector as a whole.

Although the EU's attempt to build an enduring and visible link between product, place, and quality has not been without its problems and challenges, it does highlight how the relationship between product quality and geographical origin is a complex, dynamic and multi-layered one. Notwithstanding this, it has been observed that these EU GIs do have a global resonance that extends well beyond European borders (Mancini, 2013; Heath 2017).

7 Middle East: Homemade Traditional Foods

Extending from the western Mediterranean Sea to Afghanistan and Pakistan (FAO/WHO, 2005), several middle-eastern countries share a similar social, political, economic, cultural and religious ancestry with similarities in their food control systems and food safety programs (Al-Kandari & Jukes, 2009). Notably, some countries in this region import almost 90% of their food (FAO/WHO, 2005). Examples of traditional foods consumed in the middle east include hummus; yogurt and clarified butter (samn or ghee); labneh (concentrated yogurt) dishes such as Shuwaa, Khouzi, Kibbeh, Mansaf, Masquof, and Shawarma; and desserts such as Halva. Common barriers to keeping homemade foods safe are proper storage temperature during food processing. Consequently, common sources of foodborne illness in home prepared food are cooked food being stored at ambient temperature for several hours as well as inadequate heating before serving food (Karabudak, Bas, & Kiziltan, 2008). In one study assessing food handling/preparation of homemade foods, a majority of homemade food was stored at home and only 14% of the food was stored in the refrigerator (Sudershan, Gavaravarapu, Jasti, Rao, & Polasa, 2008). Traditional food safety risks can therefore arise in the home as well as in commercial and industrial settings.

Whilst, steps toward improving food safety systems in the Middle East have been made, efforts to control the safety of locally produced and imported food still require further improvement (FAO, 2004). Healthcare facilities, clean drinking water, and use of sanitation facilities have been found highly deficient extremely low in rural areas of some of the less-developed Middle Eastern countries. This might be an important factor in high morbidity and mortality patterns. Not all of the population in the region has access to safe water. Improvement of the capacity of food control laboratories to enhance food control services is also necessary (FAO, 2004). Poor personal hygiene practices are a very important issue. With the exception of major exporters and large epidemics, low investment in food safety infrastructure such as laboratories, quarantine facilities and slaughterhouses is common.

In the Middle East in recent years, countries which regard food safety as key in agricultural development and economic sustainability have worked to strengthen their infrastructure in modern food safety and develop new systems for the regulation and oversight of food safety. This includes the development of a modern, farm to fork, risk based approach to food safety, the modernisation of food legislation, development of food inspection services; food monitoring and surveillance systems; foodborne disease surveillance systems; food safety education and training and development of new food safety authorities to enhance consumer protection and develop food trade including trade in traditional foods. Because countries are obliged to demonstrate equivalence of national SPS requirements with those of more developed countries, this can hinder access to developed markets and increase costs for exporters because of the difficulties for developing countries of meeting the level of international standards.

8 Africa: Developing Food Safety Systems

Some of the 54 countries making up the continent of Africa continue to rely on subsistence farming, whilst others have formal and high level food safety infrastructure allowing access to export markets and a safe domestic food supply (Anelich, 2013). Among the food regulatory challenges facing African countries is the low level of development of food regulatory systems generally across the continent. These systemic shortcomings affect traditional foods. High post-harvest losses, due to lack of technology and improper storage facilities, such as cold storage because of a weak energy supply will have an impact on accessing markets before food deteriorates. This will in turn inhibit investment by large food businesses in the affected countries and will hinder development of local agri-food businesses including those producing traditional foods (Anelich, 2013).

The top EU food imports from sub-Saharan Africa between 2013 and 2017 were cocoa beans, cocoa paste and powder, tropical fruit including fresh or dried, nuts and spices, coffee, tea and citrus fruits (EC, 2018). Aflatoxins are toxic secondary metabolites produced notably by *Aspergillus flavus* and *A. parasiticus* that frequently invade foods and feedstuffs before and after harvest. The presence of the four major aflatoxins include aflatoxin B1, B2, G1 and G2, with aflatoxin B1 contamination resulting from fungal contamination of dietary staples including cereals, spices, oilseeds, groundnuts, have been reported widely (Makun, Dutton, Njobeh, Gbodi, & Ogbadu, 2012). Aflatoxin contamination persists in processed products because toxins are heat stable. The biggest challenge in regulating mycotoxins in Africa is the lack of enforcement of legislation partly attributed to the presence of informal food market systems in many countries where raw agricultural produce from farms and storage barns are sold directly to consumers without being screened for mycotoxins nor are they subjected to inspection for spoilage. Other factors that worsen the aflatoxin burden in Africa are public ignorance of the existence of the toxins; and introduction of contaminated food into the food chain due to shortage of food supply caused by drought, wars and other socioeconomic and political insecurity (Wagacha & Muthomi, 2008).

Most countries in Africa are members of the Codex Alimentarius Commission (CAC) and members of the WTO and, whilst all African countries have food standards, these range from few and rudimentary standards, to antiquated standards to Codex standards which are regarded as the minimum standards that countries should comply with, for conducting international and regional trade, and for assuring a safe and nutritious domestic food supply (Anelich, 2013). Capacity building to develop science-based food standards is required in keeping with WTO, SPS and TBT principles. African governments must build or strengthen already existing regional laboratories to monitor mycotoxins in foods and feeds on regular basis and to ensure that they are in compliance with set standards. Africa must reinforce its food quality control agencies with professionals who can demonstrate that they have both the academic and technical capacity for mycotoxin management. An effective mycotoxin surveillance and food quality control system which ensures that all food and feed destined for human and animal consumption does not contain mycotoxins above the regulated limits should be in place across the continent. This approach will facilitate harmonisation of food standards, and consequently food safety standards, at regional and international levels facilitating trade.

9 China: Diffuse Production, Expanding Population and Modern Regulation

Chinese food culture is characterised by its diversity due to a long history, the vast size of China, and the fact that China is a multinational country with more than 50 distinct ethnic groups. In terms of history, chemical analyses of ancient organics absorbed into pottery jars from the early Neolithic village of Jiahu in Henan province in China revealed that a mixed fermented beverage of rice, honey, and fruit (hawthorn fruit and/or grape) was being produced as early as 7000 BC. This prehistoric drink paved the way for unique cereal beverages of the proto-historic second millennium BC, preserved as liquids inside sealed bronze vessels of the Shang and Western Zhou Dynasties. These findings provide direct evidence for fermented beverages in ancient Chinese culture, which were of considerable social, religious, and medical significance, described in the Shang Dynasty oracle inscriptions (McGovern et al., 2004).

Eating behavior, once formed, has continuity. When people move to other regions or countries, they will continue their traditional eating habits, tastes, and cooking methods. People eat special foods to celebrate important events or festivals. Specific foods will be served for specific social events in China, for example, moon cakes for the mid-Autumn festival and dumplings for the Spring festival. The traditional foods for celebrating one's birthday in China are noodles and peaches. In Chinese culture, food is also used as a symbol or a means to impart social information. Children peanuts (also known as the longevity fruit) are a symbol of longevity and oranges and chestnuts are symbolic of good luck (Ma, 2015). The diverse food traditions in China have been subject to modernisation, in the form of regulation, on grounds of public health, food safety, and nutrition. In addition, a series of well documented food safety issues in recent years has resulted in a loss of consumer confidence in the safety and authenticity of the food supply.

China's agricultural production is scattered among thousands of smallholdings. As a result, one processor may source raw materials from thousands of farmers. These diverse and distributed links create fragility across the food chain. It is difficult, but necessary, to ensure food safety throughout the production chain. A series of food safety incidents in dairy and meat products as well as food additives has affected the Chinese food industry in recent years. Chinese food policy has historically prioritised the provision of adequate quantities of affordable food to meet rising domestic demand, often to the detriment of the environment and public health (Zhang, Xu, Oosterveer, & Mol, 2016) There are about 1100 million kg of food, worth more than 21 billion Yuan, produced in China every day. With that in mind, it is perhaps unsurprising, given the sheer volume of food being produced to feed an enormous and expanding population, that there has been an increase in the risk of food safety accidents occurring (Liu, 2014).

The Food Safety Law (FSL) of the Peoples' Republic of China came into effect in 2009 and represented the first legislative effort to regulate from production through to consumption (Zhang & Xue, 2016). The law created a state-level food safety commission to oversee implementation of the law and included a Food Safety Risk Surveillance System (monitoring of foodborne diseases, food contamination and other risk factors contained in food), the Food Safety Risk Assessment System (risk assessment of biological, chemical and physical hazards in food and food additives) and the Food Safety Standards System, including general standards on foodborne pathogens, pesticide residues, veterinary drug residues, heavy metals, contaminant and mycotoxin limits as well as food additives, food labelling and detection methods.

In 2015, the Standing Committee of China's National People's Congress revised the 2009 Food Safety Law. The revisions to the Food Safety Law are wide-ranging, and addressed fragmented and unclear institutional responsibilities (Xue & Zhang, 2013), weak enforcement and inadequate penalties for those contravening the law (Jia & Jukes, 2013). The Law gave the authorities greater levels of power to prosecute violations and impose stricter penalties. It has been argued that China's current food safety regulations are one of the most stringent globally (Zhang et al., 2016). However concerns remain regarding the capacity of the revised legislation to regulate large numbers of small supply chain actors (Zhang et al., 2016). Nevertheless there remains is a crisis of confidence amongst Chinese consumers, though in recent years the Chinese government has actively sought to improve risk monitoring. Supervision of food safety practice needs to be better and enforcement of modern, rationalized laws used to eliminate illegal activities. Food safety risk communication and public education also need to be developed (Kendall et al., 2019).

10 South East Asia: Agricultural Intensification and International Trade

Agricultural intensification, defined as the increase in the productivity of crops and livestock per unit of input, has grown rapidly in Southeast Asia driven by large population growth, strong economies, and a shift in consumer demand. Agriculture is the primary economic activity in Southeast Asia, and the share of agriculture to GDP, along with net agricultural output per capita, has more than doubled in the last few decades (Lam, Pham, & Nguyen-Viet, 2018).

The food production, processing, and marketing system in Southeast Asia ranges from small-scale to large-scale, with products passing through multiple tiers of handlers and middlemen in the market chain. Facilities and infrastructure remain inadequate, and there is lack of knowledge and expertise on new or modern technologies and practices, especially among small-scale processors, and there is a shortage of sound, cost-effective methods for identifying specific food safety problems. The responsibility for ensuring food safety is based on a multi-agency approach due to historical or political reasons, and there is lack of coordination among agencies. In addition, specific food safety policies are either nonexistent, inadequate or of low priority in most of these countries (Othman, 2007).

Notwithstanding this, the Association of Southeast Asian Nations (ASEAN) has been one of the most prolific blocs in the world in terms of GI registrations with over 150 were registrations in the past decade. For example, Thailand has already registered GIs for 71 of its domestic products. This follows the authorities' desire to implement the "one province, one GI" scheme whereby each of the 77 provinces in the country should have at least one Geographical Indication attached to it. Two Thai products that have recently been added to the list are Pla Kulao Khem Tak Bai fish and Phon Yang Kham beef. In addition, Sung Yod rice recently became registered as a PGI by the European Union, increasing its potential of export to the European market. In negotiating a free trade agreement with the EU, Vietnam also proposed a list of 41 domestic GIs of which the EU agreed to protect 39 in its market. Provided Vietnam succeeds in establishing a systematic quality product check, and reduces the circulation of counterfeits and imitation goods in its domestic market, this agreement will likely assist in attempts to export further Vietnamese food products to the EU market.

11 Latin America: Regional Coordination and Authenticity

MERCOSUR was created in 1991 under the Treaty of Asunción signed by Argentina, Brazil, Paraguay, and Uruguay. Venezuela is a member but had its membership suspended in 2016. Associate countries are Bolivia, Chile, Colombia, Ecuador, Guyana, Peru and Suriname. Observer countries are New Zealand and Mexico. MERCOSUR comprises nearly 250 million people living in an area four times larger than the European Union (EU) and covering almost 13 million km^2. MERCOSUR trading activities account for 75% of all trade in South America, with Brazil and Argentina being the largest economies. One of the principal objectives of MERCOSUR was to be the establishment of a single common market in Latin America.

MERCOSUR has three principal decision-making institutions: the Common Market Council (CMC, Consejo del Mercado Comun), the Common Market Group (GMC, Grupo Mercado Comun) and the Trade Commission (CCM, Comision de Comercio del Mercosur). Since food is among the most important commodities traded internationally, the harmonization of national food standards and reduction of technical barriers to trade was considered a priority for MERCOSUR. Harmonization of food law is carried out by the Technical Regulations and Conformity Assessment, under the responsibility of the Food Commission.

The member countries examined the Codex Alimentarius standards and recommendations for the purpose of the MERCOSUR harmonization process. In some cases they adopted these standards, while in others they used the underlying principles, introducing changes needed to reflect the situation, issues and existing resources in MERCOSUR. Harmonization progressed slowly with four countries involved, Argentina, Brazil, Paraguay and Uruguay each with different laws, habits and interests. Nevertheless, some regulations are now harmonized for issues where technical standards may represent trade barriers. This interest in Codex activities on the part of the four countries could provide a mechanism by which the experiences of these countries and of MERCOSUR can reach a wider audience and have a stronger influence on the setting of international food standards (FAO, 1995).

The EU is concerned about food safety aspects of meat production in all Mercosur countries. Traceability, antibiotics, and hormones used as growth promoters, which are illegal to use in the EU, are the main concerns for EU farmers and consumers. Furthermore, a U.S. ban on Brazilian beef coupled with various on-going scandals in Brazil, have increased the EU's concerns about the safety of Brazil's meat exports (USDA, 2017).

Examples of traditional foods emanating from this region include Serrano Cheese, an artisanal cheese made from raw milk in Campos de Cima da Serra, a region located in the south Brazil using traditional process that dates back more than 200 years. The region has native grasslands, high altitude and well defined seasons The techniques and expertise required to produce Serrano Cheese are maintained by rural families, with only minor amendments to the process, which has been handed down from generation to generation. Production uses milk from free-range cows, cheese is produced on family farms using wooden tools. Cheese is consumed after about 15–30 days of ageing, when its main characteristics have become an intense taste, a semi-hard mass, and a yellowish colour. However the production process does not meet the Brazilian food safety regulations, which require at least 60 days of maturation for raw milk cheeses. Before this period, the sale of cheeses made from raw milk is prohibited.

Taking raw milk cheese as an example, the U.S. Centers for Disease Control and Prevention (CDC) has reported that whilst unpasteurized dairy products are consumed by only a small percentage of US dairy consumers, these unpasteurized products cause 840 times more illnesses and 45 times more hospitalizations than pasteurized dairy products. Researchers used publicly available outbreak data to look at the correlation between the consumption of raw milk and cheese and bacterial illnesses such as Shiga toxin–producing *E. coli*, *Salmonella* spp., *Listeria monocytogenes*, and *Campylobacter* spp. Most of the U.S. illnesses analyzed were linked to *Salmonella* spp. and *Campylobacter* spp. The authors of the CDC study believe that outbreak-related illnesses will increase steadily as unpasteurized dairy consumption grows, likely driven largely by salmonellosis and campylobacteriosis (Costard, Espejo, Groenendaal, & Zagmutt, 2017). A recent outbreak of food poisoning involving children and staff members at a Swiss boarding school who consumed Tomme, a soft cheese produced from raw cow milk was caused by bovine *Staphylococcus aureus* strains exclusively associated with very high within-herd prevalence of mastitis and has been described as a major contaminant in Swiss raw milk cheese. The raw milk used for Tomme production was contaminated with

Staph. aureus and levels further increased due to growth of the organism and physical concentration effects during the cheese-making process. In the light of this outbreak and the possible occurrence of other foodborne pathogens in bovine milk, consumption of raw milk and soft cheese produced from raw milk constitutes a health risk, particularly when young children or other members of sensitive populations are involved (Johler et al., 2015). Meanwhile, in the EU Regulation (EU) No. 853/2004 laying down specific hygiene rules for food of animal origin includes additional labelling requirements for raw milk, specifying that in addition to the requirements of Regulation (EU) No. 1169/2011, the labelling must clearly show: (1) in the case of raw milk intended for direct human consumption, the words 'raw milk'; (2) in the case of products made with raw milk, the manufacturing process for which does not include any heat treatment or any physical or chemical treatment, the words 'made with raw milk'.

Notwithstanding the above, Thome da Cruz and Menasche (2014) have noted that besides adapting production and expertise to food safety regulations, the valorisation of traditional foods such as Serrano cheese depends firstly on understanding the risk perceptions related to these products, and secondly on recognising and legitimating traditional forms of knowledge. Consumers' perceptions about Serrano Cheese indicated a range of precautions and knowledge of the quantity and quality of cheeses that can be consumed, and suggested that that rather than taking into account only quantitative data analysis when evaluating the production structural requirements and the quality of final foodstuffs—the approach used by conventional methods of risk analysis—cultural factors such as local eating habits; the production scale of traditional, artisanal products; and qualitative elements such as perceptions, values and layperson knowledge. The argument is made that this should be afforded an equal level of importance as the scientific risk analysis approach, which consists of risk assessment, risk management and risk communication, a process that allows risks to be scientifically estimated by technical studies. However, harmonised risk analysis principles facilitate international trade in foods (FAO/WHO, 2007). It might not be possible to fully preserve the very characteristics that make traditional foods unique, diverse and desirable in so far as the purist's version of these products that adhere strictly to traditional methods of production may fall foul of international food safety standards that are required to be met in order to be suitable for export. Viewed in this way, a consumer outside of Brazil might be able to buy a version of Serrano cheese that has been 'santised' for mass consumption both in the literal and figurative sense, but she may still have to make the pilgrimage to Campos de Cima da Serra in order to experience 'the real thing'.

12 The Future: Traditional Foods in a Global Marketplace

Turning to consider how traditional foods will be regulated going forward, it is important to recognise that whereas the making of traditional products and the cooking of traditional recipes has been regulated at the local level through social and cultural norms and practices developed over time, as well as local laws and regulations providing for food safety, traditional foods must also comply with formal food safety rules when they travel away from their location of origin. However, these formal requirements, when satisfied, can lead to a host of economic opportunities and advantages, and even special treatment in some jurisdictions, owing to their continued proven safe usage.

The Novel foods Regulation EU 2015/2283 came into force in the EU in January 2018 and replaced the original Novel Foods Regulation EC 258/97. This section considers developments in this regulatory area.

Article 1 of in the General Food Law 178/2002 states:

> This Regulation provides the basis for the assurance of a high level of protection of human health and consumers' interest in relation to food, taking into account in particular the diversity in the supply of food including traditional products, whilst ensuring the effective functioning of the internal market.

After a number of years of debate and consultation, the European Parliament and Council enacted the Novel Food Regulation EC 258/97, the regulation applied to novel foods and ingredients that had not been "previously used to a significant degree for human consumption" within the EU before May 15, 1997. These foods included foods derived from new sources (e.g. oil rich in omega-3 fatty acids from krill) or food obtained through the application of new technologies (e.g. phytosterols).

At first, relatively few new foods sources were approved under the Regulation. The regulatory procedure was rigorous and expensive and effectively blocked access to the lucrative EU market for a range of foods from exporting developing countries even though they could demonstrate a history of safe use. Noni juice, Chia seeds and Baobab are examples of the few successful applicants from outside the EU that were granted approval under the Regulation. Within the EU, Benecol, a margarine containing stanol esters that happened to be the first 'functional food' introduced in Europe in 1995, was exempted from the novel foods regulation. However, other sterol foods without a proven history of safe use had to obtain regulatory approval (Zawistowski & Jones, 2015). In 2000, the European Commission (EC) approved sterol esters as novel food ingredients and authorized their use in yellow fat spreads for lowering plasma cholesterol levels in humans. In 2004, the EC extended approval for a number of foods enriched with plant sterols, stanols, and their esters including milk based beverages; yogurt-type products; and other fermented milk-type drinks. Applications for uses of plant sterols in meat products, cooking and frying fats, and animal fats, bakery products other than rye bread, juices, and nectars were not approved.

Debate continued and a recasting of the Novel Foods Regulation in 2008 was rejected by the European Parliament due to divergence of opinion on the issue of animal cloning. The new Novel Foods Regulation (EU 2015/2283) eventually came into force in January 2018. In the new Regulation, the definition of "novel foods" has been made more detailed and a centralised assessment and authorisation procedure introduced.

Under the new regime, traditional foods from third countries (countries outside of the EU) derived from primary production with a history of safe use can be classed as a type of novel food for the purposes of EU regulation. These include traditional foods made from plants, microorganisms, fungi, algae and animals (e.g. insects). History of safe use means that the safety of the food in question has been confirmed with compositional data and from experience of continued use for at least 25 years in the customary diet of a significant number of people in at least one third country.

To take an example, in the European Union consumption of edible insects is a niche market however in Asia, Africa, Latin America, and Australia insects have been traditionally and safely consumed. Insects are an alternative source of protein for a growing world population compared to conventional animal breeding and notwithstanding any cultural aversion, consumption of insect-based foods is expected to grow in Europe in the coming decades. Consequently, the new Regulation on Novel Foods is expected to lead to an increasing offer of traditional insect products, as well as many other previously unknown or unavailable traditional food products, on the European market.

In this way, the Novel Foods Regulation may yet prove to be an example of a successful attempt to reconcile the two sometimes divergent aims of celebrating third country traditional food products (and the cultures they originate from) by allowing them access to the European market, while also ensuring modern, science-based standards of food safety are satisfied, by recognising certain traditional foods as 'traditionally safe'.

13 Conclusion

Traditional foods are products that have a special connection to a people and/or a place that has endured over generations. In an increasingly globalised food system, the producers of these products can capitalise on the association that their product has with the region it is made in, what it is made from, and the way it is made. However, the fact that something has traditionally been produced and eaten somewhere does not exempt it from the system of science-based risk assessment and reporting that characterises modern food safety regulation; a regulatory environment with international, regional, national, and local dimensions. Food, be it traditional or in some way novel, must firstly be safe. Moving beyond the basic requirement for safety, traditional foods can also become associated with enhanced quality in the minds of consumers, however this is not always necessarily the case. Nonetheless, the value of the intellectual property that vests in a traditional food products is recognised and geographical indications of food products are now a common and often contentious feature of international trade negotiations.

Although quality is highly subjective, notions of authenticity and products made according to tried and tested recipes in the places that they have been made for generations do appear to resonate with a group of modern food consumers. These consumers, spread throughout the world's richer nations, inhabit a globalised food system where food is viewed less as a staple and more as a consumer product that can carry with it certain values and associations that can be expressed through the act of purchase and consumption. Viewed in this way, traditional foods can also appear to the consumer as something special or unique, in contrast to increasingly homogenised food products that are produced by an increasingly globalised agri-food industry. Because of this, traditional foods can also become a way for a community to protect and promote its traditional identity in an increasingly commodified and commercialised world.

However, although there are benefits to be gained from launching a 'successful' traditional food onto the world stage, there are also possible disadvantages, and the global popularity of products that celebrate an authentic, artisanal production method, or a connection to geographically limited area, can also operate to disrupt traditional ways of life and destroy the underlying attributes that make a traditional product unique and popular in the first place.

Although this and many other issues that impact on how traditional foods are and will be regulated have only been obliquely referred to in this chapter, it is hoped that the frameworks and issues identified serve to stimulate further discussion and debate around what traditional foods are, what the traditional foods of the future might be, and how they should be regulated so as to ensure safety, while also preserving and promoting their unique characteristics.

References

Albayrak, M., & Gunes, E. (2010). Traditional foods: Interaction between local and global foods in Turkey. *African Journal of Business Management, 4*(4), 555–561.

Al-Kandari, D., & Jukes, D. J. (2009). A situation analysis of the food control systems in Arab Gulf Cooperation Council (GCC) countries. *Food Control, 20*, 1112–1118.

Allanson, P. (1992). Farm size and structure in England and Wales 1939–89. *Journal of Agricultural Economics, 43*(2), 137–148.

Almli, V. L., Verbeke, W., Vanhonacker, F., Næs, T., & Hersleth, M. (2011). General image and attribute perceptions of traditional food in six European countries. *Food Quality and Preference, 22*(1), 129–138.

Anelich, L. E. (2013). African perspectives on the need for global harmonisation of food safety regulations. *Journal of the Science of Food and Agriculture, 94*(10), 1919–1921.

Askegaard, S., & Madsen, T. K. (1998). The local and the global: Exploring traits of homogeneity and heterogeneity in European food cultures. *International Business Review, 7*(6), 549–568.

Atkins, P., & Bowler, I. (2000). *Food in society*. London: Arnold.

Balogh, P., Békési, D., Gorton, M., Popp, J., & Lengyel, P. (2016). Consumer willingness to pay for traditional food products. *Food Policy, 61*, 176–184.

Belletti, G., Marescotti, A., & Touzard, J. M. (2017). Geographical indications, public goods, and sustainable development: The roles of actors' strategies and public policies. *World Development, 98*, 45–57.

Bertozzi, L. (1998). Tipicidad alimentaria y dieta mediterránea. In A. Medina, F. Medina, & G. Colesanti (Eds.), *El color de la alimentación mediterránea.Elementos sensoriales y culturales de la nutrición* (pp. 15–41). Barcelona: Icaria.

Bowler, I. (Ed.). (1992). *The geography of agriculture in developed market economies*. Harlow: Longman Scientific and Technical.

Boza, S., & Munoz, J. (2016, December). *Traditional food products and trade: Exploring the linkages*. World Trade Institute Working paper no. 17/2016.

Bramley, C., & Kirsten, J. F. (2007). Exploring the economic rationale for protecting geographical indications in agriculture. *Agrekon, 46*(1), 69–93.

Brown, S., Kozinets, R. V., & Sherry, J. F. (2003). Teaching old brands new tricks: Retro branding and revival of brand meaning. *Journal of Marketing, 67*, 19.

Cerjak, M., Haas, R., Brunner, T., & Tomić, M. (2014). What motivates consumers to buy traditional food products? Evidence from Croatia and Austria using word association and laddering interviews. *British Food Journal, 116*, 1726–1747.

Champredonde, M., & González, J. (2016). Added value or valorization? Reflections from Apellations of origin in Latin America. *RIVAR, 3*(9), 139–163.

Costard, S., Espejo, L., Groenendaal, H., & Zagmutt, F. J. (2017). Outbreak-related disease burden associated with consumption of unpasteurized cow's milk and cheese, United States, 2009–2014. *Emerging Infectious Diseases, 23*(6), 957–964. https://doi.org/10.3201/eid2306.151603

Crawford, C. W. (1954). *The long fight for pure foods*. USDA Yearbook 211, Washington, DC.

DeSoucey, M. (2010). Gastronationalism: Food traditions and authenticity politics in the European Union. *American Sociological Review, 75*(3), 432–455.

Dixon, J. (1999). A cultural economy for studying food systems. *Agriculture and Human Values, 16*, 151–160.

EC. (2012, December). Regulation (EC) No 1151/2012 of the European Parliament and of the Council of 21 November 2012 on quality schemes for agricultural products and foodstuffs L343 (pp. 1–28).

EC. (2012, October). *Value of production of agricultural products and foodstuffs, wines, aromatised wines and spirits protected by a geographical indication (GI)*. Final Report.

EC. (2018, September 30). AgriFood and Trade Statistics. Retrieved from https://ec.europa.eu/agriculture/sites/agriculture/files/trade-analysis/statistics/outside-eu/regions/agrifood-sub-saharan-countries_en.pdf

Echols, M. A. (2001). *Food safety and the WTO: The interplay of culture, science and technology*. London: Kluwer Law International.

Errázuriz, C. (2010). Indicaciones geográficas y denominaciones de origen. *Revista Chilena de Derecho, 37*(2), 207–239.

European Federation of Origin Wines (EFOW). (2016). *International rules applicable to origin wines*. International Agreements. Retrieved from http://efow.eu/internationalrules-applicable-to-origin-wines/

FAO. (1995). Food, Nutrition and Agriculture—15—Food safety and trade. Food standards and Codex Alimentarius in the context of MERCOSUR. Rome: FAO.

FAO. (2004, May 5–7). *Food safety and quality in Europe: Aspects concerning in particular quality, nutritional balance, the importance of agricultural land and cultural heritage (Terroirs)*. 24th Regional Conference for Europe, Item 6, Montpellier, France.

FAO/WHO. (2005). *Practical actions to promote food safety*. Rome: Food and Agricultural Organization of the United Nations.

FAO/WHO. (2007). Food and Agriculture Organization of the United Nations World Health Organization, Food safety risk analysis: A guide for national food safety authorities. Food and nutrition paper 87. Retrieved October 1, 2018, from http://www.fao.org/docrep/012/a0822e/a0822e00.htm

Fine, B., & Leopold, E. (1994). *The world of consumption*. London: Routledge.

Galli, F. (2018). Traditional foods: Definitions and nuances. In A. Cavicchi., & C. Santini (Eds.), Case Studies in the Traditional Food Sector 1st Ed. (pp.3–24). Cambridge: Woodhead Publishing.

Gervais, D. (2010). Reinventing Lisbon: The case for a protocol to the Lisbon agreement (geographical indications). *Chicago Journal of International Law, 11*(1), 67–126.

Goodman, D., & Watts, M. (1997). *Globalising food: Global questions and global restructuring*. London: Routledge.
Guerrero, L., Guardia, M. D., Xicola, J., Verbeke, W., Vanhonacker, F., Zakowska-Biemans, S., … Hersleth, M. (2009). Consumer-driven definition of traditional food products and innovation in traditional foods. A qualitative cross-cultural study. *Appetite, 52*(2), 345–354.
Heath, C. (2017). How would geographical indications from Asia fare in Europe? In I. Calboli & N. W. Loon (Eds.), *Geographical indications at the crossroads of trade, development and culture: Focus on Asia* (pp. 186–211). Cambridge, UK: Cambridge University Press.
Herrmann, R., & Marauhn, T. (2009). Special section: Geographical indicators. *Estey Centre Journal of International Law and Trade Policy, 10*(1), 12–195.
Higgins, D. (2018). EU Policy on Geographical Indications. In Brands, Geographical Origin, and the Global Economy: A History from the Nineteenth Century to the Present (Cambridge Studies in the Emergence of Global Enterprise, pp. 218–248). Cambridge: Cambridge University Press.
Ilbery, B., & Kneafsey, M. (1998). Product and place: Promoting quality products and services in the lagging regions of the European Union. *European Urban and Regional Studies, 5*(4), 330–341.
Ilbery, B., & Kneafsey, M. (1999a). Niche markets and regional speciality foodstuffs in Europe: Towards a research agenda. *Environment and Planning A, 31*, 2207–2222.
Ilbery, B., & Kneafsey, M. (1999b). *Regional imagery and the promotion of quality products and services in the lagging regions of the European Union*, Final Report. Coventry: Coventry University, Department of Geography.
International Trade Centre (ITC). (2009). *Guide to geographical indications. Linking products and their origins*. Geneva: ITC.
Jia, C., & Jukes, D. (2013). The national food safety control system of China: A systematic review. *Food Control, 32*(1), 236–245.
Johler, S., Weder, D., Bridy, C., Huguenin, M. C., Robert, L., Hummerjohann, J., & Stephan, R. (2015). Outbreak of staphylococcal food poisoning among children and staff at a Swiss boarding school due to soft cheese made from raw milk. *Journal of Dairy Science, 98*(5), 2944–2948. https://doi.org/10.3168/jds.2014-9123
Karabudak, E., Bas, M., & Kiziltan, G. (2008). Food safety in the home consumption of meat in Turkey. *Food Control, 19*, 320–327.
Kendall, H., Kuznesof, S., Dean, M., Chan, M. Y., Clark, B., Home, R., … Frewer, L. (2019). Chinese consumers' attitudes, perceptions and behavioural responses towards food fraud. *Food Control, 95*, 339–351. https://doi.org/10.1016/j.foodcont.2018.08.006
Kühne, B., Vanhonacker, F., Gellynk, X., & Verbeke, W. (2010). Innovation in traditional food products in Europe: Do sector innovation activities match consumers' acceptance? *Food Quality and Preference, 21*, 629–638.
Lam, S., Pham, G., & Nguyen-Viet, H. (2018). Emerging health risks from agricultural intensification in Southeast Asia: A systematic review. *International Journal of Occupational and Environmental Health*. https://doi.org/10.1080/10773525.2018.1450923
Levitt, T. (1983). The globalization of markets. *Harvard Business Review, 61*, 92–93.
Lewis, D., & Bridger, D. (2001). *The soul of the new consumer*. London: Nicholas Brealey.
Liu, X. (2014). International perspectives on food safety and regulations—A need for harmonized regulations: Perspectives in China. *Journal of the Science of Food and Agriculture, 94*(10), 1928–1931.
Lowe, P., Cox, G., MacEwan, M., O'Riordan, T., & Winter, M. (1986). *Countryside conflicts: The politics of farming, forestry and conservation*. Aldershot: Gower/Maurice Temple Smith.
Ma, G. (2015). Food, eating behavior, and culture in Chinese society. *Journal of Ethnic Foods, 2*(4), 195–199. https://doi.org/10.1016/j.jef.2015.11.004
MacMaolain, C. (2015). *Food law: European, domestic, and international frameworks*. Oxford: Hart Publishing.
Mancini, M. C. (2013). Geographical indications in Latin America value chains: "A branding from below strategy" or a mechanism excluding the poorest? *Journal of Rural Studies, 32*, 295–306.

Makun, H. A., Dutton, M. F., Njobeh, P. B., Gbodi, T. A., & Ogbadu, G. H. (2012). Aflatoxin contamination of foods and feeds: A special focus on Africa. In A. A. Eissa (Ed.), *Trends in vital food and control engineering* (pp. 187–234). London: InTech. https://doi.org/10.5772/24919

Marsden, T., Murdoch, J., Lowe, P., Munton, R. C., & Flynn, A. (1993). Constructing the countryside. London: UCL Press.

McGovern, P. E., Juzhong, Z., Jigen, T., Zhiqing, Z., Hall, G. R., Moreau, R. A., ... Wang, C. (2004). Fermented beverages of pre- and proto-historic China. *Proceedings of the National Academy of Sciences of the United States of America, 101*(51), 17593–17598. https://doi.org/10.1073/pnas.0407921102

Miele, M. (2001). *Creating sustainability*. Wageningen: University of Wageningen.

Miller, D. (1998). *A theory of shopping*. Oxford: Berg.

Miller, D., Jackson, P., Rowlands, M., Thrift, N., & Holbrook, B. (1998). *Shopping, place and identity*. London: Routledge.

Molnár, A., Gellynck, X., Vanhonacker, F., Gagalyuk, T., & Verbeke, W. (2011). Do chain goals match consumer perceptions? The case of the traditional food sector in selected European Union countries. *Agribusiness, 27*, 221–243.

Morgan, K., & Murdoch, J. (2000). Knowing networks? Knowledge, power and innovation in the food chain. *Geoforum, 31*(2), 159–173.

Morris, C., & Young, C. (2000). "Seed to Shelf ", "Teat to Table", "Barley to Beer" and womb to tomb: Discourses of food quality and quality assurance schemes in the UK. *Journal of Rural Studies, 16*(1), 103–115.

Mulholland, R. (2000, September 18). The Roquefort revolutionary. *The Independent on Sunday*, pp. 26–29.

Murdoch, J., Marsden, T., & Banks, J. (2000). Quality, nature and embeddedness: Some theoretical considerations in the context of the food sector. *Economic Geography, 76*(2), 107–125.

Murdoch, J., & Miele, M. (1999). "Back to nature": Changing worlds of production in the food sector. *Sociologia Ruralis, 39*(4), 465–483.

Nygard, B., & Storstad, O. (1998). Deglobalisation of food markets? Consumer perceptions of safe food; the case of Norway. *Sociologia Ruralis, 38*, 35–53.

O'Connor, B. (2004). *The law of geographical indications*. London: Cameron May.

OECD. (1995). *Niche markets as a rural development strategy* (p. 13). Paris: OECD.

Othman, M. N. (2007). Food safety in southeast Asia: Challenges facing the region. *Asian Journal of Agriculture and Development, 4*(2), 83–92.

Parrot, N., Wilson, N., & Murdoch, J. (2002). Spatializing quality: Regional protection and the alternative geography of food. *European Urban and Regional Studies, 9*(3), 241–261.

Pieniak, Z., Verbeke, W., Vanhonacker, F., Guerrero, L., & Hersleth, M. (2009). Association between traditional food consumption and motives for food choice in six European countries. *Appetite, 53*, 101–108.

Quinones-Ruiz, X. F., Penker, M., Belletti, G., Marescotti, A., & Scaramuzzi, S. (2016). Why early collective action pays off: Evidence from setting protected geographical indications. *Renewable Agriculture and Food Systems, 32*, 179-192.

Rai, R. K. (2009). The TRIPS article 23 extension stalemate continues: A way-ahead for the developing countries. *Journal of Intellectual Property Rights, 14*, 405–422.

Rangnekar, D. (2003). *Geographical indications. A review of proposals at the TRIPS council: Extending article 23 to products other than wines and spirits*. [Issue Paper No. 4]. Geneva: UNCTAD-ICTSD.

Rangnekar, D. (2010). The law and economics of geographical indications: Introduction to special issue of the journal of world intellectual property. *The Journal of World Intellectual Property, 13*(2), 77–80.

Rudawska, E. D. (2014). Customer loyalty towards traditional products—Polish market experience. *British Food Journal, 116*, 1710–1725.

Sgarbi, J., & Menasche, R. (2015). Valorização de produtos alimentares tradicionais: os usos das indicações geográficas no contexto brasileiros. *Cuadernos de Desarrollo Rural, 12*(75), 11–31.

Shaw, A. (1999). "What are 'they' doing to our food?" Public concerns about food in the UK. *Sociological Research Online, 4*(3). Retrieved from http://www.socresonline.org.uk/socresonline/4/3/shaw.html

Sudershan, R. V., Gavaravarapu, S. R. M., Jasti, P. R., Rao, M. V. V., & Polasa, K. (2008). Food safety related perceptions and practices of mothers—A case study in Hyderabad, India. *Food Control, 19*(5), 506–513.

Sylvander, B. (1995). "Specific quality products"; an opportunity for rural areas. *LEADER Magazine, 3*(3), 28–41.

Sylvander, B. (1998). *Raw milk or not raw milk? Procedural working out of a local agreement and innovations systems*. Systems and Trajectories of Innovation. Berkeley, CA: Institute of International Studies, University of California.

Takaki, R. (2012). *Strangers from a different shore: A history of Asian Americans*. Updated and rev. Boston, MA: Little, Brown and Company.

Thevenot, L. (1998). *Innovating in "qualified" markets: Quality, norms and conventions*. Systems and Trajectories of Innovation Conference. Berkeley, CA: Institute of International Studies, University of California.

Thome da Cruz, F., & Menasche, R. (2014). Tradition and diversity jeopardised by food safety regulations? The Serrano Cheese case, Campos de Cima da Serra region. *Food Policy, 45*, 116–124.

Tregear, A., Kuznesof, S., & Moxey, A. (1998). Policy initiatives for regional foods: Some insights from consumer research. *Food Policy, 23*, 383–394.

USDA. (2017). Potential agricultural trade implications of the EU-MERCOSUR FTA Gain Report BR 1712. Retrieved from https://www.fas.usda.gov/data/brazil-potential-agricultural-trade-implications-eu-mercosur-fta. October 1, 2018

Verbeke, W., Guerrero, L., Almli, V. L., Vanhonacker, F., & Hersleth, M. (2016). European consumers' definition and perception of traditional foods. In *Traditional foods* (pp. 3–16). Boston, MA: Springer.

Verbeke, W., & Roosen, J. (2009). Market differentiation potential of country-of-origin, quality and traceability labeling. *The Estey Centre Journal of International Law and Trade Policy, 10*(1), 20–35.

Wagacha, J. M., & Muthomi, J. W. (2008). Mycotoxin problem in Africa: Current status, implications to food safety and health and possible management strategies. *International Journal Food Microbiology, 124*, 1–12.

Whatmore, S. (1994). Global agro-food complexes and the refashioning of rural Europe. In A. Amin & N. Thrift (Eds.), *Globalization, institutions and regional development*. Oxford: Oxford University Press.

Whatmore, S., & Thorne, L. (1997). Nourishing networks: Alternative geographies of food. In D. Goodman & M. J. Watts (Eds.), *Globalising food: Agrarian questions and global restructuring* (pp. 287–304). London: Routledge.

WHO. (2015). *WHO estimates of the global burden of foodborne disease*.

World Trade Organization (WTO). (2005). Issues related to the extension of the protection of geographical indications provided for in Article 23 of the TRIPS Agreement to products other than wines and spirits. Compilation of Issues Raised and Views Expressed [PDF document]. Retrieved from https://www.wto.org/english/tratop_e/trips_e/ta_docs_e/5_2_wtgcw546_e.pdf

World Trade Organization (WTO). (2016). Annex 1C. Agreement on Trade-Related Aspects of Intellectual Property Rights [PDF document]. Retrieved from https://www.wto.org/english/docs_e/legal_e/27-trips.pdf

Xue, J., & Zhang, W. (2013). Understanding China's food safety problem: An analysis of 2387 incidents of acute foodborne illness. *Food Control, 30 (1)*, 311–317.

Zawistowski, J., & Jones, P. (2015). Regulatory aspects related to plant sterol and stanol supplemented foods. *Journal of AOAC International, 98*(3), 750–756.

Zhang, L., Xu, Y., Oosterveer, P., & Mol, A. P. (2016). Consumer trust in different food provisioning schemes: Evidence from Beijing China. *China Journal of Cleaner Production, 134*, 269–279.

Zhang, W., & Xue, J. (2016). Economically motivated food fraud and adulteration in China: An analysis based on 1553 media reports. *Food Control, 67*, 192–198.

Zografos, D. (2010). *Intellectual property and traditional cultural expressions*. Cheltenham: Edward Elgar Publishing.

Chapter 10
Traditional Foods, Sensory Excitements and Pleasure

Mohammad Shafiur Rahman

1 Introduction

Quality is usually defined as the degree of excellent to the consumers, and it is to satisfy the needs and expectations of the consumers and customers (Peri, 2006). Safety is the first element of quality followed by sensory perceptions and expectations. All traditional foods need to satisfy safety, and should have desired taste, flavor and promote health. Usually, texture and aroma confirmed the uniqueness of traditional products in terms of their sensory characteristics (Berodier, Lavanchy, et al., 1997). Similarly, aroma or odor was identified as a key attribute influencing the overall liking of the traditional product (Kalviainen, Roinonen, & Tuorila, 2003). Traditional foods are usually considered taste good and preference to the consumers. However, in reality it could be perceived either good for health (i.e. natural products, without chemical modification and additives) or bad for health (i.e. high fat, salt and sugar contents; and high microbial and other contaminants) (Cayot, 2007). Safety is important issue of traditional foods, for example traditional cheeses for raw milk. In many countries, these products are forbidden for export. It is important to characterize the typical sensory properties of traditional products, especially when Protected Geographical Indication (PGI) or Protected Designation of Origin (PDO) labels are used (Cayot, 2007).

Traditional food products are specially distinguished and known for their unique sensory properties, and these are associated and linked to a certain local area, region, or country (Guerrero et al., 2009). Traditional foods are prominent and noticeable to a region's individuality, history, culinary heritage and local values. These are perceived as new foods when they are introduced into a new territory (Januszewska & Viaene, 2012). Traditional foods and dishes are a part of the national legacy, and

M. S. Rahman (✉)
Department of Food Science and Nutrition, College of Agricultural and Marine Sciences, Sultan Qaboos University, Muscat, Oman
e-mail: shafiur@squ.edu.om

M. Al-Khusaibi et al. (eds.), *Traditional Foods*, Food Engineering Series,
https://doi.org/10.1007/978-3-030-24620-4_10

today these play important roles in the recognition of a country or a region. In many cases, sensory characteristics are not clearly defined and chemical compositions are not well established for majority of traditional dishes, and this caused difficulty in identifying authentic products and quality control. However, traditional recipes and steps in preparation are generally well known and preserved (Korosec, Golob, Skvarca, & Bertoncelj, 2012). This chapter focusses on the sensory aspects of traditional foods.

2 Consumers Diversity

Mak, Lumbers, Eves, and Chang (2012) reviewed the factors affecting food consumption by tourist. They identified five socio-cultural and psychological factors: cultural/religious influences, socio-demographic factors, food-related personality traits, exposure effect/past experience, and motivational factors. The motivational factors can have five main dimensions: (1) symbolic (i.e. visit a destination by its cultural factors include trying local traditional food in order to explore the local food culture), (2) obligatory (i.e. essentiality of food consumption factors, such as health concern, and physical need for sustenance), (3) contrast (i.e. seek contrast from the daily routine), (4) experience, and (5) extension, and pleasure (i.e. sensory experience and appeal, and togetherness) (Mak et al., 2012). Kim, Eves, and Scarles (2009) identified nine motivational factors underlying the consumption of local traditional foods. These are exciting experience, escape from routine, health concern, learning knowledge, authentic experience, togetherness, prestige, sensory appeal, and physical environment. Therefore, traditional foods play an important role in economic gain from tourists.

Consumers differ in neophilic (i.e. variety-seeking) and food neophobic (i.e. less reluctant to try new things) and they may perceive their own traditional foods differently than traditional foods from other countries. Neophilic people are active searchers or variety seekers. They are likely to try novel foods (Pliner & Melo, 1997) and tend to give higher scores on willingness-to-try scales (Olabi, Najm, Baghdadi, & Morton, 2009). However, their choices could be in addition to liking or sensory preference (Mustonen & Tuorila, 2010). Neophobic people are afraid of new foods and drinks and they have a tendency to avoid new products (Januszewska & Viaene, 2012). Januszewska and Viaene (2012) studied the relationship between variety-seeking and food neophobia in application to liking of traditional products. The neophobic segments showed a strong negative attitude toward the consumption of new foods. The variety-seeking consumers gave significantly higher liking scores than neophobes for the majority of traditional products. The odor liking score was the best sensory discriminating descriptor between the neophilics and the neophobics and their willingness to buy scores on three tradition foods depended on the types of traditional products. These two segments of the population on education level, income, or gender. Wu, Raab, Chang, and Krishen (2016) studied the factors affecting Chinese tourists' food consumption in the United States. It was identified that the concerns for food safety (negatively) and dining table manners could aggravate Chinese tourists' attitudes toward consuming unfamiliar local foods. However, communication and the food's sensory appeal are not significant in predicting their attitudes toward local foods, and the food neophobia (i.e., a food-related personality trait) moderates Chinese tourists' decisions in relation to local food consumption.

Wang, De Steur, Gellynck, and Verbeke (2015) studied the Chinese consumers' choice of traditional food and European food. A new Food Choice Motive of mainland Chinese consumers encompasses six dimensions: health concern, time or money saving, sensory appeal, availability and familiarity, mood and food safety concern. Availability, familiarity, mood, food safety affected positively, while money saving negatively affected the attitude toward traditional foods. Sensory appeal and attitude were positively linked to purchase intentions, while mood was negatively linked to the purchase intention toward traditional food.

Vanhonacker et al. (2013) studied the innovations in traditional foods and their impact on perceived traditional character and consumer acceptance. In general, consumers were open towards innovations in traditional food products. The acceptance was related to a reinforcement of the traditional character (e.g. a label that guarantees the origin of the raw material), benefits that either outweigh the importance of the traditional character, or improve negative attributes associated with the traditional character (e.g. reduction of fat content). In general, consumers are open towards innovations in traditional food products.

3 Traditional Spices to Enhance Vegetable Consumption

There is a growing interest in the value and use of healthy foods (Kramer & Twigg, 1970). It is important to intake sufficient fruits and vegetables, which contains essential components of a healthy diet. It contains sources for vitamins, dietary fibers, and other health functional bioactive compounds and reduce the risk of many diseases, such as cancer, stroke, and cardiovascular diseases (Feng et al., 2018). Vegetables have advantage over fruits due to low level of sugars. Vegetables lack in natural liked sweet and savory attributes with innately disliked bitter attributes (Poelman, Delahunty, & de Graaf, 2015). These are disliked mainly due to strong bitter attributes, which suppresses endogenous sweetness, enhances disliked vegetable aromas (Dinehart, Hayes, Bartoshuk, Lanier, & Duffy, 2006; Sharafi, Hayes, & Duffy, 2013). A substantial portions of these unpleasant tastes are due to the phytonutrients, such as naringin (bitter flavonoid) (Drewnowski & Gomez-Carneros, 2000), glucosinolates in broccoli (Schonhof, Krumbein, & Bruckner, 2004), allyl isothiocyanate, dimethyl trisulfide, dimethyl sulfide, and methanethiol in cauliflower (Engel, Baty, Le Corre, Souchon, & Martin, 2002). In order to maintain the health functionality of vegetables, it is better option to masking instead of removing (Sharafi et al., 2013). Salt is capable of masking bitter attributes

(Breslin & Beauchamp, 1997). The alternative option is to use herbs and spices to improve sensory properties of vegetables since salt has high concern on the issue of hypertension (Feng et al., 2018; Manero et al., 2017; Nikolaus, Ellison, Heinrichs, Nickols-Richardson, & Chapman-Novakofski, 2017). Each vegetable contains its distinctive characteristics and flavors, thus the selections of the spices and herbs could be varied (Feng et al., 2018). At present, approximately 87% of the US population fails to meet the vegetable intake recommendations (Feng et al., 2018). Therefore, tradition spices and herbs based on ethnicity could play an important role to suppress the bitter or unpleasant taste and could increase the healthy consumptions of vegetables.

4 Sensory Evaluation of Traditional Foods

The sensory methods of particular traditional foods must be simplified and uniform and need to be developed with considering the typicality of the products (Etaio et al., 2010). This would help quality evaluation of specific traditional foods. Januszewska and Viaene (2012) pointed that the same product could be evaluated differently in terms of sensory characteristics across European regions. Traditional foods often possesses special sensory characteristics and a sharper odor and taste than commercial products (Barcenas, Perez de San Roman, Perez Elortondo, & Albisu, 2001). The sensory perceptions of traditional products have cognitive structure and personality of the respondents and these pre-dispositional attributes called traits (Engel & Blackwell, 1982).

First, the evaluation of sensory characteristics of traditional foods consists of establishing sensory profiles. The complexity of the tasks needs longer training periods together with more complex statistical analyses. Sensory evaluations fist need to perform a sorting test on a large number of traditional products using inexperienced panel. It needs to follow a free description of the criteria used to sort the products. A qualitative descriptive analysis could be performed by trained panel. In addition, dynamic measurements such as time–intensity measurements need to be performed (Cayot, 2007). It is important to consider the cultural background with age, gender, smoking habits or disease of the panel members to modulate the olfactory capacity (Thomas-Danguin et al., 2004).

The desired sensory flavor, texture and taste of the traditional foods are the main reasons of their preference by the consumers. The key aroma compounds and desired texture are important for traditional foods. The marketed preference of the traditional foods are due to high quality as consumer perceived, continued consumption of these preferred products (Appledorf, Pennington, Wheeler, & Koburger, 1974). Appledorf et al. (1974) compared 25 health food products with traditional foods. The only three health foods' (odor of apple butter and pizza, and color of ketchup) sensory attributes found to be superior as compared to the traditional foods. All other traditional foods on the other hand were individually preferred for odor, color, texture or flavor.

It is important to understand the relationship of consumption of familiar and transitional foods, and degree of linking and intake (Laureati, Pagliarini, Calcinoni, & Bidoglio, 2006). In an Italian elderly people aged between 57 and 98, it was shown that the basic criteria associated with food choice: simple cooking, tradition and sensory aspects. The most important factor associated with food preference was referred to the sensory attributes (i.e. sight, smell, taste, texture and hearing). Among these, most important one was the taste followed by smell and appearance. The first courses preferred by elderly people were found to be those classified as traditional Italian preparations, while the lowest liking rating were found to be those classified as simple-cooked/unfamiliar and complicated-cooked/unfamiliar dishes (Laureati et al., 2006). They also explored the most frequent reasons supplied by elderly people as an explanation of their choices. It includes childhood (for example, "...since I was a little girl I used to help my mother cook these favorite dishes") and youthful (for example, "...I like Lasagne because they remind me of celebrations and make me happy.") memories and knowledge of the most favorite dishes. In addition, elderly people often suggested local tradition as familiarity ("...I prefer these because they are traditional dishes, of course I appreciate them... I have eaten them so many times that I could not refrain from eating them...") with and knowledge of a specific dish as a reason for their choices. The simple-cooked/familiar courses as dishes from the Italian tradition and past-experiences were also considered, thus confirming that the concept of tradition may result in multiple interpretations (Rappoport, Peters, Huff-Corzine, & Downey, 1992). The protocols of sensory analysis of traditional olive oil (Carbonell-Barrachina, 2007), cottage cheese *Strukljі* (boiled rolls) (Korosec et al., 2012) and Spanish confectionery *turron* (Carbonell-Barrachina, 2007) are presented in the literature.

5 Traditional Food Products

5.1 Meat Products

Considering the oriental meat products, *sukuti* is one of the most valuable traditional meat products in Nepal, where refrigeration is very expensive method of preservation (Halwai, 2005). In traditional method, meat strips (e.g. buffalo meat) are hung over the fireplace in kitchen to dry at low heat and smoke of burning woods. The process is continued until meat becomes adequately dry. Sometimes air and sun drying could be used. In the recent days, pre-spicing (i.e. table salt, rock salt, chili powder, meat masala, and vinegar) the lean meat strips and pre-cooking on its own juice are used followed by conditioning. Crispness is the major textural attributes of this snack meat (Halwai, 2005).

5.2 Traditional Cheeses

There are different types of traditional cottage cheeses in Europe and other countries. In seventeenth century, the cottage cheese *Strukljі* has arisen from Slovenian cuisine, and it can be found on traditional holiday menus in the Slovenian regions and typical versions are available in restaurants and public canteens, and tourist farms. Strukljі are prepared mostly either filo (or phyllo) pastry or leavened dough, and served as a savory side dish with meat or sweetened as a dessert. In addition, walnuts and raisins are often added to the cottage cheese filling (Korosec et al., 2012). Korosec et al. (2012) measured sensory attributes (i.e. by descriptive method) of ten samples of cottage cheese Strukljі collected from different sources in order to preserve the authenticity and quality control. They developed the guidelines for training of sensory assessors of this dish, principal sensory parameters and attributes, and criterial of assessment and analysis techniques. The parameters and attributes are presented in Table 10.1.

The principal sensory parameters were discussed and determined during initial sessions as similarly used by Etaio et al. (2010) for red wines. The flavor was determined by Flavor Profile Method (Meilgaard, Civille, & Carr, 2007; Stone & Sidel, 2004). The intensities of the flavor attributes were evaluated according to a 5-point scale with anchor words: not present, threshold, slight, moderate, strong (Keane, 1992). The sensory parameters were scored on 5-point verbal scale regarding distinctiveness as 1: untypically, and 5: very typically. Spider diagram showed least differences in the appearance, while the highest variability was determined in taste. *Strukljі* from a tourist farm were assessed as the least typical in nearly all parameters. Less variability was observed in the samples produced in kitchens (such as inns and restaurants) with higher frequency of production. In addition, selection of ingredients affected noticeably the sensory quality of the final product (Korosec et al., 2012).

The intensity of perceived aroma attributes was reported with 5-point intensity scale (i.e. 0: not present, 1: threshold, 2: slight, 3: moderate, 4: strong). The flavor attributes were grouped to positive and negative to the product based on the

Table 10.1 Principal sensory parameters defining the quality of cottage cheese *Strukljі* and their attributes

Sensory parameter	Attributes
Overall appearance	Surface color, roughness, evenness, gloss, secretion of starch
Appearance of cross section	Color of cross section, amount of filling, linkage of filling with dough
Odor	Intensity and balance of odor attributes, absence of off-odors
Taste	Intensity and balance of perceived salt, sweet, sour, bitter taste
Texture	Adhesivity, smoothness, coarseness
Flavor	Intensity, persistence and type of aroma attributes, balance of aroma attributes, mouth feeling, absence of off-flavors

Source: Korosec et al. (2012)

ingredient filo pastry and leavened dough (Table 10.2). Some attributes were common for both the ingredients. The positive attributes were cooked curd, cooked dough, butter, crumbs, walnuts, milky, lemon, raisin, spicy (sweet), lemon zest, and rum; while the negative ones were flour, rancid, fish, fusty, sour, egg, plastic, yeasty, honey, cardboard, soapy, bitter, condensed milk, and oily mouth coat. Among positive flavor attributes, cooked curd (i.e. slight to moderate intensity) was perceived in all samples, while spicy sweet aroma was detected only when leavened dough was used. In addition, walnut flavor was distinctive as moderate to strong. Samples from industrial showed most intense negative aroma attributes of flour and fish, while samples from tourist farm showed remarkable intensity of plastic. These could be from preparation and freezer (Korosec et al., 2012).

Cayot (2007) discussed the sensory analysis of traditional cheese. Volatile fractions were analyzed to characterize five commercial Camembert-type cheeses among which three were made from raw milk (Kubickova & Grosch, 1997; Peres, Begnaud, & Berdague, 2002). More than 100 volatile compounds were identified. A collaborative study was developed under the COST action 901 and the program FLAIR sensory analysis of cheese (Cayot 2007). The objective was to harmonize the training of the tasting panel and to develop a common method for characterizing hard and semi-hard cheese leading to a common lexicon. The sensory characteristics (i.e. surface, mechanical, geometrical, kinaesthetic and oral sensations) were studied for different cheeses with Protected Designation of Origin (PDO) labels, such as Appenzeller for Switzerland, Parmigiano-Regiano and Fontina for Italy, Mahon for Spain and Comte for France (Lavanchy et al., 1993).

A complete methodology guide are available for the sensory (i.e. smell, aroma and taste) characterization of these types of cheese (Berodier, Lavanchy, et al., 1997; Berodier, Stevenot, & Schlich, 1997). In the case of *Comte* cheese, 91 flavor attributes were identified considering 99 samples from 20 plants. It was then identified 28 key attributes to characterize the cheese and could confirm the PDO level to

Table 10.2 Positive and negative flavor attributes of cottage cheese *Struklji* regarding the type of dough

Struklji from Filo Pastry		*Struklji* from leavened dough	
Positive	Negative	Positive	Negative
Cooked curd	Flour	Cooked curd	Rancid
Cooked dough	Rancid	Cooked dough	Sour
Butter	Fish	Butter	Fried egg
Bread crumbs	Fusty	Bread crumbs	Plastic
Milky	Sour	Milky	Yeasty
Lemon	Fried egg	Lemon	Soapy
	Plastic	Walnut	Honey
	Cardboard	Raisin	Condensed milk
	Soapy	Spicy (sweet)	Oily mouth coat
	Bitter	Lemon zest	
	Oily mouth coat	Rum	

Source: Korosec et al. (2012)

preserve sensory quality of such products. Similarly, descriptive analysis was proposed by Retiveau, Chambers, and Esteve (2005) for French cheeses. They developed general lexicon considering 43 samples of French cheeses and using five-assessor panel; and 31 attributes were established, and classified as seven categories. The five highly trained assessors completed 120 h of training in all aspects of sensory techniques and a minimum of 1000 h of general sensory testing. In addition, they received 6 h of orientation to the reported project.

5.3 Olive Oil

Olive oil is an important cooking and salad oil used extensively in Europe, America and many other countries. It is an integral part of the cultural and culinary heritage of the Mediterranean countries. There are three types of olive oil available: extra virgin, virgin and olive oils. Extra virgin olive oil is directly produced from olives with any chemical treatments. Climate conditions or inappropriate processing conditions, virgin olive oil could have high degree of acidity and/or defective color, flavor or aroma. These oils are termed as "lappante virgin oils" and these must pass through a refining process to adjust these defects. The refining techniques could be caustic-refining, bleaching, and deodorizing, adsorption, and membrane filtration. These oils are termed as "refined olive oil" with chemical structure of olive oil without any distinctive aroma or flavor. These oils are marketed by adding certain quantity of extra virgin, virgin or ordinary virgin olive oil for aroma and flavor, and categorizes commercial "olive oil" (Carbonell-Barrachina, 2007).

In 1956, the International Olive Oil Council (IOOC) was created to adopt rules and standards for authenticity and multidisciplinary activities related olive oil production (IOOC, 2018). They provide general basic vocabulary [COI/T.20/Doc. no. 4; 18 June 1987] and general methodology for the organoleptic assessment of virgin olive oil [COI/T.20/Doc. no.13/Rev. 1; 20 November 1996]. It includes prior knowledge required, standard procedures and criteria of taste panel to perform the sensory analysis of virgin olive oils. In addition, they provide organoleptic assessment of virgin olive oil [COI/ T.20/Doc. no. 15/Rev. 1; 20 November 1996] to determine criteria needed to assess the flavor characteristics and classification method of virgin olive oil (Carbonell-Barrachina, 2007) (Table 10.3).

Table 10.3 Classification of virgin olive oil on flavor characteristics [COI/T.20/Doc. no. 15/Rev. 1; 20 November 1996] IOOC (2018)

Grade	Median defect (*X*)	Fruity attribute (*Y*)
Extra virgin	$X = 0$	$Y > 0$
Virgin	$0 < X \leq 2.5$	$Y > 0$
Ordinary virgin	$2.5 \leq X \leq 6.0$ $X \leq 2.5$	– $Y = 0$
Lampante virgin	$X \geq 6.0$	–

Source: Carbonell-Barrachina (2007)

5.4 *Spanish* Turron

Turron is typical Spanish confectionery products made from toasted almonds, sugar, inverted sugar and honey and manufactured in a traditional way. The key ingredients, almonds and honey must be from Eastern Spain. In 1996, these products received qualifying Protected Geographical Indications (PGI), ratifying the denomination of Spanish origin. Spanish Ministry of Agriculture, Fisheries and Food in 1996 established what should be the percentages of raw materials. In the case of PGI products, it is needed to increase knowledge on the raw material and technological causes for particular sensory attributes and it can establish regulatory assurance (Carbonell-Barrachina, 2007).

Sucrose, honey, and water are placed in a boiler pan and homogenized with an anchor impeller; and concentrated to approximately 80° Brix. At this stage, toasted almonds are added and the whole mass is manually homogenized. This product is called Alicante *turron*. Jijona *turron* can be produced using further processing steps. The homogenized mass is cooled to solid mass and crushed in a stone miller, and the suspension is heated at 50–60 °C for about 1 h with gentle stirring in a special hemispherical boiler pan (i.e. approximately 0.6 m diameter), called bioxet. The first heating process is followed by second heating at 70–80 °C until desired consistency is achieved. The total holding time in the boixet is approximately of 120 min (Chiralt, Galotto, & Fito, 1991; Lluch, Galotto, & Chiralt, 1992). The deformable semi-solid soft mass from boixet becomes hard (i.e. breakable non-flowing semi-solid) when cooled at room temperature. This product is called Jijona *turron* (Carbonell-Barrachina, 2007). In Spain, *turron* is commercially classified according to its almond percentage as shown in Table 10.4.

Jijona *turron* is the most traditional and popular type of *turron* in Spain. The most important quality attributes of Jijona *turron* are its hardness, oiliness, consistency, crumbliness, adhesiveness, almond content, honey aroma, almond aroma,

Table 10.4 Minimum requirements of almond content in different commercial categories of soft and hard *turron* according to the Technical-Sanitary Regulation and Regulating Council of the Specific Denominations of Jijona and Alicante *turron* (RCSDJAT) (Spain)

Type	Category	Almond (%) (BOE, 1996)	Honey (%) (RTS, 1982)	Specific name
Hard	Supreme	60	10	Alicante *turron*
	Extra	46	–	–
	Standard	40	–	–
	Popular	34	–	–
Soft	Supreme	60	10	Jijona *turron*
	Extra	50	–	–
	Standard	44	–	–
	Popular	30	–	–

Note: Other RCSDJAT conditions are satisfied
Source: Carbonell-Barrachina (2007)

and sweetness. It is important that consumers and manufacturers perceived these attributes similar way, and honey and almond could be selected accordingly. Carbonell-Barrachina (2007) studied the effects of different almond cultivars, and toasting conditions on volatile compounds and sensory responses. Total 58 compounds were detected in toasted almonds. These included ketones, aldehydes, pyrazines, alcohols, aromatic hydrocarbons, furans, pyrroles, terpenes, and linear hydrocarbons. However, pyrazines together with furans and pyrroles significantly contributed to the characteristic toasted aroma of almond. It is important to optimize the toasting conditions for each almond cultivars (Vazquez-Araujo, Enguix, Verdu, Garcia-Garcia, & Carbonell-Barrachina, 2008). Concentration of sugars and honey in the boixet heating play an important role to create texture and flavor. Heating process decreased the characteristic honey volatile compounds (i.e. acetoin, benzaldehyde, phenylacetaldehyde, and linalool oxide), while Maillard and Strecker degradation reactions increased furfuryl alcohol, furfural and 5-methyl furfural (Martinez & Chiralt, 1995; Martinez, Chiralt, & Fito, 1997; Martinez, Fito, & Chiralt, 1996; Vazquez-Araujo et al., 2008).

5.5 Traditional Olive and Fermented Olive

Lopez-Lopez, Moreno-Baquero, Rodriguez-Gomez, Garcia-Garcia, and Garrido-Fernandez (2018) studied the sensory properties of traditional Spanish-style table olives and fermented with probiotic *Lactobacillus pentosus* TOMC-LAB2. The attributes were based on gustatory (acid, salty, and bitter) and kinesthetic sensations (hardness, fibrousness, and crispness) with two additional queries: (i.e. overall score, liking and buying predisposition). The consumer's perception trends was based mainly on the salty perception and overall score. In both products, overall score, and buying predisposition were driven by appearance, odor, salty (negatively, i.e. salt had a marked adverse effect), hardness, and crispness. This indicated that probiotic table olives with low salt could be commercialized similar to the traditional table olives.

5.6 Omani Halwa

Halwa is a traditional candy in Oman consumed on a day-to-day basis as well as at formal social and religion occasions. There are three types of Omani *halwa*: white, yellow, and black *halwa* (Al-Shamsi, Rahman, Abdullah, & Claereboudt, 2013). Similar to any food product, consumer segmentation is important for traditional food products. The consumer segmentation based on preferences could be used to develop integrated communications and managements of segmented customers. It is easier and effective if used segments (i.e., coherent) rather than all populations of customers or consumers (i.e., heterogeneous). This also helps retailers to engage communications and contacts of homogeneous customers, thus making easier the decision process for customers, as well as easier marketing of the products (Bock & Uncles, 2002; Gaedeke, 2001; Granzin, Olsen, & Painter, 1998; Griffith, 2010; Wang, 2004; Whelan & Davis, 2006).

Al-Shamsi et al. (2013) conducted a consumers' preference mapping and their behavior with regard to traditional *halwa* using a structured survey questionnaire. The total numbers of participant was 152. They determined factors affecting preferences for Omani *halwa* and their behavior and this would be beneficial to the producers and retailers. In the first step, 22 important attributes (i.e., rank) preferred by the consumers were determined and listed as most important to least important (i.e. 1–10). Five types of consumer groups were identified through hierarchical cluster analysis and nonmetric multidimensional scaling (MDS), and the importance of the attributes was listed for each group (Figs. 10.1 and 10.2). The distribution of populations for selecting *halwa* in each cluster were as follows: cluster GR 1 (26%), cluster GR 2, (14%), cluster GR 3 (21%), cluster GR 4 (18%), and cluster GR 5 (20%), which indicate that nearly similar segments of the consumers belong to each cluster. In general, color, sweetness, appearance, mouthfeel, and solubility were the top five most important attributes considering the whole population. Many attributes were also common for different groups. For examples, color of *halwa* was found to be a common important attribute within each group's top three preferences; sweetness was also within the top five important attributes for four groups; and appearance and mouthfeel were within the top five important attributes for three groups. In addition, solubility was within the top five for two groups.

The second step was used to determine their levels of preferences (scale: 1–9) considering 22 attributes (Table 10.5). Similar to above, five types of groups were also identified. Al-Shamsi et al. (2013) explained the purpose of the two-step questionnaire by providing an example. Color could be considered most important by two consumers (i.e., rank = 1 out of 10 for both), but the intensity level of color preference could be different; for example, one may prefer a very black color (preference level = 9), while another could prefer a yellow color (preference level = 1).

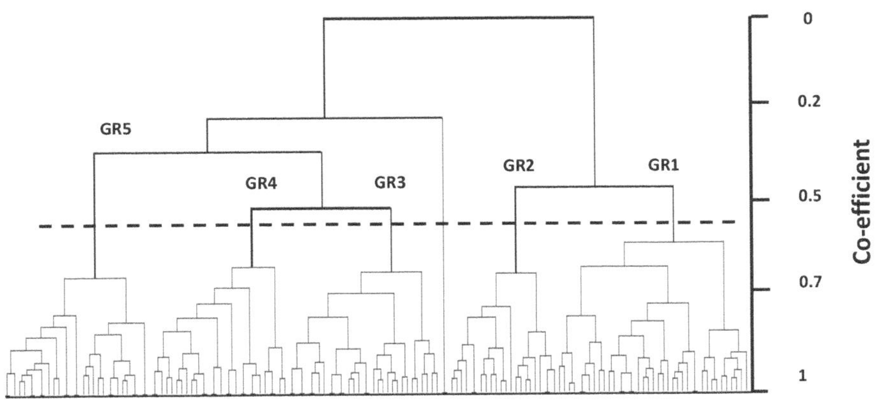

Fig. 10.1 Dendogram of the hierarchical ascending classification on the consumers based on the important of attributes (rank) (Source: Al-Shamsi et al., 2013)

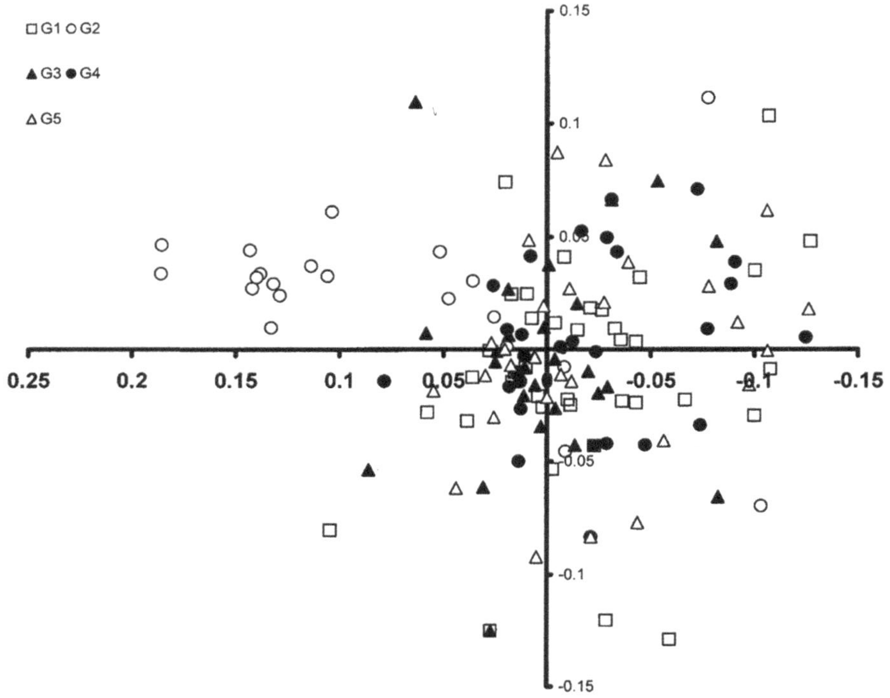

Fig. 10.2 Projection of consumers in a biplot representing five clusters based on the important of attributes (rank) (Source: Al-Shamsi et al., 2013)

The five groups could be considered based on the preference levels are: (1) consumers preferred black color with high sweetness and wrinkled surface (13%), (2) consumers preferred dark-brown color with high-medium sweetness and coarse-wrinkled surface (14%), (3) consumers preferred dark color with low-medium sweetness and uniform surface (21%), (4) consumers preferred brown color with medium sweetness and uniform-crease surface (33%), and (5) consumers preferred dark color with high-medium sweetness and uniform-crease surface (20%) (Fig. 10.3). These cluster analysis could help to develop marketing strategies of traditional Omani *halwa* (Fig. 10.4) (Al-Shamsi et al., 2013).

Rahman et al. (2012) also developed classification of Omani *halwa* available in the local market considering phsyico-chemical (i.e. moisture, protein, fat, ash, carbohydrate, color L, a, b values, pH and fatty acid profiles) and Instrumental Textural Profile Analysis (TPA) (i.e. hardness, springiness, fracture ability, cohesiveness, springiness, resilience, gumminess, chewiness). Four classes of *halwa* are available in the market as shown in Fig. 10.5. The four classes of halwa could be characterized as soft-resilient (G1), soft-springy-cohesive (G2), soft-springy (G3) and hard-chewy (G4) with their physicochemical properties and TPA attributes (Figs. 10.5 and 10.6). The parameters ash, color and pH values did not correlated with any TPA attributes, whereas all TPA attributes were significantly correlated

Table 10.5 Definition of the attributes and scales used for determining consumer preference level for each attributes

Attributes	Definition	Scale (1–9)[a]		
		1	5	9
Color	Color of halwa	Yellowish	Brown	Black
Appearance	Appearance of halwa	Uniform	Crease	Wrinkle
Glossiness	Degree of glossy or shiny surface	High	Medium	Low
Saffron	Saffron flavor	Low	Medium	High
Cardamom	Cardamom flavor	Low	Medium	High
Grittiness	Amount of crushed nuts	Small	Medium	Large
Sweetness	Sweetness of halwa	Low	Medium	High
Solubility	How solids are dissolving in the saliva	Slow	Medium	Quick
Chewiness	Number of chews required in preparing sample for swallowing	High	Medium	Low
Shear force	How difficult to make a punch on the flesh	Low	Medium	High
Elasticity	Degree to which the sample pushes back against the teeth during chewing	High	Medium	Low
Moth feel	Overall feeling during chewing and swallowing	Smooth	Lumpy	Crumbly
Gumminess	Degree to which sample sticks to itself in a dense, gummy mass at specific chews	Low	Medium	High
Hardness	Force necessary to attain a given deformation during biting	Low	Medium	High
Cohesiveness during biting	Degree of deformation undergone by a sample before its fractures, splits, cracks or breaks on first bite with molars or incisors	Low	Medium	High
Cohesiveness during chewing	Degree to which chewed sample holds together in one mass	Low	Medium	High
Springiness	Degree or rate at which a product physically springs back or returns to its original size/shape after partial compression	Low	Medium	High
Adhesiveness	Degree to which sample sticks to mouth surfaces	Low	Medium	High
Residual feeling in the mouth	Degree of residue feeling in the mouth	Present	Medium	Absent
Resilience	Degree of product fight to regain its original position	Low	Medium	High
Pastiness	Amount of soft, smooth mass that does not release moisture during chewing	Low	Medium	High
Oiliness during chewing	Degree of oily feeling during chewing	Low	Medium	High

Source: Al-Shamsi et al. (2013)
[a]Verbal words were used for 1, 5, 9 scale level

with specific types of fatty acid. As expected, specific TPA attributes did correlate with specific chemical composition since some components make the structure while could break the structure. TPA hardness was correlated with the moisture and sugar contents, while adhesiveness was correlated with the moisture, sugar and non-

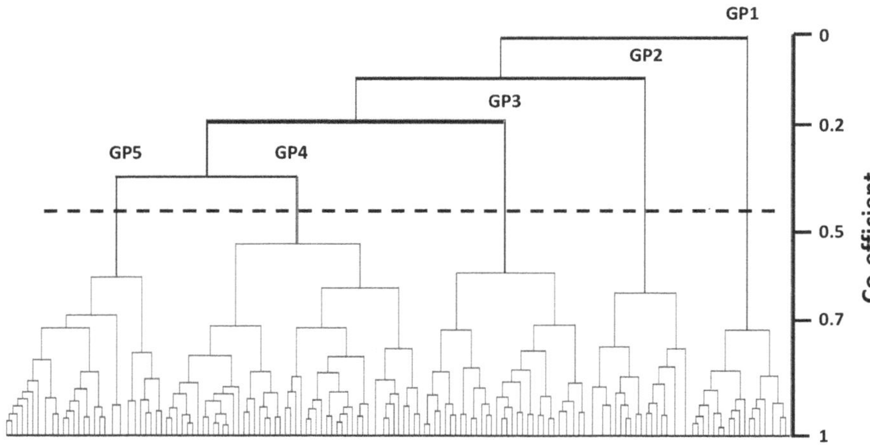

Fig. 10.3 Dendogram of the hierarchical ascending classification on the consumers based on the intensity of the preference level (intensity) (Source: Al-Shamsi et al., 2013)

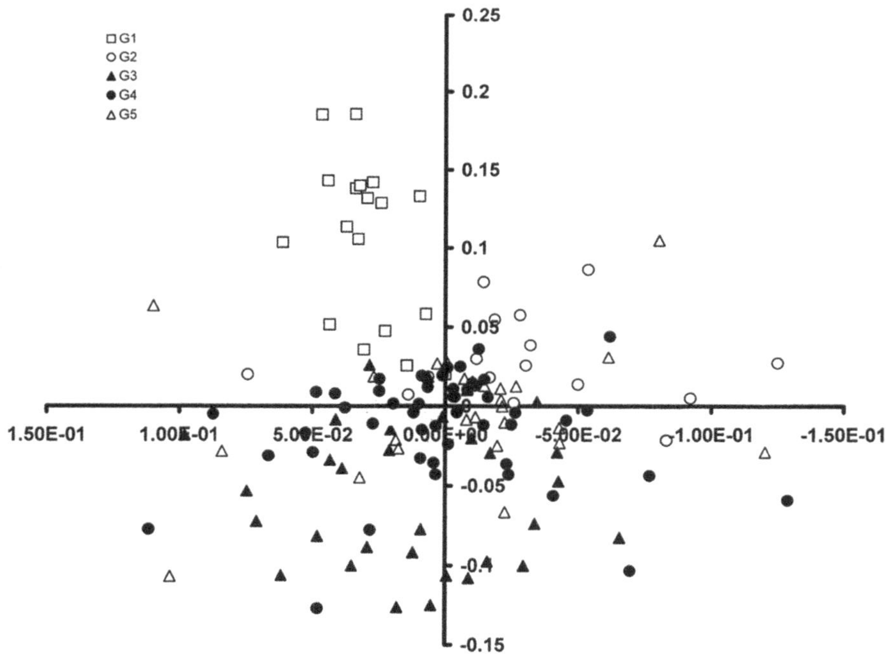

Fig. 10.4 Projection of consumers in a biplot representing five clusters based on the intensity of the preference level (intensity) (Source: Al-Shamsi et al., 2013)

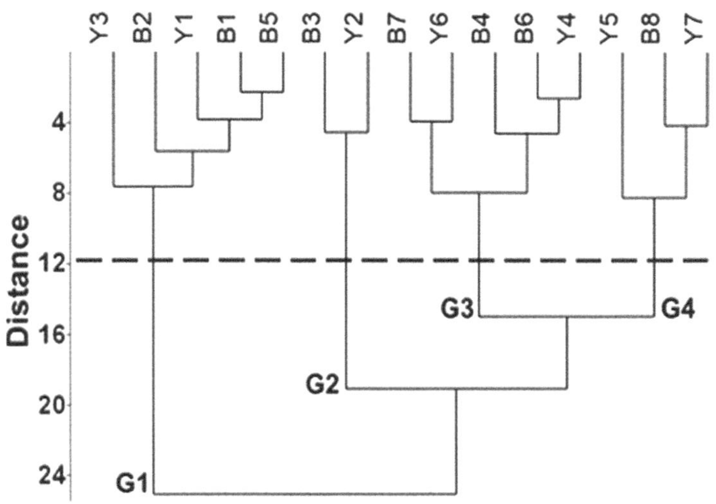

Fig. 10.5 Dendogram of the cluster analysis for the 15 halwa samples from local market (Yellow *halwa*: Y1–Y7, Brown *halwa*: B1–B8) (Rahman et al., 2012)

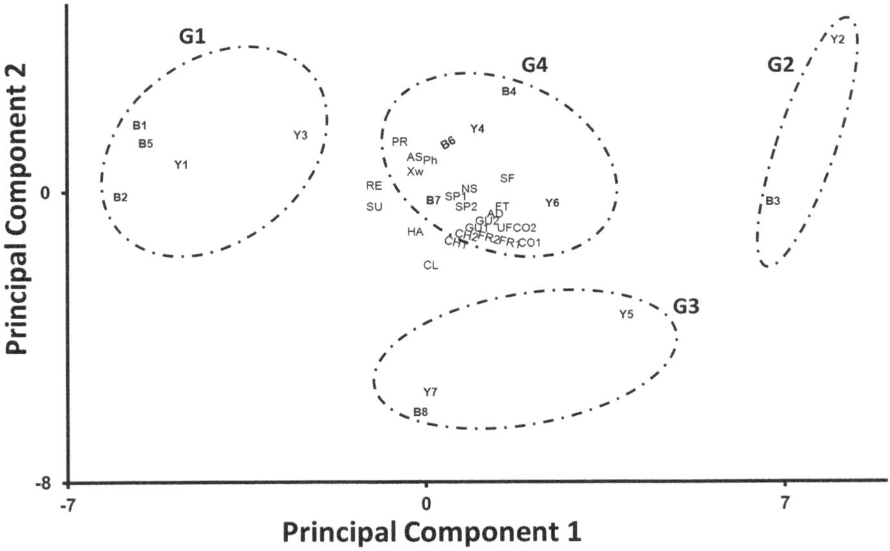

Fig. 10.6 Biplot of different *halwa* (from local market) physicochemical properties and instrumental TPA on two principal components with variables loading on the principal components (Yellow *halwa*: Y1–Y7, Brown *halwa*: B1–B8, *HA* hardness, *AD* adhesiveness, *SP1* springiness1, *FR1* firmness1, *FR2* firmness2, *CO1* cohesiveness1, *CO2* cohesiveness2, *SP2* springiness2, *RE* resilience, *GU1* gumminess1, *GU2* gumminess2, *CH1* chewiness1, *CH2* chewiness2, *XW* moisture, *PR* protein, *FT* fat, *SU* sucrose, *NS* non-sucrose carbohydrate, *AS* ash, *CL* color, *SF* saturated fat, *UF* unsaturated fat) (Rahman et al., 2012)

sugar carbohydrate contents. Water and sugar plasticize the structure, thus reduced hardness as well as stickiness. Springiness was correlated with sugar and non-sugar carbohydrate and firmness and chewiness were correlated with moisture and the total fat, saturated and unsaturated fat contents. Sugar and types of fats break the plasticity of the starch network, thus structure lost its ability to return its original shape. Cohesiveness-1 was correlated only with the protein, total fat, saturated and un-saturated fat contents, while cohesiveness-2 was correlated with sugar, total fat, saturated and unsaturated fat. Proteins make more network structure with gelatinized starch, thus cohesiveness increased. The resilience was correlated with the total fat, saturated and unsaturated fatty acids, while gumminess was only correlated with the moisture content (Rahman et al., 2012). Intermediate moisture (IM) date fruits at *tamr* stage of maturity are consumed as traditional in the Arab world. Singh, Guizani, Al-Alawi, Claereboudt, and Rahman (2013) used Instrumental Texture Profile Analysis (TPA) of nine batches of date flesh (from local market) with different quality levels to relate with its physico-chemical properties (i.e., color, mass, length, width of whole date fruit as well as moisture, crude fiber, fructose, glucose, sucrose, and pectin contents). It was observed that instrumental TPA showed correlations with physico-chemical properties. The available date fruits in the market can be classified into three different groups namely hard-resilient, soft-springy and firm-adhesive (Figs. 10.7 and 10.8). Similarly sensory texture (i.e. chewiness, elasticity, gumminess, hardness, cohesiveness during biting, springiness, adhesiveness, resilience and sweetness) and physico-chemical properties (moisture content, crude fiber, length of whole date fruits, pectin, sucrose., fructose, glucose, sugar index, color, mass, total phenols, and pH) of IM date fruits were used to classify available date fruits in the local market (Singh et al., 2015). PCA recognized three groups of date fruits as hard-chewy, soft-(medium-chewy) and soft-(non-adhesive) (Fig. 10.9). Sensory texture attributes were correlated with physico-chemical properties.

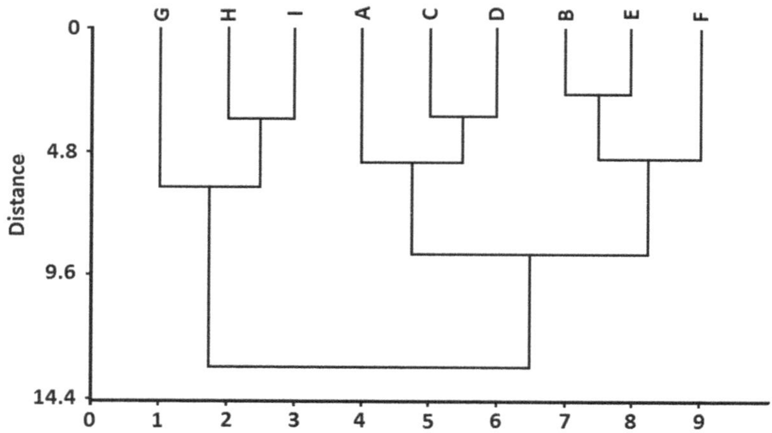

Fig. 10.7 Dendogram of the cluster analysis for different date samples [GR 1: group 1 (G, H, I); GR 2: group 2 (D, A, C); GR 3: group 3 (B, E, F)] (Singh et al., 2013)

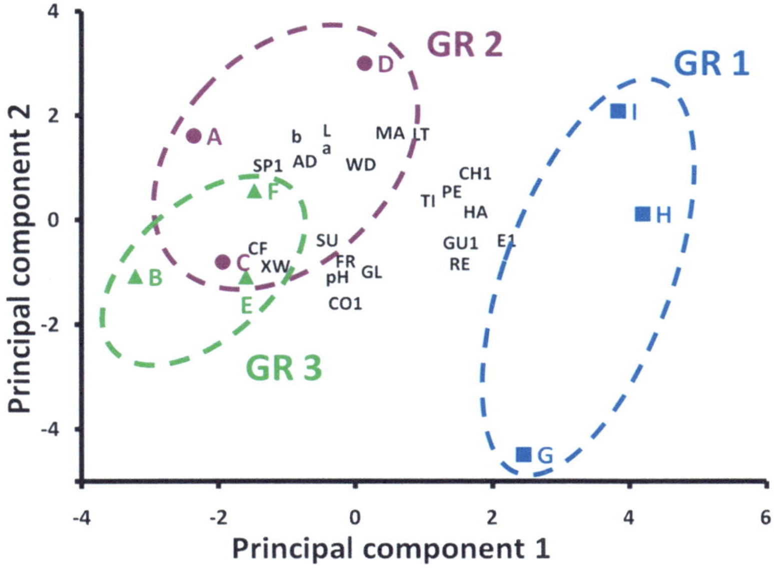

Fig. 10.8 Bi-plot of principal component 2 and principal component 1 [■: group 1 (GR 1);●: group 2 (GR 2);▲: group 3 (GR 3); *HA* hardness, *CO1* cohesiveness, *RE* resilience, *SP1* springiness, *GU1* gumminess, *CH1* chewiness, *AD* adhesiveness, *E1* elasticity 1, *XW* moisture content, *MA* mass, *LT* length, *WD* width, *CF* crude fiber, *FR* fructose, *GL* glucose, *SU* sucrose, *PE* pectin, *TI* toughness index] (Singh et al., 2013)

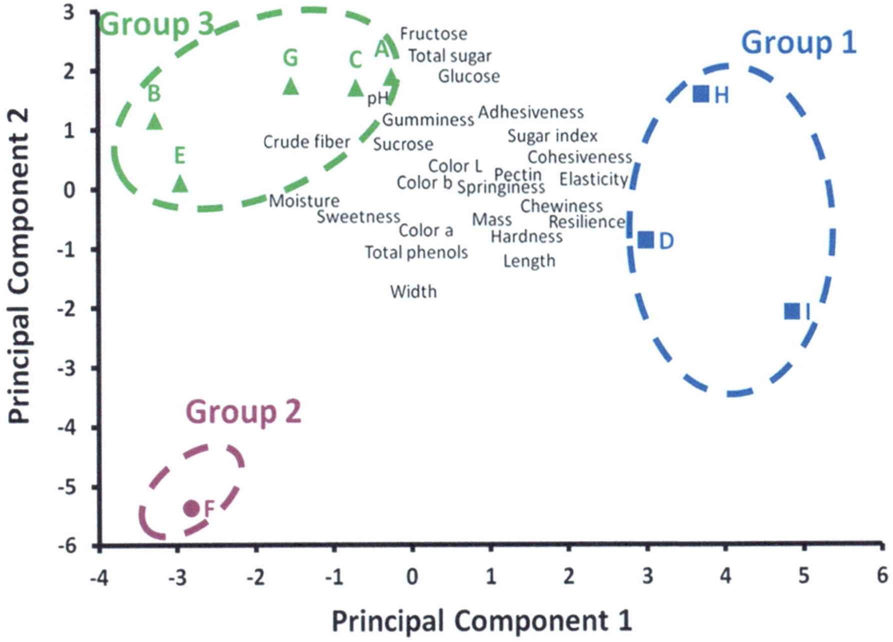

Fig. 10.9 Bi-plot of principal component 2 and principal component 1with physico-chemical properties and different date fruits from the market [Group 1 (D, H, I); Group 2 (F); Group 3 (G, A, C, B, E)] (Singh et al., 2015)

6 Conclusion

Sensory aroma, flavor, taste and texture are the key attributes for selecting traditional foods. It is important to document or develop standard of sensory profiles (i.e. methods as well as attributes) for each types of traditional food products. Traditional sensory attributes could be incorporated in developing healthy food products and their consumptions. It is also important to explore the types of consumer segments before introducing the traditional food products. Country's tourist industry could be enhanced through traditional foods since many tourists are interested to experience other culture through their foods.

References

Al-Shamsi, Q., Rahman, M. S., Abdullah, A., & Claereboudt, M. R. (2013). Consumers' preference mapping and their behavior with regard to traditional halwa using a structured survey questionnaire. *Journal of Food Products Marketing, 19*(4), 311–328.

Appledorf, H., Pennington, J., Wheeler, W. B., & Koburger, J. A. (1974). Sensory evaluation of health foods—A comparison with traditional foods. *Journal of Milk Food Technology, 37*(7), 392–394.

Barcenas, P., Perez de San Roman, R., Perez Elortondo, F. J., & Albisu, M. (2001). Consumer preference structures for traditional Spanish cheeses and their relationship with sensory properties. *Food Quality and Preference, 12*, 269–279.

Berodier, F., Lavanchy, P., Zannoni, M., Casals, J., Herrero, L., & Adamo, C. (1997). Guide d'evaluation olfacto-gustative des fromages *à* pate dure et semi-dure. *LWT Food Science and Technology, 30*, 653–664.

Berodier, F., Stevenot, C., & Schlich, P. (1997). Descripteurs de l'arome du fromage de Comt*é*. *LWT Food Science and Technology, 30*, 298–304.

Bock, T., & Uncles, M. (2002). A taxonomy of differences between consumers for market segmentation. *International Journal of Research in Marketing, 19*, 215–224.

Breslin, P. A. S., & Beauchamp, G. K. (1997). Salt enhances flavour by suppressing bitterness. *Nature, 387*, 563.

Carbonell-Barrachina, A. A. (2007). Application of sensory evaluation of food to quality control in the Spanish food industry. *Polish Journal of Food and Nutrition Sciences, 57*(4), 71–76.

Cayot, N. (2007). Sensory quality of traditional foods. *Food Chemistry, 102*(2), 445–453.

Chiralt, A., Galotto, M. J., & Fito, P. (1991). Changes in rheological properties and particle-size distribution during the manufacture of "Xixona turron". *Journal of Food Engineering, 14*, 117–128.

Dinehart, M. E., Hayes, J. E., Bartoshuk, L. M., Lanier, S. L., & Duffy, V. B. (2006). Bitter taste markers explain variability in vegetable sweetness, bitterness, and intake. *Physiology & Behavior, 87*, 304–313.

Drewnowski, A., & Gomez-Carneros, C. (2000). Bitter taste, phytonutrients, and the consumer: a review. *The American Journal of Clinical Nutrition, 72*, 1424–1435.

Engel, E., Baty, C., Le Corre, D., Souchon, I., & Martin, N. (2002). Flavor-active compounds potentially implicated in cooked cauliflower acceptance. *Journal of Agricultural and Food Chemistry, 50*, 6459–6467.

Engel, J. F., & Blackwell, R. D. (1982). *Consumer behavior*. Chicago, IL: The Dryden Press.

Etaio, I., Albisu, M., Ojeda, M., Gil, P. F., Salmeron, J., & Elortondo, P. F. J. (2010). Sensory quality control for food certification: A case study on wine. Method development. *Food Control, 21*, 533–541.

Feng, Y., Tapia, M. A., Okada, K., Lazo, N. B. C., Chapman-Novakofski, K., Carter Phillips, C., & Lee, S. (2018). Consumer acceptance comparison between seasoned and unseasoned vegetables. *Journal of Food Science, 83*(2), 446–453.
Gaedeke, R. M. (2001). Consumer consumption and perception of salmon. Implications for wild salmon marketing programs. *Journal of Food Products Marketing, 7*(1–2), 67–79.
Granzin, K. L., Olsen, J. E., & Painter, J. J. (1998). Marketing to consumer segments using health-promoting lifestyles. *Journal of Retailing and Consumer Services, 5*(3), 131–141.
Griffith, D. A. (2010). Understanding multilevel institutional convergence effects on international market segments and global marketing strategy. *Journal of World Business, 45*, 59–67.
Guerrero, L., Guardia, M. D., Xicola, J., Verbeke, W., Vanhonacker, F., Zakowska-Biemans, S., & Hersleth, M. (2009). Consumer-driven definition of traditional food products and innovation in traditional foods. A qualitative cross-cultural study. *Appetite, 52*, 345–354.
Halwai, A. (2005). Moisture sorption isotherm of Nepalese dried snack meat (Suuti). In A. B. Datta, K. M. Kundu, & G. P. Sinha (Eds.), *Proceedings of Asia Pacific Drying Conference-2005, 13–15 December 2005, Kolkata, India* (pp. 496–504). New Delhi: Allied Publishers.
IOOC. (2018). International Olive Oil Council. Retrieved from http://www.internationaloliveoil.org
Januszewska, R., & Viaene, J. (2012). Sensory evaluation of traditional products by variety-seekers and food neophobics. *Journal of Culinary Science & Technology, 10*, 192–210.
Kalviainen, N., Roinonen, K., & Tuorila, H. (2003). The relative importance of texture, taste and aroma on a yogurt-type snack food preference in the young and the elderly. *Food Quality and Preference, 14*(3), 177–186.
Keane, P. (1992). The flavor profile. In R. Hootman (Ed.), *ASTM manual on "descriptive analysis testing" (Manual 13)* (pp. 5–14). Philadelphia, PA: ASTM.
Kim, Y. G., Eves, A., & Scarles, C. (2009). Building a model of local food consumption on trips and holidays: a grounded theory approach. *International Journal of Hospitality Management, 28*, 423–431.
Korosec, M., Golob, T., Skvarca, M., & Bertoncelj, J. (2012). Protocol for sensory assessment of traditional dish cottage cheese Struklji. *Croatian Journal of Food Technology, Biotechnology and Nutrition, 7*, 5–8.
Kramer, A., & Twigg, B. A. (1970). *Quality control for the food industry*. Westport, CT: AVI Publishing.
Kubickova, J., & Grosch, W. (1997). Evaluation of potent odorants of Camembert cheese by dilution and concentration techniques. *International Dairy Journal, 7*, 65–70.
Laureati, M., Pagliarini, E., Calcinoni, O., & Bidoglio, M. (2006). Sensory acceptability of traditional food preparations by elderly people. *Food Quality and Preference, 17*(1), 43–52.
Lavanchy, P., Bérodier, F., Zannoni, M., Noël, Y., Adamo, C., Squella, J., & Herrero, L. (1993). L'evaluation sensorielle de la texture des fromages à pàte sure et semi-dure. *LWT Food Science and Technology, 26*, 59–68.
Lluch, M. A., Galotto, M. J., & Chiralt, A. (1992). Scanning electron microscopy studies of a typical Spanish confectionery product "Xixona turron". *Food Structure, 11*, 181–186.
Lopez-Lopez, A., Moreno-Baquero, J. M., Rodriguez-Gomez, F., Garcia-Garcia, P., & Garrido-Fernandez, A. (2018). Sensory assessment by consumers of traditional and potentially probiotic green Spanish-style table olives. *Frontiers in Nutrition, 5*, 1–13. (Article 53).
Mak, A. H. N., Lumbers, M., Eves, A., & Chang, R. C. Y. (2012). Factors influencing tourist food consumption. *International Journal of Hospitality Management, 31*(3), 928–936.
Manero, J., Phillips, C., Ellison, B., Lee, S. Y., Nickols-Richardson, S. M., & Chapman-Novakofski, K. M. (2017). Influence of seasoning on vegetable selection, liking, and intent to purchase. *Appetite, 116*, 239–245.
Martinez, N., & Chiralt, A. (1995). Glass transition and texture in a typical Spanish confectionery product: Xixona turron. *Journal of Texture Studies, 26*, 653–663.
Martinez, N., Chiralt, A., & Fito, P. (1997). Transport phenomena in the phase inversion operation of "Xixona turr6n" manufacture. *Journal of Food Engineering, 32*, 313–324.
Martinez, N., Fito, P., & Chiralt, A. (1996). Influence of conditions of manufacture and storage time on the textural characteristics of Xixona turr6n. *Food Control, 7*, 317–324.

Meilgaard, M., Civille, G. V., & Carr, B. T. (2007). *Sensory evaluation techniques* (4th ed.). Boca Raton, FL: CRC Press.

Mustonen, S., & Tuorila, H. (2010). Sensory education decreases food neophobia score and encourages trying unfamiliar foods in 8-12-year-old children. *Food Quality and Preference, 21*, 353–360.

Nikolaus, C. J., Ellison, B., Heinrichs, P. A., Nickols-Richardson, S. M., & Chapman-Novakofski, K. M. (2017). Spice and herb use with vegetables: liking, frequency, and self-efficacy among US adults. *American Journal of Health Behavior, 41*, 52–60.

Olabi, A., Najm, N. E. O., Baghdadi, O. K., & Morton, J. M. (2009). Food neophobia levels of Lebanese and American college students. *Food Quality and Preference, 20*, 353–362.

Peres, C., Begnaud, F., & Berdague, J. L. (2002). Fast characterization of Camembert cheeses by static headspace—Mass spectrometry. *Sensors and Actuators B, 87*, 491–497.

Peri, C. (2006). The universe of food quality. *Food Quality and Preference, 17*(1), 3–8.

Pliner, P., & Melo, N. (1997). Food neophobia in humans: Effects of manipulated arousal and individual differences in sensation seeking. *Physiology & Behavior, 61*(2), 331–335.

Poelman, A. A. M., Delahunty, C. M., & de Graaf, C. (2015). Vegetable preparation practices for 5 to 6 y old Australian children as reported by their parents; relationships with liking and consumption. *Food Quality and Preference, 42*, 20–26.

Rahman, M. S., Al-Shamsi, Q., Abdullah, A., Claereboudt, M. R., Al-Belushi, B., Al-Maqbaly, R., & Al-Sabahi, J. (2012). Classification of commercial Omani hlawa by physic-chemical properties and Instrumental Texture Profile Analysis (TPA). *Italian Journal of Food Science, 24*(3), 292–304.

Rappoport, L. H., Peters, G. R., Huff-Corzine, L., & Downey, R. G. (1992). Reasons for eating: an explanatory cognitive analysis. *Ecology of Food and Nutrition, 28*, 171–189.

Retiveau, A., Chambers, D. H., & Esteve, E. (2005). Developing a lexicon for the flavour description of French cheeses. *Food Quality and Preference, 16*, 517–527.

Schonhof, I., Krumbein, A., & Bruckner, B. (2004). Genotypic effects on glucosinolates and sensory properties of broccoli and cauliflower. *Nahrung/Food, 48*, 25–33.

Sharafi, M., Hayes, J. E., & Duffy, V. B. (2013). Masking vegetable bitterness to improve palatability depends on vegetable type and taste phenotype. *Chemosensory Perception, 6*, 8–19.

Singh, V., Guizani, N., Al-Alawi, A., Claereboudt, M., & Rahman, M. S. (2013). Instrumental texture profile analysis (TPA) of date fruits as a function of its physico-chemical properties. *Industrial Crops and Products, 50*, 866–873.

Singh, V., Guizani, N., Al-Zakwani, I., Al-Shamsi, Q., Al-Alawi, A., & Rahman, M. S. (2015). Sensory texture of date fruits as a function of physico-chemical properties and its use in dates classification. *Acta Alimentaria, 44*(1), 119–125.

Stone, H., & Sidel, J. L. (2004). *Sensory evaluation practices* (3rd ed.). San Diego, CA: Elsevier Academic Press.

Thomas-Danguin, T., Rouby, C., Sicard, G., Vigouroux, M., Barkat, S., & Brun, V. (2004). Sensory analysis and olfactory perception: some sources of variation. Deitler, K. D.; Delwiche, J (Eds.), In *Handbook of flavour characterization* (pp. 65–91). New York: Marcel Dekker.

Vanhonacker, F., Kuhne, B., Gellynck, X., Guerrero, L., Hersleth, M., & Verbeke, W. (2013). Innovations in traditional foods: Impact on perceived traditional character and consumer acceptance. *Food Research International, 54*(2), 1828–1835.

Vazquez-Araujo, L., Enguix, L., Verdu, A., Garcia-Garcia, E. A. A., & Carbonell-Barrachina, A. A. (2008). Investigation of aromatic compounds in toasted almonds used for the manufacture of *turrón*. *European Food Research and Technology, 227*(1), 243–254.

Wang, L. (2004). An investigation of Chinese immigrant consumer behavior in Toronto, Canada. *Journal of Retailing and Consumer Services, 11*, 307–320.

Wang, O., De Steur, H., Gellynck, X., & Verbeke, W. (2015). Motives for consumer choice of traditional food and European food in mainland China. *Appetite, 87*, 143–151.

Whelan, S., & Davis, G. (2006). Profiling consumers of own brands and national branding human personality. *Journal of Retailing and Consumer Services, 13*, 393–402.

Wu, K., Raab, C., Chang, W., & Krishen, A. (2016). Understanding Chinese tourists' food consumption in the United States. *Journal of Business Research, 69*(10), 4706–4713.

Index

M. Al-Khusaibi et al. (eds.), *Traditional Foods*, Food Engineering Series,
https://doi.org/10.1007/978-3-030-24620-4

The manufacturer's authorised representative in the EU is Springer Nature Customer Service Centre GmbH, Europaplatz 3, 69115 Heidelberg, Germany. If you have any concerns regarding our products, please contact ProductSafety@springernature.com

Printed and bound by CPI Group (UK) Ltd, Croydon, CR0 4YY
30/09/2026
02225448-0009